Sir Edward Hale's Regiment.
1689.

The First Captain's Colour.

SEE PAGE 293.

Private Soldier 1712

Grenadier 1751.

Private Soldier 1792

Officer and Private 1815.

Officer 1826.

Officer 1840.

Sergeant and Private 1855 Crimea

Captain 1858.

Private 1864.

Officer and Corporal 1879. Afghanistan.

Private 1881

Private 1892

Historical Records

OF THE

14TH REGIMENT,

NOW THE PRINCE OF WALES'S OWN

(WEST YORKSHIRE REGIMENT,)

FROM ITS FORMATION, IN 1685,

TO 1892.

Edited by

CAPTAIN H. O'DONNELL.

Devonport:

A. H. SWISS, 111 AND 112 FORE STREET;

OR CAN BE OBTAINED OF THE

Adjutant, 1st or 2nd Battalion of the Regiment.

PREFACE.

THE dearth of historic facts in the Regiment, and the many legendary fables extant, suggested the desirability of preparing a properly authenticated History, in which the Records published fifty years ago should be corrected and amplified—with the aid of the many sources of information, official and other, now more readily accessible—and the narrative brought down to the present time. This has now been done, it is hoped, successfully. Every care has been taken to make the History as accurate as possible. The assistance of all past and present members of the Regiment, within reach, has been solicited to enable all facts, within memory, to be corroborated and established without doubt.

A good deal of miscellaneous information—illustrations of the regimental story—which could not properly be included in the Historical Records, has been brought together in a separate Appendix Volume *(Appendix B.)* To some this Appendix may appear unnecessary and out of place in such a book as a Regimental History—appearing to belong to a decade, as it were, in the History; but this must be taken as a beginning only—the idea being periodically to revise and add to the book, so that the future generations shall be kept in touch with all phases of the life of their forerunners in the Corps, and thus to unite the past, the present, and the future of the Old Fourteenth in one bond of brotherhood. The Record proper will show the external and soldierly qualities; the Appendix, the social and civil life of the Regiment.

The Committee have to thank Mr. H. MANNERS CHICHESTER for his unwearying research into all matters requiring verification or elucidation, and for the literary assistance he has rendered throughout. The composition has been almost entirely his own.

To Mr. S. M. MILNE, of Calverley House, Leeds, the Committee have to express their best acknowledgment for the interesting chapters he has contributed on the Equipment, Badges, Costumes, and the Colours of the Regiment, and for his valuable assistance in other ways—not the least of which has been his personal superintendence of the production of the coloured plates of Regimental Costume.

To others—among whom must be numbered SIR EVERARD DOYLE, for his portrait of Colonel Welbore Ellis Doyle, and SIR HARRY VERNEY, for permission to photograph the trophies at Claydon House—their best thanks are due for the assistance they have given in supplying information for the work.

Mr. A. H. SWISS, the publisher, has devoted the utmost care and attention throughout the entire work. The Coloured Plates of Costume have been executed by MESSRS. GOODALL and SUDDICK, Leeds, from drawings by Mr. P. W. REYNOLDS. The Portraits of Celebrities were engraved by the TYPOGRAPHIC ETCHING COMPANY, London.

It is the hope of the Committee that the book will, in every way, meet with the best approval of its readers.

H. O'DONNELL,

Captain and Adjutant,

1/14 Regiment.

Aldershot,

1st September, 1893

CONTENTS.

APPENDIX A.

APPENDIX B; OR, VOLUME II.

ILLUSTRATIONS.

FULL PAGE.

ILLUSTRATIONS IN TEXT.

ILLUSTRATIONS IN THE APPENDICES.

ERRATA.

Page 30—footnote—paragraph 2, line 6: *for* "Munzo's" *read* "Munro's."

Page 32, top line: *for* "death" *read* "retirement."

Page 48, line 16: *for* "Burnett" *read* "Burnet."

Page 64, line 3: *for* "Parry" *read* "Perry."

Page 66, last line (footnote): *for* "1817" *read* "1813."

Page 152, line 26: *for* "Pratters" *read* "Protheroe."

Page 160, line 4: *for* "Marioch" *read* "Manoel;" paragraph 3, line 2: *for* "Limerick" *read* "Birr."

CORRIGENDA.

*** A slight error on pages 35 and 36 requires correction. The series of printed annual Army Lists (unofficial and official) in the British Museum commences with the year 1754. From the earliest of these lists (1754) is taken the roll of officers at page 36. The publication of the annual Army List "by authority" commenced in 1756, not in 1754.

HISTORICAL RECORDS

OF THE

FOURTEENTH REGIMENT OF FOOT.

1685.

AMONGST the Horse and Foot raised for the King's service at the time of the Duke of Monmouth's Rebellion, in the summer of 1685, was a company of one hundred musketeers and pikemen, recruited at Canterbury and in the neighbourhood by Sir Edward Hales, baronet, of Woodchurch, in the county of Kent. Companies were likewise raised by other loyalists: — Boynton, Robert Middleton, Henry Vaughan, Richard Brewer,* William Broom, John Gifford, Thomas Gifford, Mark Talbot, John Chappell, and Rowland Watson. Two of these companies had their rendezvous at Chatham and Rochester; others were formed at Sittingbourne and Faversham. The companies were collected at Canterbury, and formed into a regiment, of which Sir Edward Hales was appointed colonel; — Boynton, lieutenant-colonel; and Robert Middleton, major; by commissions dated 22nd June, 1685, and which afterwards became the FOURTEENTH REGIMENT OF FOOT.

Whilst the formation of the regiment was in progress, the rebel bands from the West country were routed at Sedgmoor on 6th July, and the unfortunate Monmouth was taken and beheaded. King James II. then decided to increase the small standing army already existing, and paid out of his Civil List; and among the corps selected to be retained was Sir Edward Hales' regiment, the establishment of which was fixed at ten companies, of sixty men each.

The regiment was in the camp formed on Hounslow Heath in the summer of the same year, which proved a source of untiring interest to the sight-seeing Londoners, although Evelyn relates that, amidst all

* Afterwards colonel of the 12th Foot.

the feasting and revelry, were "divers jealousies and discourses as to "the meaning of the array."* After having been reviewed by King James, the regiment marched to Gravesend, detaching two companies to Jersey, one to Guernsey, and two to Windsor.

1686-87.

On 1st January, 1686, the establishment of the regiment was fixed at the following numbers and rates of pay :—

SIR EDWARD HALES' REGIMENT.

STAFF :	Pay per diem. £	s.	d.
The Colonel, as Colonel	0	12	0
" Lieut.-Colonel, as Lieut.-Colonel	0	7	0
" Major, as Major	0	5	0
" Chaplain	0	6	8
" Chirurgeon, iv *s.*; Mate, ii *s.* vi *d.*	0	6	6
" Adjutant	0	4	0
" Quarter-Master and Marshall	0	4	0
	£2	5	2
COLONEL'S COMPANY :			
The Colonel, as Captain	0	8	0
" Lieutenant	0	4	0
" Ensign	0	3	0
2 Sergeants, at xviii *d.* each	0	3	0
3 Corporals, at xii *d.* each	0	3	0
1 Drummer	0	1	0
50 Soldiers, at viii *d.* each	1	13	4
Total for one company	2	15	4
Nine companies more, at the same rate	24	18	0
Total per day	£29	18	6

Per annum - £10,922 12s. 6d.

In the summer of 1687 a camp was again formed on Hounslow Heath, "the commanders profusely vying with each other in the "magnificence of their tents."† The regiment was there, and had a grenadier company added to its establishment. After passing in review before the King and the Royal Family, it marched away to Plymouth, where it spent the winter.

* *Evelyn's Diary*, vol. ii. † *Ib.*

REGIMENTAL ROLL OF OFFICERS BY COMPANIES, 1687.

(This is the earliest Roll that can be found.)

Captains :

SIR EDWARD HALES *(Colonel)*	THOS. WELD
G. BARCLAY *(Lieut.-Colonel)*	GEO. LATTON
JOHN GIFFORD *(Major)*	RICH. BREWER
JOHN CHAPPELL	THOS. GIFFORD
ROWLAND WATSON	GEO. AYLMER

PETER SHAKERLY* *(Grenadier Company).*

Lieutenants :

THOS. BUTLER	BRYCE BLAIR
ROBT. SEATON	WILLM. CAREW
RICH. BOUCHER	NICHOLAS MORGAN
GAVAN TALBOT	EDWARD GIFFORD
JAS. NICHOLSON	AUGUSTUS GIFFORD

WILLIAM FIELDING, FRANCIS SANDERSON, } *Grenadier Company.*

Ensigns :

DUDLEY VAN BURGH	CLIFFORD BREXTON
AUSTEN BELSON	GEO. BLATHWAYT
THOS. HEYWARD	EDWARD HALES†
PHILIP OVERTON	EDWARD POPE .
DUDLEY VAN COLSTER	CÆSAR GAGE.

Chaplain - - NICHOLAS TRAFPS.

Chirurgeon - - JOHN RIDLEY.

Adjutant - - JAS. NICHOLSON.

Quarter-Master - EDWARD SYNG.

---

* Sacheverel ?

† Eldest son of Sir Edward Hales, colonel of the regiment. Killed at the battle of the Boyne, fighting for King James.

1688-89.

In June, 1688, the regiment marched from Plymouth to London, and took the duty at the Tower until the middle of August, when it was relieved by the Royal Fusiliers, and marched to Canterbury, and thence, in September, to Salisbury.

Meanwhile King James was blindly pursuing the arbitrary policy that ere long was to cost him his crown. Amidst these proceedings, Sir Edward Hales, the colonel of the regiment, became a prominent figure. The King claimed the prerogative of dispensing with the oaths required by law on appointment to office. Sir Edward, who was a lord of the admiralty, deputy-governor of the Cinque ports, and held other posts besides, had recently avowed himself a Roman catholic, and for that reason was legally disqualified to take the oath and ineligible to hold a commission in the army. He was prosecuted at Rochester assizes, on the information of a discharged servant, and was convicted of a breach of the penal laws, but moved the case into the Court of King's Bench and obtained judgment in his favour, eleven of the judges siding with the King against the law. Macaulay supposes the prosecution, like the judgment, to have been a sham designed to secure a convenient precedent.* The landing in the West of England, in November, 1688, of the Prince of Orange, with a powerful force, to support the Protestant interest, put an end to further usurpations on the part of King James. Finding that his troops collected at Salisbury were indisposed to fight in the cause of arbitrary government, he returned to London and attempted an escape to France, attended by Sir Edward Hales and Quarter-Master Syng of the regiment. They were apprehended on board a custom-house vessel at Faversham, and Sir Edward was afterwards confined in the Tower of London. The King made a second attempt, and reached France in safety.

The Prince of Orange directed the regiment to be quartered at Bishops Waltham, in Hampshire, and appointed William Beveridge, an officer of one of the English regiments in the pay of Holland, as

---

* Macaulay, *History of England*, vol. vi.

colonel in the place of Sir Edward Hales, by commission dated 31st December, 1688.

On 13th February, 1689, the Prince of Orange and his consort, the Princess Mary, were crowned King and Queen. In March, the news came that King James had landed in Ireland, at the head of a French army, and had entered Dublin. On 3rd April, 1689, the first Mutiny Act, legalizing the maintenance of a standing army within the British Isles, *subject to the consent of Parliament*, and providing for the punishment of mutiny and desertion—which before were not legally punishable in time of peace—became law. It was at first passed for six months only, but with a few exceptions has been renewed annually ever since. The Orange succession was opposed in Scotland, as well as in Ireland, and the regiment, with other corps, were in consequence moved into the North. It was inspected at Berwick-on-Tweed, by the commissioners appointed for remodelling the army, on 14th June, 1689; and in August, on the news of the defeat of the Williamite forces under General Hugh Mackay, at the pass of Killiecrankie, was sent on from Berwick to Edinburgh.

1690-91.

The regiment was in Scotland during the operations against the insurgent clans, most of whom, early in 1691, lost all hope of success, and tendered their submission to the Government of King William. Probably its services were confined to the South, as the regiment is not named among those employed under General Mackay and Sir Thomas Livingstone. Enquiries have shewn that no rolls of the regiment at this period (nor for the period of its next sojourn in Scotland in 1716-22) exist among the old military records preserved at Holyrood.

1692.

On 25th February, 1692,* Beveridge's regiment (14th Foot) sailed from Leith, with Leven's (25th Foot) and Lowther's regiments, to join the army in Flanders, under the personal command of King William, serving against the forces of Louis XIV. of France. Scarcely had it

---

* *London Gazette*, 29th February-3rd March, 1692.

reached the seat of war and gone into garrison in West Flanders ere rumours of an anticipated French descent, for the purpose of replacing King James on the English throne, hurried it home again. It landed at Greenwich early in May, and with other regiments was quartered round about London, in readiness for the expected invasion.

All fears of the latter were, however, dispelled by Admiral Russell's great victory over the French fleet off La Hague (Cape Barfleur) on 19th-21st May, 1692, which gave the English the command of the sea. It was proposed to follow up the success by a grand descent on the French coast so soon as the fleet had refitted. Fourteen regiments, one of which was Beveridge's, were encamped in readiness on Portsdown Hill. Luttrell* describes the busy preparations making in July: The troops were to have with them six months' provisions; a supply for four months more was to follow. They were to take with them sixty pieces of cannon and twenty mortars, an abundance of other munitions of war, including thirty thousand stand of spare small arms. A sum of £40,000 was sent down from London for current expenses. Each soldier received 20s. (a large sum in these days) "by way of encouragement." On 26th July, 1692, every man was embarked.

The sequel is related by Macaulay, in the fourth volume of his history:†

"The transports sailed, and in a few hours joined the naval "armament in the neighbourhood of Portland. On the 28th, a "general council of war was held. All the naval commanders, with "Russell at their head, declared that it would be madness to carry "the ships within range of the guns of St. Maloes, and that the town "must be reduced to straits by land before the men-of-war in the "harbour could, with any chance of success, be attacked from the sea. "The military men declared with equal unanimity that the land-forces "could effect nothing against the town without the co-operation of "the fleet. It was then considered whether it would be desirable to

---

* Narcissus Luttrell, *Relation of State Affairs*, vol. ii., p. 516.

† Macaulay, *History of England*, vol. iv., p. 290.

"make a descent on Brest or Rochfort; Russell and the other flag-"officers, among whom Rooke, Shovel, Van Almonde, and Evertsen, "pronounced that the summer was too far spent for the enterprise.* "We must suppose that an opinion in which so many distinguished "admirals, English and Dutch, concurred, was in conformity with "what were then the recognized principles of the art of maritime war. "But, why these questions could not have been discussed a week "earlier, why fourteen thousand troops should have been shipped and "sent to sea before it had been considered what they could do, or "whether they could do anything, we may reasonably wonder. The "armament returned to St. Helens, to the disgust of the whole nation."

From St. Helens, Beveridge's and some other regiments, under the command of the Duke of Leinster, proceeded to the Downs, and then to Ostend, where they landed and encamped a little way out of the town. Reinforced by a detachment from the army in Flanders, they marched to Furnes, which the French evacuated on their approach. After throwing up some defences there, they proceeded to Dixmude, where they were employed repairing the fortifications. Some regiments then returned home, but Beveridge's passed the winter in cantonments in Flanders.

On 14th November, 1692, Colonel Beveridge was killed in a duel by one of his officers.† The colonelcy of the regiment was then conferred on Lieutenant-Colonel John Tidcomb, from the 13th Foot.

1693.

In May, 1693, Tidcomb's (late Beveridge's) regiment joined the Confederate Army under King William, which encamped at Parck,

---

* It must be remembered that the "old style" of reckoning was in use in England at that time and for sixty years afterwards, although "new style" was in use on the Continent. Ascending to the modern reckoning, the date of the council of war was the 7th August.

† Luttrell, who calls the regiment a Scotch corps, states that Colonel Beveridge's antagonist was a Captain Van Brook—evidently the Van Burgh whose name appeared among the killed at Landen the year after. A dispute arose at dinner, and swords were drawn after the captain had been abused and violently shaken by the colonel.—*Return of State Affairs*, vol. ii., p. 621.

near Louvain, to check the designs of the French on Brabant. D'Auvergne, chaplain of the Scots Guards, and the historian of these campaigns, shows the regiment as forming part of the second line of the army. After taking part in sundry movements, the regiment was in position at LANDEN on 19th July, 1693, when King William's forces were attacked by a French army, of very superior numbers, under command of Marshal Luxembourg. The battle that ensued—variously known as Landen, from a stream of that name close to the left flank of the army; and Neerwinden, from a village that formed one of the principal points of attack—was the first general action in which the regiment engaged. It has been rightly said that although the region round about was the battle-ground of the greatest powers of Europe for many ages, it has witnessed but two more terrible days—the day of Malplaquet and the day of Waterloo.* Tidcomb's regiment was in Major-General Thomas Erle's brigade, which consisted of Tidcomb's (14th Foot), Stanley's (16th Foot), Erle's (19th Foot), and Collingwood's and Graham's (afterwards disbanded). Thrice the spirited attacks of the French were repulsed, but their superior numbers eventually enabled them to carry the village of Neerwinden with fresh troops. The King then ordered a retreat, which was executed with difficulty and attended with heavy loss.

The chief events of the day have been related with picturesque minuteness by different writers, from Sterne to Macaulay; but of the movements of individual regiments, so far at least as regards Erle's brigade, there appear to be no official or other details extant. Erle himself left a sick bed, at Mechlin, to take command of his brigade, and was badly wounded at its head; and the loss of officers in the several regiments of the brigade—twenty-four officers killed and wounded—testifies that they were not backward in the fight.

Tidcomb's (14th Foot) had Captains Van Burgh, Cassim, and Henriosa, and Lieutenant Worley killed; Lieutenant Nicholson died of wounds; Captains Devaux and Stanwix; Lieutenants Campbell, Forbes, and Pettitpierre; and Ensigns Ravisson and Perrott wounded; and Colonel Graham taken prisoner. The loss in non-commissioned

---

* Macaulay, *History of England*, vol. iv., p 416.

officers and men has not been recorded for any of the regiments engaged.

In the autumn, when the army separated for winter quarters, Tidcomb's regiment marched into garrison at Bruges, and parties were sent home to England to recruit.

1694.

A Board of general officers was assembled this year at Gemblours Camp, Flanders, by order of King William, to fix the precedence of regiments of the army, a subject which had been the source of many disputes. At the recommendation of the Board, English regiments were directed to reckon their seniority according to dates of formation, Scotch and Irish regiments from the dates when first placed on the English establishment. Tidcomb's regiment thus took rank as the FOURTEENTH REGIMENT OF FOOT. The practice of calling regiments by their colonels' names, however, continued in vogue for nigh sixty years longer.

When the army took the field again in the spring of this year, the regiment was left encamped at Ghent with some other corps, under Brigadier-General Sir David Collier, to form an escort for the artillery, which was conveyed by water to Mechlin (Malines). The regiment joined the army in camp near Louvain on 4th June, and on 6th June was reviewed by King William, who expressed to Colonel Tidcomb his high appreciation of its appearance. After participating in several movements, it formed part of the splendid force of Confederate troops assembled in camp at Mont St. André, near the village of Ramillies, under the personal command of King William. It was one of the corps which attempted by a forced march to pass the enemy's fortified lines and penetrate into French Flanders; but, by extraordinary exertions, the French outstripped them, and so prevented the intended invasion. The regiment was next encamped near Rousselaer, with the army covering the siege of Huy. Detachments of the enemy were very active in the neighbourhood of the camp, and on one occasion when the waggons conveying bread to the army were attacked, a detachment of Tidcomb's, forming part of the guard, was attacked, and Captain Sacheverel was mortally wounded; the only British officer killed by the enemy in this campaign.

Having to keep the field during the wet and cold of the early autumn, the troops erected huts of wood and straw. On 1st October the tents of Tidcomb's regiment were accidentally set on fire and destroyed. A similar mishap had befallen the Coldstream Guards a few days before. When Huy surrendered, the army broke up to go into winter quarters, the regiment returning to Bruges in the second week in October.

1695.

In May, 1695, the regiment marched from Bruges to Dixmude, where it remained several days. The Duke of Wurtemburg took command of the troops assembled at this point, and, advancing to the juncture of the Loo and Dixmude canals, encamped before the fortress of Kenoque, on which an attack was to be made, to draw off the French in that direction. Tidcomb's (14th Foot) took part in this service, and its grenadier company was engaged in driving the French from some buildings and entrenchments near the Loo canal, and repulsing the attempts to retake them. A redoubt was taken, and a lodgment effected in some works near the bridge over the canal, in which operations the regiment had several men killed and wounded. The demonstration against Kenoque had the desired effect; the great fortress of Namur was left open to the attack of the main army of the Allies. Namur, reported to be all but impregnable, was held by ten thousand French, under the famous Marshal Boufflers. It was invested by King William on 3rd July, 1695.

The attack on Kenoque was then desisted in, and Tidcomb's and other regiments were withdrawn from West Flanders and joined the covering army under the Prince of Vaudemont, at Wouterghem. From Wouterghem, Tidcomb's regiment advanced nearer Namur, pitching its tents at Templeux, about one and a half leagues from Namur, to take its turn of duty in the trenches at the since famous siege.

Tidcomb's (14th Foot), Stanley's (16th Foot), and Collingwood's, brigaded together under Lord Cutts, formed the guard of the trenches on 7th July, 1695. On the morrow an attack was ordered on the covered way of a line of defence constructed by the enemy on the hill of Bouge, to cover some of their works. The attack was to be made at the relief of guard, so to have as many troops as possible in the trenches. The following quaint account of the assault is given in the *London Gazette* of 22nd July, 1695. The date is wrongly stated as Monday, 18th July, a mistake repeated by some later writers :*

"The signal being given half an hour after six of the evening, the "several battalions marched forward with the greatest courage and un- "dauntedness that was ever seen, without taking any notice of the enemy's "fire, which was very furious, and the fusiliers in front carried their fascines† "to the very palisades, where, laying them down, they fired upon the enemy, "and the grenadiers threw their grenades into the tower (of La Bouge) and "the works, while the battalions marched close after them, with arms "shouldered, until they came so near that they presented over the palisades, "drove the enemy thence, and pursued them through a large place of arms "at the bottom of that work. The enemy making fresh fire from the counter- "scarp, and a redoubt on the other side of a hollow way on our flank, "my Lord Cutts, with three battalions from the trenches, viz., Tidcomb, "Stanley, and Collingwood, came immediately to sustain our men. With "the first of these he marched to the lower palisades, and with the Guards "again repulsed the enemy, part in the water, and the rest as they could "best escape. Brigadier Fitzpatrick moved at the same time with the "regiment of Lord George Hamilton (1st Royals),‡ the Royal Fusiliers, "Ingoldsby's (23rd Royal Welsh Fusiliers), Sanderson's,§ Lauder's, and "Maitland's (25th Foot), the first two were sent to relieve the Dutch Guards "and those with Lord Cutts at the lower palisades, and the rest were drawn "up by the tower, to sustain as the action should require, except the regiment "of Lauder, which was posted on the other side of the hollow way, to "prevent being flanked or surprised. In the meantime, Major-General "Ramsay ordered all the pikemen to carry fascines and to dig till a work "was made sufficient to resist musket-shot, where he posted the regiments "of Ingoldsby, Sanderson, Lauder and Maitland, and at the point of day "drew off the rest of the troops to the first parallel. The heat of the action "lasted about two hours, during which time we possessed ourselves of the "enemy's works, which were defended by great numbers of men."

---

* The attack was made on Monday, 8th July (19th July, new style), 1695. The 18th July, old style, was a Thursday, and new style a Sunday.

† The assaulting column had to traverse nine hundred paces of open ground from the first parallel to the covered way. It was preceded by detachments of fusiliers, each man carrying a fascine, and a working party with tools and gabions.

‡ Lord George Hamilton, afterward Earl of Orkney, colonel of the Scots Royal, as the regiment was then called.

§ Sanderson's Foot, afterwards disbanded. Its colonel, in 1702, raised a marine regiment, which became the 30th Foot, and is now the 1st East Lancashire Regiment.

The old guard did not dismount until the fight was over, when Tidcomb's returned to its camp at Templeux. Its loss was severe: Lieutenant Ravisson, killed; Captain Carew and Ensign Perrott died of their wounds; Captains Pope, Jackson, and Forbes, and Ensign Cormach, wounded. The loss in non-commissioned officers and men is not on record.

The regiment left Templeux and took its part in the line of circumvallation, and on 10th July again formed part of the guard of the trenches. It was on duty in the trenches once more on 16th July, when Captain Forbes and several private soldiers were killed. On 17th July a detachment of the grenadiers of the regiment was engaged in an attack on the counterscarp. The attack was made, in a rush, at the time the guards in the trenches were relieved. The French disputed the post with great bravery, defending the glacis for some time, but at last were driven from the work, and a lodgment effected. Tidcomb's had Lieutenant Williams, of the grenadier company, killed, and Captain Devaux wounded with the working party.

The regiment was again on duty in the trenches on 19th July and 24th July. On the following day the town surrendered, the garrison, now reduced to seven thousand men, returning into the Castle, which was forthwith besieged.

After the surrender of the town of Namur, Tidcomb's regiment quitted the lines of circumvallation, and rejoined the covering army under the Prince of Vaudemont, which, on 8th August, encamped near the village of Waterloo, and afterwards took up a position near Namur. A numerous army of French, under Marshal Villeroy, was attempting to raise the siege of the Castle of Namur. The opposing armies confronted each other on 16th, 17th and 18th August. The 19th was expected to be the decisive day, and King William's forces were under arms before dawn; but as the day broke it was seen that Villeroy had fallen back several miles. William at once sent to summon the Castle for the last time; but without success. The assault was ordered for the morrow.

It was made in four columns, each composed of troops of a different nationality—Brandenburghers, Dutch, Bavarian, and English. The

English column, under Lord Cutts, was to assault the breach in the Terra Nova cavalier. Four sections of grenadiers, each consisting of a sergeant and fifteen men, formed the forlorn hope; then followed a column of detachments of grenadiers from the various regiments of the army—in all, seven hundred men, under Colonel Evans of the 1st Guards, the grenadiers of the Guards leading; next came Courthope's regiment, with drums beating and colours flying;* other battalions which had not been in action before were to follow in reserve. At a concerted signal, the springing of a couple of powder-barrels, the troops moved out of the trenches with great spirit and impetuosity, to find themselves, a few minutes later, in disorder and out of breath, with a precipice in front of them, under a terrible fire, and a shower, scarcely less terrible, of fragments of rock and wall. Every officer of rank was killed or wounded. Cutts himself was shot in the head.† The men were falling back in confusion, until Cutts, whose wound had been hastily dressed, succeeded in rallying them, and led them, not to the place whence they had been driven back, but to the aid of the Bavarian column, which, after a gallant but unsuccessful onset, was beginning to waver. The appearance of the "Salamander," as Cutts was called, and his men changed the fate of the day. Two hundred English grenadiers, bent on retrieving, above all, the disgrace of the recent repulse, were the first to force a way, sword in hand, through the palisades, to storm a battery that had made grievous havoc among the Bavarians, and to turn the guns on the garrison.‡ Meanwhile, the Brandenburghers and Dutch had been equally successful; and, when evening closed, the Allies had a lodgment a mile long in the outworks of the castle. The advantage had been purchased by the loss of two thousand men. A detachment from the grenadier company of Tidcomb's regiment, under the command of Lieutenant Sewell, formed part of Lord Cutt's column, and had several men killed and wounded, and Lieutenant Sewell wounded.

Overtures for the surrender of the castle followed, and on 1st

* D'Auvergne, *History of the Campaign in Flanders*, vol. ii.

† Macaulay, *History of England*, vol. iv., p. 596.

‡ See Macaulay's *History of England*, vol. iv.

September, 1695, the famous fortress of Namur, which the French had boasted might be given up, but could never be taken, fell to the Allies. With the fall of Namur—an achievement that immensely enhanced the prestige of the English troops—the campaign closed for the year. Tidcomb's regiment shortly afterwards went into winter quarters in the villages near the Bruges canal.

1696-98.

Early in 1696 the invasion alarms caused by the assembling of large bodies of French troops at Calais and Dunkirk, and the plot against the life of King William, caused the Guards and other regiments to be summoned home in haste. Among the latter was Tidcomb's (14th Foot), which landed at Gravesend on 22nd March, and proceeded to Canterbury and Faversham, whence it moved, in November, to London, and took the duty at the Tower.*

---

* The following advertisements of deserters, inserted in the *London Gazette* (as was then customary) during the stay of the regiment in the Tower may be noted as offering some idea of the sizes, ages, etc. of the men composing the regiment. The uniform is not described, as in many advertisements of the same class and period :

London Gazette, 24th December, 1696.—Deserted out of Colonel John Tidcomb's regiment, Thomas Clarke, a well-set, middle-sized man, with lank brown hair, aged twenty-two, born two miles from Bradford in Somersetshire ; Will Richards, a short well-set man, in his own hair, aged 24, born near Chichester ; Gabriel Hall, a tall man, with fresh complexion, in a white bob wig, born in Leicestershire ; Saml. Hall, a tall pale-faced man, with lank hair, came from Burton-on-Trent, in Staffordshire. If all, or any, of them will return to their colours at Ratcliff, they will be kindly received and forgiven. Whoever secures them, or any, and gives notice to Mr. John Moyer, of Freeman's Court, near the Royal Exchange, shall have two guineas each and their charges.

London Gazette, 25th January, 1697.—Deserted out of Captain Thos. Morris's company, in Colonel John Tidcomb's regiment, Wm. Woodward, of middle stature, lank brown hair, very much disfigured with small-pox, a glover by trade, and born in Worcester ; Wm. Godwin, a black man, full eyed, aged about twenty-one ; Thos. Powell, a tall, lusty, scabby-faced man, a shoemaker by trade, and born at Worcester ; John Starkey, a fair man, full of the small-pox, born at Enfield, writes a good hand, aged twenty-one ; John Good, a tall, lusty man, with lank black hair, by trade a tailor. Whoever secures all, or any, of them, and gives notice to the captain, at his quarters at Bow or at the Tower of London, shall receive a guinea reward each, or if all, or any, of them return within ten days they shall be pardoned.

London Gazette, 15th July, 1697.—Deserted out of Major Pope's company, in

The strength of the regiment at this period was thirteen companies, including one of grenadiers, with a total of forty-four officers, one hundred and four non-commissioned officers, seven hundred and eighty drummers and privates, and sixty-nine servants. The cost on the estimates was £16,145 3s. 4d. for the year.†

On 9th March, 1697, negotiations were opened at the village of Ryswick, in Holland, to terminate the war which France had waged since 1688 against Holland, Germany, Spain, and England. A treaty, since known as the Treaty of Ryswick, whereby the French king agreed to restore some of his conquests, and to recognise William III. as King of England, was signed on 10th September (old style) the same year. Sweeping reductions of the army were at once insisted on by the House of Commons, and some regiments, to be retained in the service, were transferred to Ireland to escape reduction.

Tidcomb's (14th Foot) landed at Cork and Belfast in March, 1698, and was reduced to forty men per company. It remained in Ireland for the next seventeen years, during which, owing to the imperfection of the Irish Military Records, the information regarding its movements is scanty in the extreme.

1701-2.

Luttrell speaks of the regiment as at Limerick, and under orders for the West Indies, in the summer of 1701. On 8th February, 1702, King William died, and the accession of Queen Anne was

---

the Honble. Colonel John Tidcomb's regiment of foot, Wm. Hartley, a pretty, well-set man, with dark brown hair and pale face, aged about thirty. Also out of Captain Norris's company, in the same regiment, Thomas Leake, a tall, slender man, with brown hair, a shoemaker by trade, born at Taurendean. Also, out of Captain Oldy's company, in the same regiment, Wm. Currey, a short man, with short brown hair, aged twenty. Whoever secures them, or either, and gives notice to Mr. John Moyer, Freeman's Yard, Cornhill, shall receive three guineas for the first, and one guinea each for the others.

† See *Treasury Papers*, vol. lxiii., p. 22.—The total strength of the English army, including Scotch, Irish, and foreign regiments in English pay, was eighty-seven thousand four hundred and forty of all ranks, of which sixty-nine thousand three hundred and fourteen were foot. Officers' servants were at this time supernumeraries and non-combatants, mustered on the rolls for pay. When Sterne, who was born and brought up in a marching regiment at this period, makes Captain Toby Shandy take Corporal Trim as his servant, when disabled at the battle of Landen, he probably describes a common practice of the day.

speedily followed by the Declaration of War by England, Holland, and Austria against France and Spain, commencing what is known as the "War of the Spanish Succession."

1703-5.

In the autumn of 1703 the regiment furnished a draft of fifty men for Lord Mountjoy's, and another draft of like strength for Colonel Brudenell's, regiments (afterwards disbanded), on their embarkation to accompany the Archduke Charles of Austria, the claimant of the Spanish throne, to Portugal, whither he was to be conveyed by the English fleet. The regiment appears to have been in garrison in Dublin from 7th August to 31st December, 1703.

In the autumn of 1704, and again in the spring of 1705, additional drafts were sent by the regiment to the army employed in Portugal under the Earl of Galway. They were conducted thither by Captain Laffit, Ensigns Shackford and Blount, and three sergeants, whose expenses, amounting to £70 19s. 4½d., are directed to be paid by a warrant, dated 5th July, 1705. In August the same year, the regiment furnished one captain, one lieutenant, one ensign, two sergeants, and fifty rank and file towards completing the regiments of Charlemont, George, and Caulfield (afterwards disbanded) on their embarkation with the army under the famous Earl of Peterborough, which captured Barcelona, and achieved astonishing successes in Catalonia and Valencia.

1706-8.

The regiment was quartered in Dublin from March to November, 1706, and the soldiers received the additional penny a day granted by King William III. in 1699 to all regiments employed on Dublin duty. As Tidcomb's (14th Foot) had performed the duty of two regiments for some time, the allowance was extended to all detachments in consideration of the good conduct of the corps.

In 1707 the Act of Union with Scotland was passed, and the Union flag, as arranged down to 1801, became the principal colour of all British regiments of foot.

Although stationed in Ireland, the regiment was recruited in England, the enlistment of Irish recruits being forbidden at that time and long afterwards. By an Act of Parliament passed in 1708 the English counties were grouped in recruiting districts, to facilitate recruiting; but the arrangement was not followed out.

1709-14.

An Act of Parliament passed in 1710 authorized justices of the peace to impress vagrants and other persons who in their judgment had no recognized means of subsistence, for service in the land forces. Men so impressed were not to be under five feet five inches in height, and had to serve for five years; they were not sworn-in like voluntary recruits.

On 11th April, 1713, the Treaty of Utrecht was signed, ending the War of the Spanish Succession, with its many victories and colonial accessions.

On 13th June, Lieutenant-General Tidcomb, described by Dean Swift as "a wit, as well as a brave officer," died at Bath, and the colonelcy of his regiment was bestowed on Colonel Jasper Clayton, from the half-pay of a newly-raised corps, which had then lately been disbanded.

On the death of Queen Anne, 1st August, 1714, the Elector of Hanover succeeded to the English throne as King George I.

1715-16.

The Accession of the House of Brunswick was speedily followed by renewed activity on the part of the adherents of the Stuart cause. In the summer of 1715 these efforts began to assume ominous shapes. Reinforcements were brought over from Ireland, and among them Clayton's regiment, as it was now called, which landed at Saltcoats, Ayrshire, early in the summer.

In October, 1715, the Earl of Mar raised the standard of the Pretender in Braemar, and summoned the clans to arms. The royal forces, under Major-General Wightman, were encamped near Stirling, where Clayton's regiment joined them about this time, and the Duke of Argyll assumed command; but the royal forces only amounted to four thousand men, whilst Mar was reported to have a following of ten thousand.

When the rebels moved towards the Firth, the royal troops advanced to Dumblain, where they encamped on 12th November. On 15th November, the hostile forces confronted each other at day-break on Sheriffmuir. Argyll placed himself on the right, at the head of the cavalry; Major-General Wightman commanded the centre; General Whitham, the left. The clans under the command of Clanronald, Glengarry, Sir John Maclean, and Campbell of Glenlyon, led by Mar, made such a furious charge on the left wing of the royal army just as it was taking up a new alignment, that, says an account published shortly afterwards at Perth under the authority of the Earl of Mar, "in seven or eight minutes one could neither perceive the "form of a battalion or squadron of the army before us." But the highlanders were not so successful against the English right. Argyll charged them vigorously with his cavalry and pushed them across the Water of Allan, after which he returned to the field, and, being joined by three battalions under Major-General Wightman, took possession of some mud walls and enclosures to cover himself from a threatened attack of the enemy's right wing, which came up to the support of their left. But, through jealousy or some other cause, the highlanders did not renew the attack. The armies held these positions until the evening, when Argyll retired to Dumblain, and Mar to Ardoch. The loss on both sides was about equal; about eight hundred of the rebels were killed and wounded, while the loss of the royal troops was over six hundred. Both sides claimed the victory; but the advantage rested with Argyll, who returned to the field next day and carried off the wounded to Stirling, and by this action checked and delayed the rebel progress towards the south. Clayton's regiment lost one lieu-tenant and six rank and file, killed; fourteen rank and file, wounded; and Captain Barlow, Lieutenant Griffin, and several men taken prisoners.

The landing of the Pretender at Peterhead six weeks afterwards, for a while infused new life into his adherents. Reinforcements joined the royal troops, which were at Stirling, under the Duke of Argyll, and Clayton's (14th Foot) was brigaded with the Buffs, Scots Fusiliers (21st Foot), and Egerton's (36th Foot), under Brigadier-General Morrison. On 29th January, 1716, the royal troops—now mustering ten thousand men, inclusive of militia—marched from Perth, where they arrived three days later in a very exhausted state, having bivouacked, without shelter, in the snow on their way. Timorous counsels, however, prevailed in the Pretender's camp, and--instead of attacking Argyll at once—the rebels fell back towards Montrose, whither Argyll followed them, by way of Errol and Dundee, with three battalions of foot and some dragoons and militia. On 4th February the Pretender and other leaders of the rebellion escaped from Montrose to France, and the highlanders, deserted by their commanders, dispersed. Clayton's regiment, which had followed the insurgents with Argyll, was then quartered for a while at Dunkeld.

1717-21.

After the rebellion, the regiment was stationed at Fort William, which, since the days of William III., has stood beside Loch Eil, a navigable arm of the sea near the confluence of the Locky and Nevis, in Invernesshire. There it was stationed in 1717. In 1718 it marched to Perth, and, later, to Inverness, where it remained until June, 1719.

After the Earl of Mar's rebellion was put down, Charles XII. of Sweden made some preparations for a descent in favour of the Stuarts, and when that project failed, the King of Spain fitted out an expedition to place the Pretender on the British throne. The Spanish fleet was dispersed by a storm, but two ships reached the coast of Scotland in April, 1719, and four hundred Spaniards, with about one hundred Scottish and English gentlemen, landed at Kintail, on the mainland within Skye, and encamped opposite the castle of Donan, where they were joined by about fifteen hundred of the clans. To oppose this force, Major-General Wightman marched from Inverness, on 5th June, with three troops of the Scots Greys and the regiments of Montagu (11th Foot), Clayton (14th Foot), and Harrison (15th Foot). About 4 p.m. on the 10th June they arrived at Glenshiel, when the Spaniards and Highlanders fell back and took up a position amidst the romantic scenery of the Pass of the Streichell, where, an hour later, they were attacked by the King's troops, the infantry climbing from crag to crag, and the dragoons moving along the mountain road, to form the pass. The enemy made a brief stand on the summit, but a bayonet charge sent them flying. The troops passed the night in the hills. Next day the Spaniards laid down their arms, and the Highlanders dispersed. The Marquis of Tullibardine and other Jacobite leaders fled to France.

After this service the regiment marched to the Castle of Bran, near Kainloch-Benchven, Invernesshire, and in 1721 moved to Edinburgh.

1722-27.

In May, 1722, the regiment left Scotland and marched to Hungerford. In the summer of that year it was encamped, with other corps, on Salisbury Plain, where it was reviewed by King George I. on 30th August, and afterwards returned to Hungerford.

Early in 1723 the regiment marched to Reading and Windsor. Subsequently it was encamped in Hyde Park, and in the autumn moved to Bristol. In May, 1725, the regiment commenced its march from Bristol to Berwick-on-Tweed. In June, 1726, it moved from Berwick into Lancashire; and in January, 1727, marched to Canterbury, detaching four companies to Dover, Ashford, Sandwich, and Faversham. Major-General Jasper Clayton, who was lieutenant-governor of Gibraltar, proceeded thither and assumed the command of that garrison.

Defence of Gibraltar.

Gibraltar—captured by Rooke in 1704, and ceded to Great Britain at the Peace of Utrecht—was at this time menaced by the Spaniards, who had lately entered into a close confedracy with the Empire against Great Britain and France.

On 13th February, 1727, a Spanish army of about twenty thousand men, encamped at San Roque, the commander, the Conde de Las Torres, boastfully declaring that he was ready to take possession of the fortress for His Catholic Majesty whenever he should be instructed so to do. On 23rd February the Spaniards broke ground against the fortress, and the siege—the second of the three sieges the Rock has undergone since it fell into the British hands—began in earnest. By 10th March the Spanish batteries were within one hundred paces of the defences. The attack during this siege was exclusively directed against the north front and defences, from the extremity of the Old Mole to Willis's Battery. A heavy fire was kept up from the defences; but the ordnance was old and worn out, and more casualties, it is said, was caused by the bursting of guns than by the fire of the enemy, heavy as the latter was. The siege was still continuing, when, on 1st May, the *Prince Frederick*, seventy guns, with convoy, arrived, having left Portsmouth early in March. She brought the veteran Earl of Portmore, the governor, who had chivalrously declined to plead his age and infirmities as an excuse for absence from his post, a battalion of the 1st Guards (grenadiers), under command of Colonel Guise, and Clayton's regiment (14th Foot), by which the strength of the garrison, exclusive of officers, was raised to five thousand four hundred and eighty-one men, composed as follows :—

Gunners	20
Guards	672
Pearce's (5th Foot)	432
Lord Mark Kerr's (13th Foot)	434
Clayton's (14th Foot)	640
Egerton's (20th Foot)	415
Middleton's (25th Foot)	394
Anstruther's (26th Cameronians)	396
Disney's (29th Foot)	358
Bisset's (30th Foot) and Haye's (34th Foot)	388
Newton's (39th Foot)	293
Detachments from Garrison of Minorca	480

Sayer, the historian of the Rock,* thus describes the events that followed :—

"During the ensuing week, the enemy having completed four "gigantic batteries, armed with the finest brass artillery, opened a "terrific fire all along the line. So magnificently grand was this "bombardment, previously unequalled in the history of artillery, that "for some time," says an eye-witness, "we seemed to live in flames.

* Sayer's *History of Gibraltar*, p. 202, *et seq.*

"Attempts, feeble by comparison with the relentless storm of shot and
"shell that tore over the walls, were made to check the murderous fire,
"but in vain; guns were everywhere dismounted, and as quickly as
"they were replaced were again destroyed.

"In vain the men, with undaunted courage, threw themselves upon
"the ramparts, and worked to repair the shattered parapets, the heavy
"shot tore away whole tons of earth and buried the guns beneath the
"ruins. Butts filled with sand and bound with fascines were heaped
"together as some covering for the artillery, but they were no sooner
"in position than they were swept away. For fourteen days seven
"hundred shots per hour were thrown into the fortress, and ninety-two
"guns and seventy-two mortars were in constant play.

"To this formidable armament the garrison could only oppose sixty
"guns, viz., twenty-one on the Grand Battery, twenty-three on the Old
"Mole, nine at Willis's, and five in the battery on the side of the old
"Moorish Castle. There were one hundred and thirty-five mortars and
"cohorns in the place, but only a portion of these could be brought to
"bear on the enemy's lines.

"By the 20th (May) the enemy's artillery began to experience the
"effects of the heavy firing. The brass guns drooped at the muzzles,
"and the iron ordnance in many instances burst. The fire then
"slackened rapidly, until at length only nineteen guns were in play.
"Taking advantage of this respite, the Governor employed the garrison
"day and night, restoring the shattered defences, clearing the ditch
"at Landport, filling in the breaches and repairing the embrasures.
"So efficiently and rapidly were these works executed that in a few
"days thirteen new guns were ready to open, and one hundred
"mortars were brought to bear upon the most formidable of the
"enemy's batteries. While the English had thus successfully met the
"crushing bombardment which was to compel a surrender, and were
"preparing for a final struggle, the Spaniards had shot their last ball,
"and found themselves in a position of the greatest difficulty and
"danger. Their artillery and ammunition had been recklessly
"exhausted. The roads of the interior were impassable, and no
"reinforcements could reach the camp. Provisions were becoming
"scarce; the weather was inclement; and they were threatened with

"a destructive cannonade from the fortress. Impressed with the "hopelessness of continuing the siege, the Spanish generals united in "a representation to the Conde de Las Torres that, unless His Catholic "Majesty would reinforce the army to the number of twenty-five "thousand men, Gibraltar could not be taken.

"Observing the embarrassment of the Spaniards, the Governor "harassed them with a fire that almost equalled the terrible bombard-"ment of the enemy. A supply of ammunition arrived opportunely "from England, and by the end of May the guns of the garrison had "gained a complete ascendancy over the besiegers.

"By the beginning of June, the Earl of Portmore had perfected his "arrangements for the final bombardment of the besiegers' lines. "One hundred guns were in position, and countless mortars occupied "commanding positions on the heights. On 3rd June this mass of "ordnance opened upon the Spanish batteries, and so crushing was the "fire that not a single gun replied. Early in the day the trenches "were a heap of ruins, the parapets were in flames, and the magazines "blown up. . . . . Although the first day's fire drove the enemy "from the forts, the bombardment was still unremittingly kept up "until the whole line of batteries were completely destroyed. "Numerous deserters found their way into the fortress, and gave "lamentable accounts of the sufferings of the (Spanish) troops. Sick-"ness was carrying off thousands, and each day increased the horrors "of want.

"Such was the state of the besiegers, when, on 23rd June, a courier "arrived in the camp at San Roque, bringing despatches for the "Conde de Las Torres and a letter for the Earl of Portmore, which "latter was conveyed into the garrison by Colonel Lacy, of the Irish "Brigade. At 10 o'clock at night the colonel left the advanced "trenches, and on being challenged, replied that he had despatches "for the governor of the fortress, but the officer of the guard refused "to admit him, and threatened to fire on him if he did not instantly "return. Shortly afterwards, he again appeared at the head of the "trench, beating a drum, blowing a trumpet, and making signals of "truce. He was then admitted into the town, when he announced "the important intelligence that a suspension of hostilities had been "agreed to, and the preliminary articles of peace signed by the "plenipotentiaries of the several powers."

Thus ended the siege, which cost the garrison sixty-nine killed, two hundred and seven wounded. Died of wounds and sickness, forty-nine. Total, three hundred and sixty-one. Deserted, seventeen.

Clayton's Regiment had six killed, twelve wounded, one death from wounds and three from sickness. There were no desertions from the regiment.

The Spanish casualties were three hundred and sixty-one killed, eleven hundred and sixty-one wounded, eight hundred and seventy-five deserted. Five thousand died of sickness, or were invalided.*

Whilst the last days of the seige were passing, George I. died in Hanover, on 11th June, 1727, and George II. mounted the British throne.

Annexed is a roll of the officers of Clayton's regiment (14th Foot), whose commissions were renewed at this time, under the new sign manual.

---

* The *Journal of an Officer* (1727), of which there is a copy in the British Museum, relates some anecdotes of the siege. The following is shewn as the "duty" furnished daily during the siege:

	F.O.	Capts.	Subs.	Sergts.	Corpls.	Drmrs.	Men
Prince's Line Guard	1	2	2	7	6	2	100
King's Line Guard	...	1	2	4	4	2	100
Governor's Line Guard	...	...	...	1	1	...	15
Rock Guard	...	...	1	1	2	1	24
Middle Hill Guard	...	...	1	1	2	1	20
Willis' Battery Guard	...	1	1	2	2	1	42
Signal Ho. Guard	...	...	1	1	1	1	20
South Port Guard	...	...	1	1	1	...	15
Main Guard	...	1	1	3	2	2	50
Waterport Guard	...	1	2	3	2	2	50
Landport Guard	...	2	2	3	3	2	100
The Shed (at night only)	...	...	1	1	1	1	20
Europa Advance	...	...	1	1	2	1	36
Quarterly Guard	...	...	...	...	...	...	90
Orderly Sergeants	...	...	...	4	...	...	...
Castle Picquet for a Reserve	...	...	...	...	...	...	150
Picquet, Night	...	...	...	...	...	...	150

Total - - - - - - - 1191, all ranks.

ROLL OF OFFICERS OF THE REGIMENT

whose commissions were renewed under the new Royal Sign Manual at the accession of King George II.

Home Office: *Military Entry Book*, vol. xiii., 20th June, 1727.

Field Officers and Captains:

JASPER CLAYTON, *Col. and Capt.**	CHAS. STRAHAN, *Capt.*
EARL OF DUMBARTON, *Lt.-Col. and Capt.*	GEO. HEIGHINGTON, ,,
JOHN LAFOREY, *Major and Capt.*	THOS. TEMPEST, ,,
PEREGRINE THOS. HOPSON, *Capt.*†	JOHN GOUGH, ,,
EDM. WRIGHT, *Capt.*	ISAAC GIGNOUX, ,,
	GEO. MALCOLM, ,,

ROBT. MOORE, *Capt. (Grenadier Company).*

Lieutenants:

JAS. RAMSEY, *Capt.-Lt.*	AND. SIMPSON	NICHOLAS WEST
ROBERT MILLAR	JOHN SCRIVENER	WM. JAMESON
JOHN CASSEL	MATH. LOWE	JASPER CLAYTON
EZEKIEL JEFFERYS	MARK JARLAND	

JAMES HUNT, WM. JONES, } *Grenadier Company.*

Ensigns:

THOS. LYNN	JOHN BELL	DANL. COLE
ALEX. GROSERT	THOS. BAYLIES	PETER HAWKER
BASIL CARTY	JOS. MARSHALL	JOHN CAMPBELL
GEO. WARD	ED. BOOTH	

Chaplain - - JOHN LISTER.
Adjutant - - JOHN CASSELL.
Surgeon - - — LIND.
Quarter-Master - DAVID STEVENSON.

* Killed, as lieutenant-general, at the battle of Dettingen, 1743.
† Afterwards lieutenant-general; colonel, 47th Foot.

1727-41.

After the seige, Clayton's regiment remained in garrison on the Rock for fifteen years. Of this period no regimental details appear to be procurable. Down to 1739, when disputes recommenced with Spain, England, under the Walpole administration, remained free from foreign wars and complications, It is believed that small parties from the regiments in garrison at Gibraltar were at times shipped as acting marines on board cruisers employed on the Barbary coast (as in 1734) and on the coasts of Italy during the campaigns of the French and Spaniards against the Austrians; but, from the disbanding of the marine forces, at the peace of Utrecht, until the re-formation of marine regiments in 1740-41, the practice of carrying soldiers in ships of war was by no means general, and particulars of the detachments so employed are wanting.

1742-44.

Relieved from duty at Gibraltar, Clayton's regiment returned home, landing at Portsmouth in September, 1742, and after a few days rest in that garrison marched into quarters in Yorkshire, the regimental head-quarters being stationed at York. Thence the regiment moved to Berwick and Newcastle-on-Tyne.

The death of the Emperor Charles VI. at this period, was followed by the "War of the Austrian Succession," in which Great Britain supported the claims of Maria Theresa and the House of Austria against Charles VII., Elector of Bavaria, who had the support of the French. In the spring of 1743 a British force, under the Earl of Stair, was sent to Holland with this object. The British troops subsequently crossed the Rhine, and on 16th June, 1743, an army of British, Hanoverians, and Hessians, under the personal command of King George II., inflicted a signal defeat on the French, under Marshal Noailles, at Dettingen, in Bavaria. In this action, Lieutenant-General Clayton was killed *(see Appendix)*. The colonelcy of the regiment, which was still in the North of England, was then conferred on Colonel Joseph Price, from the 57th, afterwards 46th Foot, by commission dated 22nd June, 1743.

Early in 1744 a circular was issued by the Earl of Stair, commanding the forces in South Britain (England and Wales), to the commanding officers of regiments there, asking in which counties they would prefer to recruit. Most of them named the several counties in which their regiments were at the time. A General Order, dated 6th April, 1744,* followed, in which recruiting districts, determined—it would seem—by the answers to the circulars aforesaid, were assigned to various regiments, which were directed forthwith to send officers and non-commissioned officers into these districts to receive any able-bodied men as volunteers, and all such pressed men as should be turned over to them by the "commissioners appointed by the late "Act of Parliament for the more speedy and effectual recruiting of "His Majesty's land forces and marines." The act was, in point of fact, a renewal of the act authorizing the impressment of vagrants for the land service before referred to as being passed in 1710. Since the peace of Utrecht, voluntary enlistment for life, or until lawfully discharged, had been the rule.

The counties of Northumberland and Cumberland were assigned as the recruiting ground of Price's (14th Foot), and a subsequent General Order of 24th April, 1744, directed that "every volunteer, "and man impressed by the commissioners, if he be able-bodied and "fit for His Majesty's service, shall be received, even if he be under "five feet five inches. It was further enjoined that "the officer is "not to receive anyone who is not able-bodied, being too old or too "young, or having any manifest infirmity, or being too small in stature.

"The officer is to divide his recruits into three classes, according "to height: 1.—The tallest man. 2.—Men for the regiments serving "in Flanders. 3.—Men for other marching regiments."†

Later in 1744 the regiment moved from Berwick to Dunstable, and afterwards to Colchester.

1745-46.

Immediately on the receipt of the news of the loss of the battle of Fontenoy on 30th April, 1745, Price's regiment received orders to join the army under the Duke of Cumberland, in Flanders. It

* British Museum. Additional MSS., 20,005.—*Lord Stair's Order Book.*

† The regiments of horse, certain dragoon regiments, and the foot guards were to recruit at large. The counties adjoining the ports of Bristol, Holyhead, and Liverpool were to supply recruits to regiments in Ireland.

embarked at Tilbury on 15th May, 1745, landed in West Flanders, and joined the army, which was encamped on the Plain of Lessines, before the end of the month. The regiment took part in several movements: was encamped at Grammont, and afterwards on the Brussels canal, covering Dutch Brabant; but the superiority in numbers of the French enabled them to capture several fortified towns.

On 19th August, 1745, Prince Charles Stuart landed in Scotland, and the clans began to gather round him. The gravity of the situation appears not to have been fully realized at first; but at the end of August peremptory orders were sent to the Duke of Cumberland for the immediate despatch of a number of regiments from Flanders to reinforce the army at home.

Price's (14th Foot) was one of the regiments thus brought over. It formed part of the army assembled under Marshal Wade at Newcastle-on-Tyne, whence, in the second week of November, 1745, it was despatched to its old quarters at Berwick-on-Tweed. It arrived there in time to prevent the capture of the town by a part of the rebel forces on their march southwards, and, when the rebel army withdrew from Derby and recrossed the border, Price's regiment moved into Scotland and took up its quarters in the city of Edinburgh.

Prince Charles Stuart having laid siege to Stirling Castle, Lieutenant-General Hawley, who commanded the King's troops in Edinburgh, decided to attempt to raise the siege. For this purpose, Price's (14th Foot) and some other corps advanced from Edinburgh, on 13th January, 1746, under the command of Major-General John Huske. They drove a body of rebels out of Linlithgow. On the following day a division marched to Borrowstounness, and on 16th January, 1746, the troops encamped near Falkirk.

The Campaign in Scotland.

On 17th January, 1746, amidst squalls of sleet and rain, which drove in the faces of Hawley's men, and wetted their powder, was fought the battle of Falkirk Muir, of which so many discordant tales have been told. Chambers states* that "Some individuals who

---

* *History of the Rebellion of 1745-46*, revised edition.

"beheld the battle from the steeple of Falkirk used to describe its "main events as occupying a surprisingly brief space of time. They "first saw the English army enter the mist and storm-covered muir at "the top of the hill: then saw the dull atmosphere thickened by the "fast-rolling smoke, and heard the pealing sound of the discharge; "immediately after, they beheld the discomfited troops burst wildly "from the cloud in which they had been involved, and rush in far-"spread disorder over the face of the hill. From the commencement "to what they called 'the *break* of the battle' there did not intervene "more than ten minutes—so soon may an efficient body of men "become, by one transient emotion of cowardice, a feeble and "contemptible rabble. The rout would have been total, but for three "outflanking regiments.* These not having been opposed by any of "the clans, having a ravine in front, and deriving some support from "a small body of dragoons, stood their ground under the command "of General Huske and Brigadier Cholmondeley. When the High-"landers went past in pursuit, they received a volley from this portion "of the English army, which brought them to a pause, and caused "them to draw back to their former ground, their impression being "that some ambuscade was intended. This saved the English army "from destruction."

The following account of the battle is given in a private letter, by Brigadier Cholmondeley, to whose brigade Price's regiment belonged.† The letter is preserved at Somerby Hall, Lincolnshire, and is printed in the Historic MSS. Commission's Tenth Report (part 1, page 441.)

"I am sorry to renew our correspondence with so disagreeable a subject "as our scandalous affair at Falkirk, but as I am satisfied that you and "everyone else are desirous to be informed of the truth of that affair, I "send you a most exact account of what I know of it, that by Comparing "the different accounts you may attain to a true knowledge of the affair.

"On Thursday, General Huske and Brigadier Mordaunt marched to

---

* Believed to have been Barrel's (4th Foot), Price's (14th Foot), and Ligonier's (48th Foot.)

† The Honble. James Cholmondeley. He raised the 48th Foot, and commanded the 34th on the day of Fontenoy, when it won its silver laurel wreath. He lost the use of his limbs by exposure at Falkirk. He died a full general in 1775, aged sixty-seven.

"Falkirk. I marched with the 3 battallions under my command from "Borrowstounness to the same place, and we all got there about 1 o'clock. "We march'd through the Town and drew up and Incamped at a place "leaving Falkirk about 100 paces in rear of our left; our camp was very "strong, having in front a deep hollow morass ground, and upon our right "flank some enclosures with large wet ditches. We Incamped in two lines "facing the enemy, and as the ground would not allow of our extending "the Incampment any further, we had 2 regiments Incamped on the right "flank obliquely. The enemy lay behind Torwood, which was opposite "the right of our Incampment, about 2 or 3 miles from us. About 7 at "night our artillery, and Friday morning about 8, Colonel Campbell with "his Highlanders joined us, as did Cobham's Dragoons. The Highlanders "were advanced on the other side of the morassy ground that lay in our "Front, and lay there in some farm houses. About 11 o'clock the rebels "were in motion. I saw them very plain with a glass all the time they "marched to our (their?) right which was opposite the left of our Camp, "but as there were several hills between us, we could not tell exactly where "they took ground. Upon this orders were given for the men to be ready "to turn out at a minute's warning. About one, we had Information that "they were marching towards us, when the Army was immediately "ordered to stand to their arms and form in front of their Incampment, "all the Cavalry were ordered to the left to take post there and the two "lines of Infantry were ordered to face to the left and in this position we "marched them to the left near a mile and a half, but as we had hollow "roads and very uneven ground to pass over we were in great confusion. "Here we formed again, in my opinion a very good situation, but we were "no sooner formed than ordered again to take ground to the left, and as "we marched all the way uphill and over very uneven ground the men "were greatly blown.

"Our first line consisted of Ligonier's, Proyalls',* Poulteney's, Cholmon-"deley's, and Wolfe's; our second, Battereau's, Barrel's, Fleming's, "Munzo's, and Blakeney's; where Price's was formed I really cannot say.† "The Old Buffs was in reserve. Our Highlanders was left in the place "before mention'd, not only to secure our Camp, but to prevent their

---

* A misprint in the report for "Royals."

† In the only plan of the battle in the British Museum—a very well-executed one, by an officer of Battereau's—the composition of the lines, counting from the right, is shewn as follows: First line—Ligonier's (48th), Price's (14th), Royals, Pulteney's (13th), Cholmondeley's (34th), Wolf's (8th Kings), then the three regiments of cavalry. Second line—Battereau's (old 62nd, afterwards disbanded), Barrel's (4th King's Own), Fleming's (36th), Munzo's (37th), Blackeney's (27th Inniskilling). Reserve—The Buffs. It would seem probable that the similarity of the facings led Brigadier Cholmondeley to mistake the two battalions on the right of the first line for one and the same regiment. The best contemporary accounts speak of *three* battalions standing fast.

"marching a Body of Troops up and Falling on our right Flank. The "Glascow Regiment were drawn up on an Eminence in the rear of our "left Flank, and our whole Cavalry were formed upon our left; they began "the attack with spirit, which did not last long. Nevertheless they broke "a Considerable body of the Highlanders, but another Body coming upon "our left Flank our Foot gave a feint Fire and then faced to the right "about, as regularly as if they had had the word of command, and could "not be rallied until they got a considerable distance, although I don't "think they were pursued by 200 men. Barrel's regiment kept their "ground and I got my late regiment, Ligonier's, to form on their right; "Barrell's left was secured by a little Farm House. In this situation we "kept our ground, and with the assistance of the officers (who deserve the "greatest Praise for the Spirit they shewed) I got the men to be quite Cool, "as Cool as ever I saw men at Exercise, and when the Rebels were down "upon us we not only repuls'd them, but advanced and put them to flight. "During this time General Huske was rallying the other Troops that had "been broke; then I told these two battallions that if they would keep "their ground, I would go back and rally the Dragoons; they promised "they would, and kept their Word. Accordingly I went to the Dragoons "and rallied about 100 of them, and told them that I had repuls'd the "Enemy with two weak Battallions, and that if they wou'd march up I "wou'd head them, and that I wou'd order the two Battallions to march up "briskly, at the same time to give them their Fire and that they should "Fall in Sword in Hand. They were greatly pleas'd with this, and with "many Oaths and Irish exclamations swore they wou'd follow me. I "march'd them up to the two Battallions, but when we were to advance "they kept at least 100 yards behind me; with some difficulty I got them "to the top of the Hill, where I saw the Highlanders forming behind some "houses and barns (I was forced to fire a pistol among them to get them "to do this). I then returned to the two Battallions to march them up. "Here General Huske Joyn'd me, and I told him that if we cou'd get "some more Battallions to Joyn us we might drive the Enemy, but as night "was drawing on he ordered me to retire.

"My chief inducement for giving you this Minute Account is to do "justice to the officers of these two Battallions who behaved so well, that "their Stand Stopt the Rebells from pursuing our Troops which else wou'd "have been cut to pieces."

The loss in the battle was not published regimentally. The total was about two hundred of all ranks killed, wounded, and missing. The proportion of officers killed was very large, and a number of deaths were announced immediately afterwards, probably due to exposure. Among these was Lieutenant-Colonel Jeffreys, who succeeded to the lieutenant-colonelcy of Price's, from captain in the 3rd Troop of

Horse Guards, at the death of Colonel Robert Moore, in Flanders, some months before. He was replaced by Major John Grey,* of Onslow's (8th King's.)

Hawley's troops returned to Edinburgh, where the Duke of Cumberland arrived and assumed the command in chief. On 31st January, 1746, the royal forces again advanced, upon which Prince Charles raised the siege of Stirling Castle and retreated precipitately towards Inverness. Cumberland followed as far as Perth, where the severity of the weather caused a halt. The march was resumed on 20th February, but the heavy rains necessitated another halt at Aberdeen.

Early in April, the march was again resumed, and on 16th April (old style) 1746, when moving in three columns towards Inverness, the royal troops came in view of the rebel forces, about four thousand men, drawn up on Culloden Muir. Cumberland's troops deployed into three lines, Price's regiment (14th Foot) being in the centre of the first line, which was commanded by the Earl of Albemarle. The battle began with a brisk cannonade on both sides, during which the Duke of Cumberland, who was everywhere greeted with cheers and cries of "Flanders" by the soldiers, rode down the lines and made several changes in the disposition of the troops. The rebel artillery did little execution, but the royal guns, light three-pounders and six-pounders, placed in pairs in the battalion intervals of the first line, were well served, and wrought such havoc in the ranks of the clan regiments that Lord George Murray took upon himself to make an immediate attack on Cumberland's left wing. A very gallant onset was at once made with sword and targe by the centre and right of the clans, in which they broke through Barrell's and Monro's regiments in Cumberland's first line, and captured two guns, but were brought up by the bayonets of the second line. Taken in flank by the English horse, the clans were soon broken, and the attack turned into an utter rout of that portion of the rebel forces. Their left is said to have

* This officer, one of the Greys of Howick, entered the King's, as ensign, in 1709, when the regiment was with Marlborough in Flanders. He became lieutenant in 1712 and captain-lieutenant in 1727. As a captain, with thirty-four years' service, he brought the regiment out of action at Dettingen. He died a major-general, and colonel of the 54th Foot, in 1760.

retired towards the Nairn in good order, with the pipes playing and the Prince's standard displayed; but the Prince had been forced from the field by his attendants, and the day and the cause were irretrievably lost. After following the rebels some miles, Price's regiment encamped, with the rest of the infantry, near Inverness. The regiment went into action with twenty-three officers, twenty-one sergeants, eleven drummers, and three hundred and four rank and file, and had Captain Grosette and one man killed, and Captain Simpson and nine men wounded.*

In May the regiment pitched its tents in a valley near Fort Augustus, and was employed in guarding prisoners taken after Culloden. Later in the year it moved to Stirling.†

---

* The news of the victory was received with great satisfaction in London. The City voted a sum of £4,000 for distribution among the soldiers present in the battle, whereof a sum of £400 was set aside for extra gratuities to the wounded. The sums allotted to Price's regiment (14th Foot) were:

To the men of the regiment - - - - - - - - £203 2s. 6½d.
Extra to Regimental wounded men - - - - £10 18s. 1½d.

See Maclachlan's *Duke of Cumberland's Order Book*, pp. 335-36.

A large silver medal was struck and presented to the general officers and officers commanding regiments. It was of oval shape, and is stated to have been suspended by a crimson ribbon with green edges, worn round the neck.

† *Directions for observing the King's birthday, at Stirling, 29th October, 1746.* (Maclachlan's *Duke of Cumberland's Order Book*). "The Troops quartered at Stirling and St. Ninians to be formed on parade at 11 in the forenoon, from whence they are to march to the Castle and line the outer parapet right round the Castle, facing outwards. At noon 11 pieces of cannon are to be fired from the Castle, after which the Foot on the Parapet are to make a running Fire with their small-arms right round the Castle, beginning with Brigadier Price's Regiment on the right of the Great Gate and ending with Colonel Conway's posted on the left of the said Gate. As soon as the men have loaded their arms, the same number of cannon are to be fired from the Castle, then the running fire of small arms as before mentioned. That to be repeated three times by the cannon and small arms, after which they are to give three Huzas, then to march out of the Castle, back to the Parade and thence to their Quarters. The officers to be drest in their regimentals and the men to appear clean. The several guards to be reinforced at 4 o'clock, that they may be enabled to send out constant patroles to prevent the soldiers breaking of windows or committing any other disorders. As soon as the retreat has beat, the Soldiers who are not on duty to repair to their Quarters and not stir out again that night, since those who do will be taken up by the patrols and severely punished next day for disobeying orders.

"As the Magistrates intend to drink His Majesty's and the Royal Family's healths at ye Bonfire near the Cross, in the evening, 1 Captain, 2 subalterns, and

1747-50.

In June, 1747, the regiment, which had changed its quarters from Stirling to Glasgow previously, marched from Glasgow to Perth, and in September to Inverness.

The colonel, Brigadier-General Price, was in command of a brigade in Flanders, and distinguished himself at the battle of Val, or Laffeldt, 2nd July, 1747. He died at Breda in November the same year, when the colonelcy was conferred on the Honble. William Herbert, from captain and lieutenant-colonel Coldstream Guards.

The regiment remained in Scotland, and was stationed, in 1749, at Fort William, and in 1750 at Glasgow again, whence it moved to Carlisle and Newcastle-on-Tyne.

1751-52.

On 1st July, 1751, a Royal Warrant regulating the colours, standards, and clothing of the army was issued in amplification of previous Warrants of 1743 and 1747. The Warrant directed that the first, or King's Colour, of all line regiments should be the great Union as then marshalled (the crosses of St. George and St. Andrew); that the second, or Regimental Colour, be of the colour of the regimental facings, with the Union in the upper corner next the flagstaff; and that the regimental number, in Roman numbers, be placed in the centre of each within a Union wreath of roses and thistles on one stalk.

The practice of designating regiments by their NUMBERS, instead of

---

60 grenadiers are to be drawn up under arms near the Cross, and at the beating of the Retreat, in order to fire Vollies on the drinking of the said healths, after which the grenadiers are to be dismissed to their quarters and remain there.

"When the grenadiers are dismissed, sergeants' patroles are to be sent out in the Town to take up all soldiers they find out of their quarters, or making a noise in public-houses, and carry them to the guard, who shall be severely punished next day for disobeying orders. As soon as the patroles return, other Soldiers to be sent out immediately charged with these orders and continue doing so during the night. The 3 companies at Alloa to fire three times at noon, and to take signal for the firing from the cannon at Stirling Castle, and to observe the same orders of retiring to their quarters at the beating of the retreat, and against breaking windows and other disorders. A proper number of cartridges for the said Firings to be made up in time without ball, at which the officers commanding companies to make a strict scrutiny also in the men's arms to prevent their loading with ball, and to avoid accidents which might occur if not looked into. The sixth cannon to be the signal for the men to Recover and to remain with Recovered arms until the firing of the eleventh."

by their COLONELS' NAMES, now became customary, although the latter mode continued in popular use for some years longer.

The uniform of the Fourteenth Foot at this period was red coats faced with buff and buff cuffs, red waistcoats and breeches, white gaiters and cravats, buff leather belts and pouches, and three cornered black hats laced with white. The grenadiers wore mitre-shaped caps, such as Hogarth drew; and the drummers, buff coats faced with red. Cannon, in the *Historical Record of the 14th Regiment*, published in 1845, gives the facings at the period above referred to as yellow, but this appears to be a misprint, as there is no evidence that yellow facings were ever worn by the regiment.

In August, 1751, the regiment was ordered into the South of England, to furnish detachments on the Sussex coast, to aid in the suppression of smuggling, or, as the *Marching Orders* of the time phrase it, "there to be aiding and abetting the civil magistrates in "preventing owlers, smugglers, and others from running goods." Very energetic measures were at this time being enforced to put down this evil, which had attained to intolerable dimensions. After some months of this employment, the regiment was directed to call in its detachments in April, 1752, and march to Portsmouth, to embark for Gibraltar. At Gibraltar it spent the next seven years.

By the Act 24, George II., c. 23, of this year the "new style" of reckoning was adopted in the calendar. The day after Wednesday, 2nd September, was called Thursday, 14th September, 1752. Thenceforward "new style" was used in all official documents in England, Ireland, and the British Dependencies, as it had previously been for many years (when not otherwise specified) in Scotland and on the Continent. The practice of dating the year from the 25th March, the first three months being considered common to both years and written, *e.g.*, 1st January—24th March, 1749-50, or $17\frac{49}{50}$, had been discontinued in 1750, but was retained in War Office documents for some years, as in the extract on p. 31.

1753-55.

On 17th February, 1753, the Honble. Wm. Herbert was removed to the 2nd Dragoon Guards, and the 14th Regiment was given to Colonel Edward Braddock, from the Coldstream Guards.

In 1754 the publication of the annual Army List ("by authority") commenced. Annexed is the roll of officers of the 14th Foot, as it appears in the first issue.

ROLL OF OFFICERS OF THE REGIMENT, 1754.

(From a copy of the Annual Army List for 1754 in the British Museum.)*

Colonel	†Edward Baddock	17th February, 1753
Lieutenant-Colonel	John Grey	17th February, 1745-46
Major	Mark Renton	2nd March, 1750-51
Captain	John Bell	23rd March, 1742-43
,,	Edward Booth	1st August, 1744
,,	John Meard	1st August, 1744
,,	Richard Russel	22nd June, 1745
,,	Barth. Corneille	23rd April, 1746
,,	Thomas Bowyer	23rd April, 1746
,,	Jonathan Furlong	2nd March, 1750-51
Captain-Lieutenant	George James Bruce	16th March, 1752
Lieutenant	James Montresore	23rd July, 1737
,,	Thomas Baylies	20th July, 1744
,,	Francis Lind	22nd June, 1745
,,	John Hughington	1st December, 1745
,,	Charles Grame	23rd April, 1746
,,	John Deaken	2nd March, 1750-51
,,	Leathes Johnston	31st October, 1751
,,	Rob. Richard Maitland	16th March, 1752
,,	Richard Carr	17th March, 1752
,,	Robert Price	18th June, 1753
Ensign	Thomas Brisbane	23rd April, 1746
,,	Connolly Ball	11th December, 1746
,,	James Vipont	10th March, 1752
,,	Herbert Whitfield	20th March, 1752
,,	John Geathen	21st March, 1752
,,	— Stuart	29th April, 1752
,,	Hugh Antrobus	27th March, 1753
,,	Thomas Shaftoe	18th June, 1753
,,	Thomas Seykes	22nd January, 1754
Chaplain	James Dowsing	11th May, 1748
Adjutant	John Deacon	31st August, 1751
Quarter-Master	James Brucre	2nd March, 1750-51
Surgeon	Francis Lind	3rd June, 1726

---

* The publication of the *Annual Army List* (by authority) commenced in this year. The issue of the *Monthly Army List* (by authority) did not commence until sixty years later.

† An *Army List* misprint for Braddock.

On 29th March, 1754, Colonel Braddock was promoted to major-general, and soon after was sent to America, in consequence of alleged French encroachments there, as "general and commander-"in-chief of all Our Troops and Forces y^t are in America or "y^t shall be raised or sent there to vindicate our just rights and "possessions."* He fell in the disastrous attempt on Fort Duquesne (now Pittsburg, U.S.) on 9th July, 1755, and was succeeded in the colonelcy of the 14th by the veteran governor of Gibraltar, Lieutenant-General Fowke, from the 2nd, or Queen's.

1756-58.

In 1756, war between Great Britain and France commenced with an attack by the French on the island of Minorca, which had been ceded to Great Britain at the Peace of Utrecht, in 1713. General Fowke received orders to send a reinforcement to Minorca; but, instead, called a council of war, which passed a resolution against sending it. For this General Fowke was tried by court-martial, and sentenced to be suspended for nine months. The King, however, dismissed him from the army. The colonelcy of the 14th was then given to Colonel Charles Jefferies, from the 60th Royal Americans.

1759-63.

No other incident of importance is recorded in connexion with the stay of the regiment at Gibraltar. Towards the end of the year 1759 it was relieved from duty there and returned home. A Monthly Return—the earliest return of the regiment in the Public Record Office, London—is dated, possibly ante-dated, 1st December, 1759, and shows the regiment at Plymouth Dock, Lieutenant-Colonel Thomas Lister commanding, with twenty-seven officers, (exclusive of Major-General Jefferies) five staff, thirty-six sergeants, twenty drummers, and six hundred and twenty-two rank and file, including thirty-eight sick, and with two hundred and seventy-eight rank and file wanting to complete. Captain Baylies and Lieutenant Burrell are shewn recruiting; Captain Patoun, engineer at Gibraltar from August, 1759,

---

* See *King's Warrants and Orders in Council*, 1754.

and Lieutenant James Montresor,* engineer in America from 1753. The earliest muster roll,† or—rather—set of company muster rolls, of the regiment, is dated 24th June, 1760, when the regiment, having moved from Plymouth to Canterbury, was encamped on Barham Down,‡ near that city, with the 19th and 21st Foot, under Lieutenant-General Campbell. The camp was broken up in October that year, and the 14th Regiment marched into quarters in Dover Castle.

Recruiting in Ireland was first legalized at this period. Permission to recruit in Ulster was given in 1757; and, on 31st March, 1759, was extended to the whole of Ireland; but the Departmental correspondence in the Irish State Paper Office shows that it was expressly enjoined, on each occasion, that Protestants only should be enlisted. The rule requiring a recruit to declare himself a Protestant did not disappear from the recruiting regulations until the last year of the century. There is no allusion to any medical examination or test of fitness of recruits, beyond the specification of rupture as a disqualification.

On 25th October, 1760, King George II. died, and was succeeded by his grandson, George III.

The regiment was in Dover Castle in 1761, and in 1762 moved to Maidstone, and in October was guarding French prisoners of war at Sissinghurst. In December it marched to Exeter, and thence, in March, 1763, to Plymouth.

In 1763 the Seven Years War came to an end. Preliminaries of peace between Great Britain and France had been arranged at Paris in the previous autumn.

---

* This officer appears as brevet lieutenant-colonel (as well as lieutenant, 14th Foot), in the *Army List* for 1758. He was chief engineer at Gibraltar, 1748-53, and afterwards chief engineer in America. He died in 1766. Some writers have confused him with his son, Captain John Montresor, who distinguished himself at the defence of Quebec in 1760, and was chief-engineer at Bunker's Hill in 1775.

† All Regimental Muster Rolls, Monthly and Inspection Returns and the like, from the commencement of the series in 1759-60 to the beginning of the present reign, are now in the Public Record Office, London. In the 14th, as in most other corps, there are wide gaps in the earlier years.

‡ See British Museum additional MSS., Plans of Encampments.

ROLL OF OFFICERS OF THE REGIMENT, 1763.

(From the Annual Army List.)

Rank.	Name.	Rank in the Regiment.	Rank in the Army.
Colonel	*Charles Jefferyes	7th Sept., 1756	*M.-G.*, 25th June, 1759
Lieut.-Colonel	William Napier	12th June, 1760	
Major	Bartholom. Corneille	2nd August, 1760	
Captain	Jonathan Furlong	2nd March, 1750-1	
,,	George James Bruere	14th Oct., 1755	
,,	Alexander Murray	2nd August, 1760	
,,	William Blackett	29th May, 1761	
,,	Herbert Whitfield	1st Oct., 1761	
,,	Samuel Leslie	8th Feb., 1762	
Capt.-Lieut.	Joseph Vipond	1st Oct., 1761	
Lieutenant	John Geathin	17th May, 1756	
,,	Charles Ross Stewart	18th May, 1756	
,,	Samuel Lindsay	1st Dec., 1758	
,,	Joseph Goldfinch	9th Sept., 1759	
,,	William Kitchingman	11th Sept., 1759	
,,	Andrew Armstrong	14th Sept., 1759	
,,	Brabazon O'Hara	24th Nov., 1759	
,,	David Grant	2nd August, 1760	
,,	Edward Fitzgerald	20th Dec., 1760	
,,	— Browne	16th March, 1761	
,,	Thomas Evans	15th July, 1761	
,,	Richard Matthias	11th Nov., 1761	
,,	John Miller	12th Nov., 1761	
,,	John Stanton	14th Nov., 1761	
,,	James McAdam	22nd Dec., 1761	
,,	— Hautenville	3rd Feb., 1762	
Ensign	John Bruere	16th March, 1761	
,,	H. Amand Montresor	3rd Feb., 1762	
,,	— Mattier	12th April, 1762	
,,	James Urquhart	13th April, 1762	
,,	Alexander Ross	19th April, 1762	
,,	John Ness	17th May, 1762	
,,	Richard Symes	23rd June, 1762	
,,	Christopher O'Bryan	10th Nov., 1762	24th May, 1762
Chaplain	Hugh Palmer	17th Dec., 1756	
Adjutant	Brabazon O'Hara	15th May, 1761	
Qr.-Master	Thomas M'Ellis	17th May, 1762	
Surgeon	Charles Hall	20th Dec., 1758	

* Afterwards General William Napier, of Culcleugh, N.B. He died in 1780, aged 68. His descendants are the Millikin Napiers.

1764-70.

Leaving Plymouth, in March, 1764, the regiment proceeded to the neighbourhood of London, and was reviewed on Wimbledon Common, and again on 7th May, in Hyde Park, by George III., who expressed his approval of its appearance and discipline. After the review the regiment marched to Chatham and Dover.

On the death of Major-General Jefferies, in May, 1765, the King conferred the colonelcy of the regiment on Major-General the Honble. William Keppel, from the 56th Foot.

Three companies were at this time doing duty at Windsor and Hampton Court, Their good conduct attracted the notice of the King, who was pleased to direct that the badge of the White Horse, with the motto—*nec aspera terrent*, should be placed on the new pattern caps of the grenadiers and drummers, and that the alterations in the clothing specified below should be made at the next issue:

"14th October, 1765.

"Alterations in the clothing which is to be delivered in the year 1766 to "the Fourteenth Regiment of Foot, commanded by Major-General Keppel, "and which are approved by His Majesty—

"The breeches to be buff.

"The grenadiers to have black bearskin caps fronted with red; the "horse and motto to be in white metal.

"The drummers to have white bearskin caps, with red cloth fronts; "motto and horse, white metal.

"By order of the King,

"(Signed) EDWARD HARVEY,

"Adjutant-General."

Towards the end of May, 1766, the regiment went into billets at Hounslow, and on the royal birthday (4th June) was reviewed on Hounslow Heath by the King, who again expressed his approbation of its appearance and movements in the field. After the review it marched to Salisbury and places adjacent, and at the end of the month embarked at Southampton for Halifax, Nova Scotia, and was stationed in Nova Scotia and in Massachusetts for several years.

A notice, dated War Office, 10th January, 1767, contained the following: "Whereas many mistakes and inconveniences have arisen from the want "of an authoritative announcement of the commissions granted to officers "in His Majesty's Army, notice is hereby given that the said commissions "shall for the future be regularly published in the *London Gazette*."

It may be added that for years the rule was not invariably observed, and that the publication of indexes of the appointments gazetted only commenced in 1785.

1771-77.

During 1771 and part of 1772 the regimental head-quarters were at Castle William, Massachusetts Bay. A light company is believed to have been added to the regiment about this period, but no record of its formation has been found. In the summer of 1772 the regiment proceeded to the West Indies, to join a force to be employed against the refractory Caribs in the island of St. Vincent. Colonel William Dalrymple, 14th regiment, was appointed to act as major-general of the troops so employed.*

The island of St. Vincent, then known as one of the Ceded Islands, had been taken from the French by the British in 1762, and ceded to Great Britain at the peace in the following year. It contained two races of natives, known as the *red* and the *black* Caribs, the former being aborigines, the latter descendants of a cargo of African slaves, who had escaped from a vessel wrecked on the island long before. The black Caribs, the same who proved such troublesome foes and unerring marksmen twenty-three years later, were the dominant race. They claimed the island as their own by right of conquest of the aboriginal inhabitants, and, although French in their sympathies, denied the right of cession by France and the territorial authority exercised by Great Britain. The Crown had sold certain lands to English planters, who petitioned for protection against the Caribs. In 1771 the governor-in-chief of the Ceded Islands was directed to assemble a Special Commission, of which Sir William Young, the

---

* *Calendar State Papers*: *Home Office*, 1770-72, par. 147. Colonel Dalrymple afterwards became a full general, colonel 47th Foot, and lieutenant-governor of Chelsea Hospital. He died in 1807. Contemporary accounts speak of him as a man of most amiable manners and disposition.

governor of Dominica,* was president, to enquire into the matter. The commission offered certain terms, which were rejected by the Caribs, who appeared in arms and fired on all boats approaching the coast of the island. A military force of some two thousand five hundred men was assembled from the West Indies and America for the protection of the settlers.† Proclamation of the terms offered by the Special Commissioners was made on 7th September, 1772, and on 24th September the troops were put in motion. The operations lasted five months. This, Sir William Young explains, was largely due to the humane and cautious system on which the operations were carried out by General Dalrymple: "Enterprizes of mere ravage and "slaughter into the Carib country at no time entered into the plan of "the war. The general sought gradually to attain and establish military "posts throughout the country, so as to hem in the Caribs in their "several districts, prevent their communication, control their move-"ments, and finally necessitate their surrender."‡ These posts, fourteen in number, were fully established by 25th January, 1773, when the Caribs agreed to terms. A treaty, acknowledging the sovereignty of the British Crown and defining the allegiance and rights of the natives, was agreed to by the Carib chiefs, in the grand camp at Macaricau, 27th February, 1773, with which the operations ended.§ The losses of the troops appear to have been heavy: died of sickness, forty-eight; killed, twenty-six; drowned, nineteen; deserted, five; wounded, forty-three.

---

* Sir William Young subsequently published "An Account of the Black Caribs of St. Vincent" (London, 1795), in answer to the misrepresentations contained in the earlier editions of Bryan Edwards's *History of the West Indies.*

† The troops, consisting of detachments of 6th, 14th, 31st, 50th, 60th Royal Americans, and 70th Foot, with small parties of royal artillery and marines, mustered one hundred and four officers, one hundred and forty sergeants, eighty drummers, and two thousand two hundred and seventy-three rank and file.—Sir William Young.

‡ *Account of the Black Caribs*, p. 89.

§ As in the case of other colonial wars of more recent date, widely different views appear to have been upheld at home of the need of the operations and of the conduct of the Government and the settlers. See *Annual Register*, 1773, chap. vii. A copy of the treaty will be found in the fifth edition of Bryan Edwards's *History of the West Indies* (appendix to vol. i.), in which the strictures contained in earlier editions of that work have been withdrawn by the author.

ROLL OF OFFICERS OF THE REGIMENT, 1773.

(From the Annual Army List.)

RANK.	NAME.	RANK IN THE	
		REGIMENT.	ARMY.
Colonel - -	Honble. Willm. Keppel	31st May, 1765	*M.-G.*, 25th May, [1772
Lieut.-Colonel	William Dalrymple - -	27th March, 1763	
Major - - -	Jonathan Furlong - -	27th March, 1765	
Captain - -	William Blackett - - -	29th May, 1761	
,, - -	Samuel Leslie - - - -	8th Feb., 1762	
,, - -	Charles Fordyce - - -	9th Sept., 1763	8th Oct., 1761
,, - -	Edwin Mason - - - -	19th Oct., 1763	8th Dec., 1762
,, - -	Brabazon O'Hara - -	27th March, 1765	
,, - -	James Gifford - - - -	21st August, 1765	31st March, 1762
,, - -	Stanhope Harvey - - -	6th April, 1772	9th April, 1760
Capt.-Lieut. and Capt.	John Stanton - - - -	21st Feb., 1772	*Capt.*, 25th May, [1772
Lieutenant -	James Urquhart - - -	11th Jan., 1763	
,, -	John Bruere - - - -	17th April, 1763	
,, -	David Cowper - - -	1st June, 1763	
,, -	Daniel Mattier - - -	27th March, 1765	
,, -	Alexander Ross - - -	18th Sept., 1765	
,, -	William Browne - - -	10th June, 1766	24th Oct., 1760
,, -	John Batut - - - -	25th Dec., 1770	17th March, 1761
,, -	William Napier - - -	16th Sept., 1771	
,, -	John Dalrymple - - -	6th Jan., 1772	
,, -	Cornelius Smelt - - -	21st Feb., 1772	
,, -	Edwin Gower - - - -	24th July, 1772	
Ensigns - -	Andrew Laurie - - -	24th Jan., 1767	
,, - -	Henry Hallwood - - -	15th July, 1767	
,, - -	Hill Wallace - - - -	9th April, 1771	20th March, 1771
,, - -	Peter Henry Leslie - -	12th August, 1771	
,, - -	Francis Wilkie - - -	16th Sept., 1771	
,, - -	Henry Medley Kelvington	21st Feb., 1771	
,, - -	Peter Burnett - - - -	24th July, 1772	
Chaplain - -	Hugh Palmer - - - -	17th Dec., 1756	
Adjutant - -	David Cooper - - - -	24th July, 1772	
Qr.-Master -	Peter Burnett - - - -	10th June, 1762	
Surgeon - -	Charles Hall - - - -	20th Dec., 1758	

The total sick on 25th January, 1773, was three hundred and thirty-seven. The strength of the 14th Regiment engaged in the operations was twelve officers, four staff, nineteen sergeants, eleven drummers, and three hundred and forty-two rank and file. No return of the regimental casualties, which are included in the above total, has been discovered.

At the close of the war a medal was given by the Colonial Legislature to the troops, but to what grades and to what proportion in each grade cannot now be ascertained. The medal was in silver, having on one side a bust of King George III. in a suit of armour, and on the other the figure of Britannia holding out an olive branch to a kneeling Carib. It was worn with a dark red ribbon.*

When the Caribs had been brought to submission, the regiment returned to America, leaving a number of sick men and others behind.

It was at first stationed in Virginia, and was in the "Old Dominion" still when the bloody battle of Bunker's Hill was fought on Charleston Heights, above Boston, on 17th June, 1775. Two promising young officers of the regiment, who were temporarily attached to the staff of Sir William Howe, lost their lives on that occasion—Lieutenant and Adjutant Bruce, killed, and Ensign Hesketh, mortally wounded.

On 18th October, 1775, Lieutenant-General the Honble. William Keppel was removed to the 12th Dragoons, and Major-General Robert Cunninghame, from the 58th Foot, became colonel of the regiment.

On 8th December, 1775, at midnight, a detachment of one hundred and twenty men of the regiment, under command of Captain Fordyce, was despatched from Norfolk, Virginia, where the regiment was stationed, to attack an American entrenched post at a place called Great Bridge, in the neighbourhood. The detachment crossed the bridge at daybreak, the grenadiers leading, Lieutenant Batut being at the head of the first section; but on nearing the entrenchment, a heavy fire was opened upon them by an American force of superior

* See D. Hastings Irwin's *War Medals*, pp. 22, 23; also Tancred's *Historical Records of Medals*, in which the Carib medal is figured.

numbers. Captain Fordyce and twelve men were killed within a few yards of the breastwork, and Lieutenant Batut and sixteen men were wounded and taken prisoners. The rest of the detachment fell back across the bridge to a small post held by a British detachment under Captain Leslie. The Americans buried Captain Fordyce with full military honours.

The American forces increased so fast, that the royal troops were withdrawn from Virginia in 1776, and the 14th Regiment joined Sir William Howe's Army at New York, where a detachment that had been left at Halifax, N.S., also joined. Part of the regiment was quartered on Staten Island, the other part taking the field. The regiment was now very weak, owing in a great measure to its losses in the West Indies. It was therefore directed to draft the men fit for service into other corps in America, and to return to England, where it arrived in small detachments in the summer of 1777, and at once began to recruit.

1778-81.

The remains of the regiment on arrival were stationed as recruiting detachments at Shepton Mallet, Somersetshire, and at Chatham. Subsequently it removed to the neighbourhood of London, and was stationed in detachments at Charlton, Lee, Sydenham, Lewisham, etc., head-quarters being at Greenwich. Subsequently it furnished detachments at Brentford, Richmond, Hampton Court, and various other places in the course of the year 1778. In the winter of 1778-79 it appears to have been at Chatham, and in the early summer of the latter year joined the great camp formed, under command of Lieutenant-General Richard Pierson, on Coxheath, near Maidstone, to cover London in the event of an invasion. The force there collected, in addition to some cavalry and artillery, consisted of the 6th, 14th, 50th, 65th, and 69th regiments of foot and sixteen regiments of militia. The camp was broken up in November, 1779, and the 14th was distributed in small detachments along the Kentish and Sussex coasts, with head-quarters at Tenterden and afterwards at Battle.* In May, 1780, the regiment was ordered to march, by way

* *Marching Order Books*, vols. lxiii., lxiv., lxv.

of Lewes and Brighton, to Gosport, where it encamped at Stokes Bay, and furnished working parties at Fort Monckton, and guards over the French, Spanish, and American prisoners of war at Forton prison.

An admiralty letter, dated 25th August, 1780,* and addressed to the Senior Sea Officers at Portsmouth and Plymouth, mentions that it is intended to employ two regiments of foot, of seven hundred men each, viz., the 14th Foot at Gosport and the 97th Foot at Plymouth, as marines on board the squadron preparing for a cruise under Vice-Admiral Sir Francis Geary. The admirals were directed to put themselves in communication with the officers commanding these regiments for that purpose. The troops were not to be employed on board vessels carrying marines, but were to replace the marines in vessels where the marine detachments were weakest, after such marines had been transferred to other vessels to make up the complement of the latter. The regiments were not to be mixed, but to serve on board separate vessels.

It is not clear what vessels were chosen to receive them, or what proportion of either regiment was ultimately embarked; but the head-quarters of the 14th Regiment appears to have been removed from Gosport to Dorking, and the details not embarked to have been quartered there and at Reigate, while the rest of the regiment was afloat.† Admiral Geary resigned his command at this time, and was succeeded by Vice-Admiral Darby, who left Spithead immediately afterwards to cruise at the mouth of the Channel, for the protection of the homeward-bound trade. The fleet returned to Spithead in December,‡ having run out of provisions. The troops remained on board, as the Admiralty books early in the following year record an application from the commanding officers of the regiments in question, that as their men "continue to serve on board the fleet," they may be provided with slop-clothing like marines, a request which appears to have been granted.

---

* *Admiralty Records:* Sec. Dept., *Marines*, vol. xxi., p. 445.

† Regimental Returns of Officers, in Public Record Office.

‡ *Naval Chronology*, vol. iii. *Admiralty Records : Marines*, vol. xxii.

ROLL OF OFFICERS OF THE REGIMENT, 1780.

(From the Annual Army List.)

RANK.	NAME.	RANK IN THE REGIMENT.	RANK IN THE ARMY.
Colonel - -	Robert Cuninghame - -	18th Oct., 1775	*L.-G.*, 29th Aug., 1777
Lieut.-Colonel	James Rooke - - - -	13th Oct., 1779	
Major - - -	John Stanton - - - -	16th Dec., 1778	
Captain - -	James Urquhart - - -	22nd Dec., 1772	
,, - -	David Cooper - - - -	25th March, 1774	
,, - -	Alexander Ross - - -	22nd Feb., 1775	
,, - -	William Browne - - -	15th Oct., 1775	
,, - -	John Batut - - - -	15th Oct., 1775	
,, - -	William Ramsay - -	10th Dec., 1775	
,, - -	Thomas Hayter - - -	16th Dec., 1778	
,, - -	Frank Percival Elliot -	2nd June, 1779	
,, - -	Honble. — Monson -	1st Feb., 1780	
Capt.-Lient. and Capt.	Hill Wallace - - - -	24th Oct., 1776	
Lieutenant -	Peter Burnet - - - -	25th March, 1774	
,, -	Thomas Haviland - -	22nd Feb., 1775	
,, -	Thomas Parke - - -	30th Sept., 1776	
,, -	James Grant - - - -	24th Oct., 1776	
,, -	— Brownrigg - - - -	6th June, 1778	
,, -	Gavin King - - - -	7th June, 1779	
,, -	J. Bampfylde Tyndale -	1st Sept., 1779	
,, -	Hugh Perry - - - -	10th Nov , 1779	
,, -	Hercules Harrison - -	1st Dec., 1779	
,, -	William Anne J. Crawley	10th Dec., 1779	
,, -	John Alston - - - -	7th Jan., 1780	
,, -	John Whitlocke - - -	26th April, 1780	
,, -	William Minet - - -	2nd June, 1780	8th April, 1780
Ensign - -	Thomas Cockrane - -	2nd June, 1779	
,, - -	Richard Lloyd - - -	7th June, 1779	
,, - -	Charles Brandon Ludlow	31st August, 1779	
,, - -	Frederick Maitland - -	1st Sept., 1779	
,, - -	John Jones - - - -	22nd Oct., 1779	
,, - -	James Cotter - - - -	10th Nov., 1779	
,, - -	Landon Lennon - - -	1st Dec., 1779	
,, - -	Thomas Northery - -	10th Dec., 1779	
,, - -	— Napier - - - - -	7th Jan., 1780	
,, - -	— Dunbar - - - -	26th April, 1780	
Chaplain - -	John Wright - - - -	17th June, 1777	
Adjutant - -	— Brownrigg - - - -	7th June, 1779	
Qr.-Master -	Peter Burnet - - - -	10th June, 1768	
Surgeon - -	Charles Hall - - - -	20th Dec., 1758	

The 14th and 97th detachments were on board the fleet of twenty-nine sail of the line, under Admiral Darby, which, with a convoy of two hundred vessels, effected the second relief of Gibraltar on 12th April, 1781. After very great difficulties, the principal stores were landed; the 97th Regiment was put on shore as a reinforcement for the beseiged garrison; and the colliers, which formed part of the convoy, were scuttled and sunk until such time as their cargoes could be broached. Admiral Darby then returned home, reaching Spithead on 21st May, 1781. Soon after the fleet again left on a channel cruise. There are no monthly returns or muster rolls of the regiment at this period now existing; but a "Return of Officers, 14th Foot," dated 14th July, 1781, shows the head-quarters at Dorking under command of Major Urquhart, Lieutenant-Colonel James Rooke being absent on duty,* and the following officers serving on board the fleet: Captains Batut and Burnett, Captain-Lieutenant William Minet,† Lieutenant Crowley, and Ensigns Ludlow, Maitland,‡ and Lennon. Most, if not all, of the detachments on board the fleet appear to have landed at Portsmouth in August, 1781, and were ordered to join the head-quarters, which had proceeded from Dorking to Canterbury. In September the regiment marched from Canterbury to Guildford and Petersfield, and subsequently to Hilsea Barracks, where in December, 1781, it was held in readiness to embark for the West Indies as soon as the transports provided for it should be in readiness.

1782-87.

In January, 1782, the regiment, Major Urquhart commanding, embarked at Portsmouth for Jamaica. It was on board transports in the harbour of the island of St. Lucia at the time of Admiral

---

* As Governor of Goree. Afterwards lieutenant-general and colonel, 38th Regiment. Died 1802.

† Afterwards Lieutenant-General William Minet, some time lieutenant-colonel commanding 2nd battalion 30th Regiment. Died 1830.

‡ Afterwards General Fred. Maitland, colonel 58th Foot. Died 1848. His record of service shows that he commanded the acting marines of the *Union*, ninety guns, Captain J. Dalrymple.

Rodney's great victory over the French fleet under the Count de Grasse, 12th April, 1782, and afterwards mounted guard over the count when a prisoner of war on that island.

From St. Lucia the regiment went on to Jamaica. It furnished the guard of honour to receive Prince William Henry (afterwards William IV.) on his first landing as a midshipman at Spanish Town, and mounted guard over his quarters during his stay in the island.

At this time County Titles were bestowed on regiments of foot not entitled "royal," as an aid to recruiting, The difficulties increased more and more as the war went on, notwithstanding the introduction of short service enlistment—"for the continuance of the war only, "and not to exceed three years." At the end of the war all short-service men who had served three years and over were discharged, but were permitted to re-engage for *unlimited* service only, receiving five guineas bounty. Medical examination of recruits was first enjoined by a War Office order of 27th October, 1790.

Soon after its arrival in Jamaica, the regiment received an order, dated Horse Guards, 31st August, 1782, directing it to adopt the title of the FOURTEENTH, OR BEDFORDSHIRE REGIMENT OF FOOT, and to cultivate a connexion with the County of Bedford in the interests of recruiting.

On 21st December, 1786, Captain Harry Burrard,* 60th Royal Americans, became major, *vice* Urquhart, retired, and remained in the regiment until transferred to the 1st Foot Guards in 1789.

Colonel Rooke attained the rank of major-general in 1787, but remained sole lieutenant-colonel of the regiment until his promotion as colonel-commandant in the 60th Royal Americans, in 1789, when Lieutenant-Colonel Welbore Ellis Doyle, from half-pay, late 105th Foot, succeeded him.

---

* Afterwards Lieutenant-General Sir Harry Burrard, Bart., K.B. He superseded Sir Arthur Wellesley in Portugal, in 1808, and was one of the signatories of the Convention of Cintra. He died in 1813.

ROLL OF OFFICERS OF THE REGIMENT, 1785.

(IN JAMAICA.)

(From the Annual Army List.)

Rank.	Name.	Rank in the Regiment.	Rank in the Army.
Colonel	Robert Cuninghame	18th Oct., 1775	*L.-G.*, 29th Aug., 1777
Lieut.-Colonel	James Rooke	13th Oct., 1779	*Col.*, 17th Nov., 1780
Major	James Urquhart	11th May, 1785	
Captain	David Cooper	23rd March, 1774	*Maj.*, 19th Nov., 1783
,,	Alexander Ross	25th Feb., 1775	
,,	William Browne	15th Oct., 1775	
,,	William Ramsay	10th Dec., 1775	
,,	Honble. — Monson	1st Feb., 1780	
,,	William Burnett	21st Jan., 1784	16th Dec., 1782
,,	Alexander MacBean	15th Dec., 1784	24th June, 1782
Capt.-Lieut. and Capt.	Hugh Campbell	7th March, 1782	*Capt.*, 2nd Mar., 1780
Lieutenant	Thomas Parke	30th Sept., 1776	
,,	James Grant	24th Oct., 1776	
,,	J. Bampfylde Tyndale	1st Sept., 1779	
,,	Hugh Perry	10th Nov., 1779	
,,	Hercules Harrison	1st Dec., 1779	
,,	Wm. Anne J. Crawley	10th Dec., 1779	
,,	Thomas Cochrane	13th Sept., 1780	
,,	Frederick Meard	1st Dec., 1784	
,,	Benjamin Hoche	9th Feb., 1785	5th Nov., 1784
Ensign	Thos. Ger. Elrington	21st Sept., 1781	
,,	Francis Elrington	24th Oct., 1781	
,,	Richard Topp	18th March, 1782	
,,	— Perkins	3rd July, 1782	
,,	Thomas White	21st July, 1784	
,,	Wm. Atwood Oliver	31st Dec., 1784	
,,	John Langton	19th Jan., 1785	
Chaplain	John Wright	17th June, 1777	
Adjutant	James Grant	6th April, 1784	
Qr.-Master	— Langdon	4th July, 1781	
Surgeon	James Snagg	25th Dec., 1782	

1787-92.

On 4th April, 1787, Lieutenant-General Robert Cunninghame was removed to the 5th Royal Irish Dragoons, and was succeeded in the colonelcy of the 14th by Lieutenant-General John Douglas, who had been colonel of the old 21st Light Dragoons, disbanded in 1783.

On 27th August, 1789, Colonel George, (fourth) Earl of Waldegrave, replaced Lieutenant-General John Douglas, transferred to the colonelcy of the 5th Dragoon Guards. Lord Waldegrave died six months after his appointment, when Lieutenant-Colonel the Honble. George Hotham, from captain and lieutenant-colonel 1st Foot Guards, was appointed colonel.

After nine years in Jamaica, the regiment embarked for home, in H.M.S. *Dover*, forty-four guns, on 9th April, 1791, and landed at Portsmouth on 10th June following. In the autumn of the year it marched from Portsmouth to Chatham, and thence to Canterbury. On 21st November, 1791, it furnished a guard of honour at Dover, to receive T.R.H.H. the Duke and Duchess of York, on their arrival from the Continent, the Duke of York having married some weeks previously the Princess Charlotte Ulrica, Princess Royal of Prussia.

Early in 1792 the regiment returned to Chatham, and were brigaded with the Buffs under Colonel Fox. In June the two regiments with other corps, encamped on Bagshot Heath, to practice field manœuvres, under the Duke of Richmond. The regiment remained in camp three weeks, during which period it was reviewed several times by the King.* Afterwards it returned to Chatham, where it spent the remainder of the year.

---

* An amusing account of the doings in the camp, taken from the newspapers of the day, appeared in *Colburn's United Service Magazine*, in December, 1877, in an article entitled "Autumn manœuvres eighty-five years ago." Major Everett's *Historical Records of Farrington's Regiment, afterwards the 29th Foot*, contains some very interesting matter under the above date.

ROLL OF OFFICERS OF THE REGIMENT, 1790.

(RETURNING HOME.)

(From the Annual Army List.)

RANK.	NAME.	RANK IN THE REGIMENT.	RANK IN THE ARMY.
Colonel	George Hotham	18th Nov., 1789	13th Feb., 1782
Lieut.-Colonel	*Welbore Ellis Doyle	13th March, 1789	21st March, 1783
Major	Alexander Ross	13th March, 1789	
Captain	William Browne	15th Oct., 1775	
,,	William Ramsay	10th Dec., 1775	
,,	William Burnett	21st Jan., 1784	16th Dec., 1782
,,	Alexander Macbean	15th Dec., 1784	24th June, 1782
,,	George Wathen	6th April, 1785	27th Jan., 1783
,,	George Clewlow	5th August, 1785	14th Oct., 1778
,,	James Grant	6th June, 1788	
Capt.-Lieut. and Capt.	†Geo. Townsend Walker	13th March, 1789	
Lieutenant	Hugh Perry	10th Nov., 1779	
,,	Hercules Harrison	1st Dec., 1779	
,,	Thomas Cochrane	13th Sept., 1780	
,,	Frederick Meard	1st Dec., 1784	
,,	John Jones	3rd Oct., 1786	5th April, 1783
,,	Thos. Goodall Elrington	24th Sept., 1787	
,,	Francis Elrington	24th Sept., 1787	
,,	Kenneth Mackenzie	14th May, 1788	13th Dec., 1782
,,	Richard William Topp	6th June, 1788	
,,	Richard Hawkins	18th June, 1788	31st March, 1783
,,	Thos. Green Clapham	10th Oct., 1788	24th May, 1782
Ensign	Thos. Robinson Grey	18th March, 1785	
,,	Robert Campbell	29th June, 1785	
,,	John Bower	24th August, 1785	
,,	William S. Grant	24th Sept., 1787	13th Sept., 1786
,,	John Douglas	24th Sept., 1787	
,,	James Watson	25th Sept., 1787	24th June, 1783
,,	Thomas Powell	30th April, 1788	
,,	David Hunter	6th June, 1788	
Chaplain	John Wright	17th June, 1777	
Adjutant	John Douglas	9th April, 1789	
Qr.-Master	John Langdon	4th July, 1781	
Surgeon	James Duffy	15th Nov., 1789	

* See *Appendix*.

† Afterwards General Sir Geo. Townsend Walker, Bart., G.C.B. Died 1843.

1793-94.

Whilst the allied armies, under the Duke of Brunswick, were operating on the French frontier in the summer and autumn of 1792, Great Britain had maintained an attitude of strict non-intervention; but on receipt of the news of the execution of Louis XVI., the French minister was ordered to quit the Court of St. James in eight days, and immediately after, on 1st February, 1793, the French National Convention declared war against the King of Great Britain, the Stadtholder and the Court of Spain, opening the long strife of 1793-1815. On 11th February the King's proclamation for making reprisals appeared, and on 26th February the Duke of York embarked for Holland with three battalions of the foot guards. Other regiments followed, although there appears to have been difficulty in taking up transport for them. Among the first of these was the 14th Foot, which embarked at Dover, under command of Lieutenant-Colonel Welbore Ellis Doyle, on 19th March, 1793, and landed at Helvoetsluys, on the island of Voorn, on 25th March, being the first British line regiment to arrive at the scene of war.

The latest return of the regiment before embarkation that has been found is dated Dover, 1st March, 1793, and shows the strength present as one lieutenant-colonel, one major, five captains, eight subalterns, two staff, sixteen sergeants, fourteen drummers, and three hundred and fifty-eight rank and file (including forty sick), and eight sergeants, four drummers, and one hundred and seventy-four rank and file wanting to complete. Six officers were out recruiting, and it is probable that the battalion received an accession of numbers in the shape of recruits, before embarking three weeks later.

The Campaign in Flanders.

When the Duke of York landed in Holland, the left wing of the French Republican armies, under General Dumouriez, was engaged in besieging Willemstadt, on the south bank of the Meuse, with a view to a movement on Amsterdam; whilst their right wing menaced Liege. Both operations failed. The siege of Willemstadt was raised on 15th March, and when the right wing of the Republicans, which had been driven back from the line of the Meuse, attempted to renew the attack, it was utterly routed by the Austrians. Summoned to Paris, to account for his failures, Dumouriez deserted to the Austrian camp. He was succeeded by General Dampierre, by whom the scattered remains of the armies were collected in the frontier towns of French Flanders, and, eventually, in a grand retrenched camp at Famars, near Valenciennes. The theatre of war having thus been shifted from Holland to the French border, the 14th proceeded from Helvoet to Brielle, where it embarked for Antwerp with other corps. Sir Harry Calvert, afterwards adjutant-general of the forces and colonel of the regiment, who at the time was aide-de-camp to the Duke of York, has recorded in his diary its appearance on arrival.*

APRIL 9TH, 1793.—A brigade, consisting of 14th, 37th, and 53rd, under Major-General Ralph Abercromby, disembarked at the ferry opposite Antwerp, and was quartered in the neighbouring hamlets. We remarked with concern that the recruits of the brigade were, in general, totally unfit for service, being mostly either old men or boys—extremely weak and short.

APRIL 26th.—Two of the regiments of the brigade totally unfit for service, so much so, 37th and 53rd left at Ostend and Bruges. The recruits sent before embarkation are worse than I ever saw, even at the close of the American War, ... a cruel circumstance on commanding officers, and on General Abercromby.

The 14th marched from Antwerp to Ghent, proceeding thence by canal boats to Bruges, where it halted a few days. From Bruges it marched to Tournay, where supplies for the British troops had been collected on account of the facilities for canal communication. The flank companies, with those of the 37th and 53rd regiments, were placed under Major Matthews, 53rd, and detached to Marquain to watch the movements of the enemy, on which service they were employed up to 20th May.

Meanwhile, on 1st May, a general attack was made by General Dampierre on the allied position; but the Republicans were driven back to their camp at Famars. Another engagement took place on 8th May, in which the brave General Dampierre fell.

---

* *Journals and Correspondence of Sir Harry Calvert* (London, 1854), pp. 52, 53.

ROLL OF OFFICERS OF THE REGIMENT, 1793.

(From the Annual Army List.)

RANK.	NAME.	RANK IN THE REGIMENT.	RANK IN THE ARMY.
Colonel - -	Honble. Geo. Hotham	18th Nov., 1789	*M.-G.*, 28th Apl., 1790
Lieut.-Colonel	Welbore Ellis Doyle -	13th March, 1789	*Col.*, 21st Mar., 1782
Major - - -	*Alexander Ross - -	14th March, 1789	
Captain - -	William Browne - -	15th Oct., 1775	*Maj.*, 18th Nov., 1790
,, - -	William Ramsay - -	10th Dec., 1775	*Maj.*, 18th Nov., 1790
,, - -	William Burnett - -	21st Jan., 1784	16th Dec., 1782
,, - -	Alexander Macbean -	15th Dec., 1784	24th Jan., 1782
,, - -	†Edmund, Earl of Dungarven - -	3rd Feb., 1791	24th Jan., 1791
,, - -	Thomas Richard Dyer	4th May, 1791	
,, - -	George Garnier - -	14th Dec., 1791	24th Jan., 1791
Capt.-Lieut. and Capt.	Hugh Perry - - - -	2nd May, 1792	
Lieutenant -	Thomas Cochrane - -	13th Sept., 1780	
,, -	Frederick Meard - -	1st Dec., 1784	
,, -	John Jones - - - -	3rd Oct., 1786	5th April, 1783
,, -	Francis Elrington - -	24th Sept., 1787	
,, -	‡Kenneth Mackenzie -	14th May, 1788	13th Dec., 1782
,, -	Frederick William Topp	6th June, 1788	
,, -	Thomas Green Clapham	10th Oct., 1788	24th May, 1782
,, -	Thomas Robinson Grey	25th Aug., 1790	
,, -	Robert Campbell - -	16th Feb., 1791	
,, -	James Watson - - -	18th April, 1792	
,, -	John Bower - - - -	2nd May, 1792	
Ensign - -	Thomas Powell - -	30th April, 1788	
,, - -	William Stuart - - -	26th Jan., 1791	
,, - -	James Greaves - - -	23rd Feb., 1791	
,, - -	— Hickson - - - -	8th July, 1791	
,, - -	§Richard G. Elrington -	27th July, 1791	
,, - -	Stanhope W. Dormer -	28th Dec., 1791	
,, - -	John Hutchenson - -	2nd May, 1792	
,, - -	Henry Andrews - -	25th July, 1792	
Chaplain - -	John Wright - - -	17th June, 1777	
Adjutant - -	Francis Elrington - -	8th July, 1791	
Qr.-Master -	Henry Coupar - - -	4th April, 1792	*Ens.*, 6th June, 1782
Surgeon - -	John Ramsay - - -	8th June, 1791	

Agents: COX AND GREENWOOD.

* Afterwards General Alexander Ross, colonel 59th regiment, and governor of Fort George. He was aide-de-camp to Lord Cornwallis in America, and brought home the despatches of the battle of Camden in 1780. He was one of Cornwallis' most intimate friends and correspondents. (See *Cornwallis Correspondence*, 3 vols.) He died in 1807.

† Afterwards General Edmund, eighth Earl of Cork, K.P. Died in 1856.

‡ Afterwards Lieutenant-General Sir Kenneth Mackenzie Douglas, Bart. Died in 1833. *(See Appendix.)*

§ Afterwards Major-General Richard Goodall Elrington, C.B., K.H. Many years lieutenant-colonel commanding 47th Foot. Died in 1847. *(See Appendix.)*

The French attacked along the whole line of Allies, extending some nine leagues, but were everywhere unsuccessful, except at the wood of Vicognes, near St. Amand, where the Prussians were worsted until the scene was changed by the arrival of the British Guards. This was the first conflict between British and French troops in the long struggle which was destined to close on the not far distant field of Waterloo twenty-two years after. The Guards only were engaged.

The allied forces, under the Prince of Coburg, now amounted to eighty thousand troops, and as repeated disasters had taught the Republicans to keep on the defensive, it was determined to attack Valenciennes without delay. The French, as before stated, had a great intrenched camp at Famars, south and south-west of that fortress, and also held Mont D'Anzain, north of the town. So long as they retained these positions the place could not be invested. It was therefore arranged that an Austrian column, under General Clairfayt, should assault Mont D'Anzain, whilst other columns, under the Duke of York, and another Austrian general, Ferraris, attacked the camp at Famars.

On 22nd May, 1793, the army moved forward, and encamped in front of the village of Seybourg. At 10 o'clock the same night the troops were under arms, and at midnight moved off the ground in different columns. The fog was so dense that it was 5 o'clock next morning before it had lifted enough to disclose the scene around. Sir Harry Calvert* and the few others who had left any account of it describe the scene as magnificent. The whole plain was covered with columns of troops—horse, foot, and artillery—of the various nationalities, each marching on its particular objective, and all moving with the order and regularity of a review, in the sunlight of a bright May morning. The column under the immediate command of the Duke of York† was flanked by two bodies of cavalry, one British and Hanoverian, the other Imperialist. The enemy at once occupied their advanced redoubts and batteries, to dispute the

---

* *Journals and Correspondence of Sir H. Calvert*, pp. 75–77.

† British Guards and Prussian troops.

passage of the Ronelle. General Otto, who was sent with a column of troops to mask Quesnoy, crossed without much loss, and so effectually drew off the French intended to defend the passage at Maresche, that the Duke of York afterwards crossed at that place without loss. Soon after 8 o'clock the column under General Ferraris approached the enemy's advanced works in front of the village of Aulnoit. After some cannonading, the troops attacked with great spirit, carrying the works, the efforts of the French cavalry to retake them being repulsed by the Hanoverian and Austrian cavalry. The Duke of York's column having by this time crossed at Maresche, the French retired into their camp at Famars, the Allies following to within cannon-shot of the batteries. On reconnoitring the position, it was resolved to postpone the attack until dark. In the night, however, the enemy evacuated the camp, of which the army took possession next morning, and as the Duke of York's head-quarters were established at Famars, the action became known by that name. The same night the French abandoned Mont D'Anzain, which had been attacked unsuccessfully by the Austrians under Clairfayt.

Abercromby's brigade, now consisting of the 14th and 53rd regiments only, and a battalion made up of the flank companies of the 14th, 37th, and 53rd regiments, formed part of General Ferraris' mixed column of eleven thousand men. A despatch, dated Famars, 24th May, 1793, states: "I am happy to have the honour of inform-
" ing you that the combined forces under the Prince of Coburg and
" H.R.H. the Duke of York have defeated the enemy and driven
" them out of the strong camp at Famars. . . . . The troops of the
" different nations engaged displayed the greatest firmness and in-
" trepidity. The British who had an opportunity of distinguishing
" themselves were the 14th and 53rd regiments and the battalion
" formed from the grenadier and light companies, commanded by
" Major-General Ralph Abercromby. Seven pieces of cannon and
" two hundred prisoners were taken in the redoubts."

The loss in the 14th was two sergeants, and five rank and file wounded, and two rank and file, besides, wounded with the flank battalion. The Duke of York expressed his approval of the conduct of the regiment in General Orders.

The present regimental quickstep, *Ca Ira*, dates from this attack on Famars. Sir Francis Doyle, grandson of Colonel Welbore Ellis Doyle (who commanded the regiment on the occasion), thus relates the family tradition of the circumstances of its origin:

"Had he (Colonel Doyle) lived, his extraordinary personal qualities "must have secured him the highest distinction; as it is, he has left "a mark upon the British Army, which trifling in itself, ought yet "among brother-soldiers to keep his name alive. To him it is owing "that the French revolutionary tune, *Ca Ira*, has remained to the "14th Regiment since 1793 as their chosen quick-march. At the "battle of Famars, in that year, the French attacked so fiercely that "his regiment wavered for a moment. The revolutionary fever, in "truth, blazed out as a new element in war, and everywhere the "discipline learned under average drill-sergeants was at a loss how to "meet it. He, however, was not at a loss, for, dashing to the front, "he called out in a loud voice, 'Come along, my lads; let's break "the scoundrels to their own damned tune. Drummers, strike up "*Ca Ira!*' The effect was irresistible, and the enemy found them-"selves running away (it was an Irish exploit, and a bull is permis-"sible) before they could turn round."

Strangely enough, owing to the subsequent loss of the regimental records and order-books referred to on another page, the only information on the subject in the existing records is contained in the following letter, received many years afterwards, from the late Major-General Richard Goodall Elrington, C.B., K.H., who was a lieutenant in the 14th Regiment at Famars.

"Mullingar, 27th November, 1833.

"My dear Colonel,

"I have much pleasure in stating that the conduct of the 14th "Regiment from 1793 to 1795 on the Continent of Europe was "good, and frequently mentioned in General Orders for gallantry "before the enemy.

"That at the attack on the Heights of Famars, near Valenciennes, "at the opening of the campaign of May, 1793, the present quick-"step, or march of the 14th, called *Ca Ira* (a French tune), was

Dawsons. Ph. Sc.

Lieut. Colonel Welbore Ellis Doyle,
who Commanded the 14th Regiment at the
Storming of Famars 23rd May 1793.

"given to the regiment in General Orders, in consequence of that "tune having been played by the band* under fire when the men "were marching up to the attack of the batteries on the Heights "of Famars.

"That at the siege of Dunkirk, at a sortie of the enemy on 6th "September, 1793, the 14th Regiment was at the head of the "column that repulsed the sortie, and in one fire from the enemy, "when the regiment had not more than three hundred men in the "field, about fifty-six men were brought down, and notwithstanding, "the 14th continued, and made a gallant charge.

"Truly yours,

"(Signed) R. G. ELRINGTON."

After the capture of Famars, the French withdrew across the Scheldt. Three weeks later the siege of Valenciennes was commenced, the conduct of which was entrusted to the Duke of York. The 14th Regiment formed part of the besieging army, and on 23rd July, 1793, furnished a detachment for the storming of the Mons hornwork. Lieutenant-Colonel Doyle was appointed to command one of the assaulting columns, and obtained permission to place at its head a hundred volunteers of his own regiment.

"My lads," he is reported to have said, addressing the regiment, "The general has done us a great honour. We have been selected "to perform an important, and, I will not disguise from you, a "dangerous duty. We have to carry yonder work, said to be mined "underneath; we must carry it therefore in such fashion that the "enemy may not have time, as he retires, to blow us and it up "together. I want a hundred of you to follow me there. Volun-"teers, 'Recover Arms.'" The whole regiment recovered as one man. "Very good," said the Colonel; "then I'll take the hundred "men next for duty." And with the hundred men next for duty the work was carried, and so rapidly that the enemy had no time to spring their mine.

* The regimental legend has it that the drums played it.

On 26th July the following General Order appeared: "His Royal "Highness the Commander-in-Chief returns his best thanks to Major-"General Abercromby, Colonel Leigh, and Lieutenant-Colonel Doyle "for the gallantry they showed in the attack last night."

The regiment had one sergeant and three rank and file killed; one officer,* one sergeant, and fourteen rank and file wounded; and seven rank and file wounded with the flank battalion.

After a splendid defence of forty-three days, during which they endured great privations, the French garrison of Valenciennes, about four thousand men, surrendered, marching out with the honours of war, and the place was delivered up to the Duke of York.

The British troops then marched towards Cambray. Subsequently they separated from the Imperialists, taking with them a few Austrian regiments, for the purpose of attacking the great fortress of Dunkirk. On reaching Menin it was learned that the French had driven the Dutch troops out of some redoubts at Lincelles, recently taken by the Prince of Orange. The Brigade of Guards, under Major-General (afterwards Lord) Lake, was sent to the rescue, and drove the French out headlong, the Guardsmen "hustling and cuffing the Frenchmen "as they were wont to do London mobs." † The 14th was sent to relieve the Guards the same night, and held the village until it was handed over again to the Dutch.

The day after, the 19th August, the army resumed its march to Dunkirk, and on 24th August the 14th took part in driving the French outposts between the Furnes Canal and the sea back into the town, upon which occasion the troops had to force their way through strong double hedges and across deep ditches full of water. In one instance a deep ditch, surrounding a chateau, was obstructing the progress of the grenadier company of the 14th, when Lieutenant

---

* This officer is described in the despatch as Lieutenant Duer, said to be badly wounded. No officer of the name belonged to the regiment. Probably, Lieutenant Bower is meant, who was promoted to an independent company.

† Corporal Brown, of the Coldstream Guards, who afterwards published a very good account of the campaign, describes the French cavalry as looking well, but the infantry as small, undersized men, and very ragged. They had many fine women, nicely dressed, on their baggage-waggons, the guardsman adds.

Thomas Green Clapham leaped into the ditch, and stood breast-high in the water, to enable the grenadiers to cross, by stepping on his shoulders, and pursue the retiring French, which they did with great alacrity.

On 6th September the French made a sortie from Dunkirk in great force, directing their attack chiefly against the besiegers' right, on which occasion the 14th, under command of Major Alexander Ross (Lieutenant-Colonel Doyle being ill), and its light company, under Lieutenant Kenneth Mackenzie,* which was with the battalion of flank companies, particularly distinguished themselves. Loudly cheered by the Hungarian regiment of Esterhazy as they went by, the 14th passed through two Austrian regiments which were retiring pressed by the enemy, and charged the French, driving them back into the covered way. The 14th light company appears to have then been on the left, and Lieutenant Mackenzie, by pushing forward half the company, with a few Austrian soldiers who had joined it, was enabled to open a flank fire on the retreating French. But when the rest of the 14th retired, the French returned to their works, and opened a heavy fire therefrom on the light company, which lost one-third of its number. The 14th had one sergeant, one corporal, and eight privates killed; Captains Cochrane and Gardner, Lieutenants Kenneth Mackenzie, T. Powell, and R. G. Elrington,† Ensigns Smith and Williams, Volunteer McGrath, one sergeant, one corporal, and thirty-six privates wounded. Owing to its heavy loss, the flank battalion was broken up, and the companies rejoined their regiments.‡ Another French sortie, on 8th September, was repulsed by the 14th and 37th regiments. The delays in the arrival of the siege-train and of the naval squadron which was to co-operate in the attack, enabled

* Afterwards Lieutenant-General Sir Kenneth Douglas, Bart. (See *Appendix.*) He was wounded by a grape-shot in the shoulder.

† Lieutenant Richard Goodall Elrington. (See *Appendix.*) This officer received a musket-ball, which was afterwards extracted from his left, the ball having passed from one side of the body to the other after the closing of the wound.

‡ See *Philliport's Royal Military Calendar*, 1820, "Services of Major-General Kenneth Mackenzie."

Carnot to mass an immense body of troops, hurried in carriages, carts, and vehicles of every description from every part of France, in the neighbourhood of Dunkirk. All hope of bringing the operations to a successful conclusion* was abandoned, and the siege was raised, the 14th marching by Furnes and Ypres to Menin. In October, 1793, the regiment marched to Oudenarde, where it furnished a guard over two thousand French prisoners. It was sent to the front on outpost duty several times, and during a movement of the enemy on Menin and Wevelghem repulsed an attack upon the outpost at Vervicke.

1794-95.

Early in 1794 the regiment left Oudenarde for Wevelghem, and remained on outpost duty until April, when the army assembled, and was reviewed by the Emperor of Germany (afterwards Francis I. of Austria) on the heights of Cateau. Here H.R.H. Prince William of Gloucester, nephew of George III., was appointed to command the brigade composed of the 14th, 37th, and 53rd regiments, in which he was soon afterwards succeeded by Major-General the Honble. Henry Fox.†

In a general attack on the enemy's positions on 17th April, 1794, the 14th formed part of the column under Lieutenant-General Sir William Erskine, and was in the attack on the village of Prémont and the adjoining wood on its left.

The French having been driven from their positions, the siege of Landrécies commenced, the 14th being with the covering army encamped on the heights of Cateau.‡ The French, under General Chapuy, attacked the heights on 26th April, and were repulsed with great slaughter by the British cavalry. On this occasion the light

---

* Special difficulties attended the engineering operations likewise, as whenever the trenches were dug deep enough to afford cover, they filled with water.

† Brother of the Right Honble. Charles James Fox.

‡ Whilst at Tournay, the men received compensation for the loss of their kits on the retreat from Dunkirk. A sergeant was allowed £2 10s. and a private £2 2s. for a full kit, and lesser amounts in proportion for smaller losses.—*Corporal Brown.*

company of the regiment was thrown forward into a wood, where it gallantly held in check a very considerable body of French who were meditating an attack on the camp. On the fall of Landrécies, the British moved into the neighbourhood of Tournay, where, on 10th May, 1794, they were attacked by a numerous body of French, whom they defeated with severe loss. The regiment only lost one man on this occasion.

This was followed up by a combined attack on the French positions, for the purpose of driving the French out of Flanders, which afforded the regiment fresh opportunities for signalizing itself. Leaving Tournay on 16th May, it was successful in driving the French from their posts in the direction of Lille on the following day. Several Austrian columns failed to accomplish their parts in the combined movement. The British, having penetrated into the French position, were thus left unsupported in the presence of very superior numbers of the enemy. On 18th May the regiment was attacked by an overwhelming force; but, firing by alternate wings and platoons with the greatest steadiness, it held its assailants in check. Brevet Lieutenant-Colonel Browne, the veteran commanding officer of the regiment, was so much exhausted that he sat for a time on a chair in the rear of the colours. At length an aide-de-camp from General Fox brought orders for the regiment to retire, which it did in unbroken order. Surrounded by the enemy, fired upon by musketry and artillery, and threatened in rear by cavalry, it displayed astonishing firmness, moving over ground strewed with dead. The road to Lannoy, by which it had advanced in the morning, was found to be in possession of the enemy, and an abattis, with cannon, thrown across it. The first discharge struck down several of the grenadiers, whereupon General Fox, who had come up, remarked to Captain Clapham, "I fear we shall have to lay down our arms." "No, Sir!" Captain Clapham answered; "the 14th can go through them." At this juncture Corporal Gilbert Cimitière, of the grenadier company (a French immigrant personally acquainted with the locality), undertook to lead the regiment through the enclosures. Under his guidance, and followed by the French, the regiment quitted the main road. Colonel Browne, shot through the body, was carried along in a blanket by four grenadiers, but suffered so much pain that he requested them to stop, and he and they were made prisoners. The command of the regiment devolved on Captain Parry, of the light company, which was then commanded by Lieutenant Graves. This officer and Lieutenant R. G. Elrington commanded the two rear companies of the column, which formed alternately to receive the French cavalry. Though every road was barricaded, and the hedges lined with troops, the brigade fought its way through the enclosures with great gallantry and resolution, and gained the position at Templeuve, having, however, lost every gun with the column except one of the battalion guns of the 14th, under Lieutenant Phillott, Royal Artillery.* The guide of the column, Corporal Gilbert Cimitière, was afterwards rewarded with a commission.†

Captain Lewis Tobias Jones,‡ of the 14th Regiment, who published a *Journal of the Campaigns in Flanders in* 1794-95, describing this affair, says:

"Major-General Fox, with the Fourteenth, Thirty-seventh, and "Fifty-third, was engaged with the whole of the French column which "had marched from Lisle, and the different corps which had driven "the rest of the army back fell upon his flanks and rear; perhaps "there is not on record a single instance of greater gallantry or more "soldierlike conduct than was exhibited that day by these three "regiments. At length, General Fox, finding that the whole army "had left him, began to think of retreating, to effect which it was "necessary to get possession of the causeway leading to Leers; but "before that could be accomplished he was obliged to charge several

---

* Two battalion or curricle guns, light six-pounders, with a subaltern and some gunners of the artillery to work them, were then attached to each battalion.

† Corporal Cimitière was appointed to an ensigncy in the 6th West India Regiment, 1st July, 1795, and promoted to a lieutenancy without purchase in the 48th Regiment 15th June, 1796. He commanded that regiment, as major, at Albuhera (gold medal). He became a lieutenant-colonel in 1824, and retired from the Service in 1828.

‡ Afterwards Barrack-Master at Sunderland. He joined the 14th as captain-lieutenant, from lieutenant, 57th, late in the campaign. He was father of the present Admiral of the Fleet, Sir Lewis Tobias Jones, G.C.B.

"battalions of the enemy, who were astonished that such a handful "of men should presume to give them battle, and expected every "moment that they would lay down their arms; but with a degree "of intrepidity that words cannot describe, and is indeed hardly "conceivable, they gained the wished-for point, and then formed "with such regularity that the enemy could not assail them; they "secured their retreat towards Leers, and the next morning joined "General Otto's column."

The loss of the regiment on this occasion was one sergeant and thirteen rank and file killed; twenty-two rank and file wounded; Lieutenant-Colonel Browne wounded and taken prisoner; three sergeants, two drummers, and sixty-eight rank and file taken prisoners and missing, many of whom were taken prisoners in consequence of being wounded and unable to continue the retreat. Lieutenant-Colonel Browne died at Lille on the following day, much regretted by the officers and men he had commanded with distinguished bravery on many trying occasions. The conduct of the regiment was commended by H.R.H. the Duke of York.

Action at Tournay.

The regiment resumed its post in front of Tournay, and was in position on 22nd May, 1794, when General Pichegru attacked the Allies with an immense body of troops, assailing first the right, and afterwards the centre of the line. The 14th, being on the left, were not engaged during the early part of the day; but in the afternoon the enemy carried the post of Pontechin, on the high road from Courtray to Tournay, and drove the Austrian troops back on the town, on which Fox's brigade of the 14th, 37th, and 53rd—the "Fighting Brigade," as it was called in the army—was ordered to the post of danger.

As the 14th came up, under Brevet Lieutenant-Colonel Ramsay, the Duke of York complimented them, assuring them of his perfect reliance on their gallantry.

In full view of the inhabitants, who thronged the ramparts in painful suspense as to the issue, the brigade formed in front of the Austrians, and advanced at a run, wholly unsupported. Charging along the chaussée, the 14th carried the village of Pontechin, re-formed beyond the houses under a heavy fire,* and with a loud shout rushed forward to storm a battery on some rising ground near a windmill, which the French defended for a short time, but afterwards abandoned, leaving the regiment in possession of several pieces of cannon. This sudden burst of British valour, coming like a thunder-clap, amazed and confounded the French, who, after a while, gave way at all points, and were pursued till dark.

There was a stubborn contest in an orchard, where the grenadier and light companies of the 14th were engaged, and where more than one remarkable instance of personal contempt of danger occurred. A grenadier, named Ryan, refused to avail himself of the cover of a tree, saying, "They cannot touch me;" but the next moment he fell forward, apparently dead, when Captain Clapham turned him over, and said, "Ryan, you are only shot through the face; you will do well yet." "Is that all?" replied the grenadier; and, jumping up, he commenced reloading his firelock, adding, "I'll have another rap at them." With difficulty he was persuaded to go to the rear. He served many years afterwards, with deep furrows in his cheeks.

The 14th Regiment had one sergeant and four rank and file killed; Captain Cochrane, major of brigade, died of his wounds; one sergeant and twenty-eight rank and file wounded, and five men missing. Lieutenant-Colonel Ramsay† had his horse shot under him, and received four musket-balls through his hat.

The following General Order was published, dated Tournay, 23rd May, 1794:—

"His Royal Highness the Commander-in-Chief desires to express

* While the troops were re-forming outside the village, a hare ran across the line. A man named Tovey knocked it down with his musket, and placed it in his haversack, with surprising coolness, although under so heavy a fire that it was difficult to form the men up, so frequent were the casualties among them, thus exemplifying that distinguished feature in the character of the British soldier—coolness and collectedness in the midst of danger.

† Afterwards Lieutenant-General William Ramsay. Died, 1817.

"his most particular thanks to Major-General Fox; to the 14th "Regiment, under the command of Lieutenant-Colonel Ramsay; to "the 37th Regiment, commanded by Captain Lightbourne; to the "53rd Regiment, commanded by Major Wiseman, and to the de-"tachment of artillery attached to them, under command of Captain "Trotter, for that display of intrepidity and good conduct, which "reflects the greatest honour upon themselves, at the same time that "it was highly instrumental in deciding the important victory of the "23rd inst.

"His Royal Highness much laments the loss they have sustained, "but flatters himself they feel it, in some measure, compensated by "the credit they have gained."

The following is Captain Jones' account of the action:—"The "Duke of York detached seven Austrian battalions, and the Second "Brigade of British Infantry (14th, 37th, and 53rd Regiments), under "Major-General Fox, who, though they had lost so many men only "four days before, anxiously wished to get into action. Nothing "could exceed their spirit and perseverance; they stormed the "village of Pontechin, and, after firing a few shots, rushed with "fixed bayonets into the heart of the enemy, and turned the fate "of the day once more in favour of the Allies. The charge was "conducted with such skill and activity that it immediately threw "the enemy into confusion, and forced them to give way. At this "time the artillery came into action, and directed their fire so well, "and followed it up with such activity, the enemy could never be "rallied to renew the attack, although they had fresh troops con-"stantly coming up, but continued to lose ground till dark. Such "a battle has seldom been fought; the enemy was in action, under "a constant fire of cannon and musketry, over twelve hours, and left "twelve thousand dead on the field, five hundred taken, and seven "pieces of cannon.

"The loss of the Allies in this memorable action amounted to "about four thousand men; one hundred and ninety-six were British, "and all, except three, from General Fox's brigade. It is a fact, "although it appears incredible, that a single British brigade, and "that brigade less than six hundred men, on that great day absolutely won the battle, for had it not come up, the Allies would have been beaten."

Forty-two years afterwards the 14th Regiment received royal permission to place upon its colours the word "TOURNAY," in commemoration of its distinguished services on that memorable day—a day of defeat, if the truth be told, for all but "the Fighting Brigade."

Despite the bravery and success in individual encounters of the troops, the Allies were driven from the Austrian Netherlands (Belgium), and a series of retrograde movements followed, with little fighting. An attempt was then made to defend Holland; but Dutch sympathies were with the French Republicans. In August, 1794, the 14th was with its brigade, in camp near Antwerp; then in position near Breda; afterwards in camp beyond Bar-le-Duc. There a redistribution of the army took place, and Fox's brigade from the Second, became the Fourth Brigade, composed of the 14th, 37th, 38th, and 53rd regiments. For a time the 14th Regiment was in garrison at Nimeguen, and when that town was evacuated, proceeded to Linden Castle, the army being in cantonments along the far side of the Waal, to defend the passage of that river. In November there was another redistribution of brigades, and Fox's was formed of the 14th, 38th, 63rd, and 80th regiments. Major-General (afterwards Sir David) Dundas succeeded the Duke of York at this time. Towards the end of December the Waal became frozen over, and a body of French crossed on the ice, but were defeated on 30th December, 1794. The frost afterwards became more intense—the winter was one of the coldest of the century—and on 4th January, 1795, another body of French passed the river on the ice. At this time the 14th, under the command of Lieutenant-Colonel the Honble. Alex. Hope, who had been promoted into it just before, was at Linden Castle, whence it advanced to take part in a combined attack on the enemy. On 7th January it crossed the Rhine on the ice at Rhenen, and proceeded to Buren Castle. On the following morning Major-General Lord Cathcart advanced with the light companies, thirty Uhlans, and a detachment of the 27th Regiment to reconnoitre; and the 14th and 27th regiments were afterwards brought up to attack the enemy at Gueldermalsen. The 14th formed on the ice on the left of the dyke, and the 27th across the enclosure on the right, supported by the picquets and Uhlans, and afterwards by a squadron of light dragoons and the field-pieces, escorted by the grenadiers of the 14th, under Lieutenant Elrington, who marched in front of the guns. Advancing in this order, the troops drove the French before them until they arrived at Gueldermalsen, where a protracted resistance was made. Lieutenant Elrington, with the 14th grenadiers, charged the French artillery at the bridge, bayonetting the gunners at the guns, and carrying the post with great gallantry. The British battalion guns cleared the street, and the soldiers rushed forward, fighting from house to house, until they had passed the village, when they were assailed by the enemy in force. The 14th defended the streets, and the 27th the churchyard; the 28th, coming up most opportunely on the right, threw in a flanking fire, which compelled the enemy to retire.* The brigade remained in the village during the night. It was retired the next morning, and the 14th, 27th, and 28th were thanked in General Orders for their distinguished conduct. Lieutenant R. G. Elrington was thanked by name for his gallantry in the attack of the bridge and gun, and was soon after promoted in the 115th (Hanoverians). The 14th had twelve men killed, and Lieutenant-Colonel Honble. Alex. Hope, Captain Perry, and twenty men wounded. Colonel Hope was permanently deprived of the use of his right arm, and partly of his right side; Captain Perry died of his wounds.

After this the regiment marched to Culenburg, and was on duty on the bank of the Leek for about a week, without house, tent,

---

* Every man of the 14th was proud of the reputation the regiment had acquired, with which he identified himself; even the recruits possessed the same *ésprit de corps*. After the capture of Gueldermalsen, a young soldier, named Sullivan, struck the butt of his musket against a cask, when the musket went off and the ball passed through his body. He called to Lieutenant Graves, and said, "I hope, sir, you will let my friends know that I have always behaved as a good soldier," and immediately expired.

or cover of any sort, the weather all the time being phenomenally severe.

A retreat into Germany was now found to be inevitable, and was attended with great suffering and mortality. Retiring before an enemy whom they had never directly encountered without proving victorious, in the midst of a winter of exceptional severity, with fever rife in their ranks, and in the midst of a population that closed its doors against them and often proved more hostile than their French pursuers, their losses were very heavy, but no regimental or other returns of them appear to be extant. On one occasion after a long march, the 14th found themselves on a lonely heath, on a dark night, exposed to severe frost and a snowstorm, the men's limbs being so benumbed with cold that fatal effects were feared. Happily a farm-house and barn were discovered on the heath, and served as a refuge. At last the troops arrived at their destination, and rested awhile in comfortable quarters about Bremen, which they quitted with regret. The 14th formed part of a division of regiments under Major-General Gordon, which left Bremer Lee on 9th April, 1795. It arrived at Harwich on 7th May. From Harwich the regiment marched to Hitchin and the villages adjacent, greeted heartily on its route in every town through which it passed. Nearly every officer and man bore marks of bullets on his clothing or accoutrements; the colours, riddled with shots, were borne by the two senior subalterns (Lieutenants Stewart and Graves), so many had been the casualties among the officers.

In June the regiment was in camp at Warley, Essex, and in July received orders to march to Nutshalling (Nursling) Camp, near Southampton. As the regiment passed through Dartford on its way thither, the band struck up the republican air *Ca Ira*, whereupon the regiment was stoned by the populace, eager to express their detestation of republican doctrines, an inconvenient display of patriotic zeal which ceased when matters were explained.

The regiment afterwards embarked for Quiberon, to support the French Emigrants under M. Sombreuil, but, being detained by contrary winds, was directed to disembark and return to Southampton.

1795-97.

An armament was at this time fitting out under the command of Major-General Sir Ralph Abercromby for the reduction of the French West India islands, and to act against the insurgents in the islands of St. Vincent and Grenada. The 14th formed part of the expedition, which sailed in a vast fleet of Indiamen, transports, and merchantmen, under convoy of a naval squadron, under Admiral Christian, but was shattered and dispersed in the great storm of November, 1795, still remembered as "Admiral Christian's storm." Several vessels foundered; others were wrecked, with great loss of life, along the south-western coast; more returned to port. The expedition was refitted, and again put to sea, but, again encountering heavy gales, put back to Portsmouth a second time. The *Calypso* (transport), with part of the 14th on board, was nearly run down in a heavy gale by H.M.S. *Charon*, 44 guns, which carried away her mainyard, but, like some other individual transports, she continued her voyage, and reached Barbadoes, without convoy, in eleven weeks. Several of the regiments which returned to port had their destination changed; but the portion of the 14th that put back re-embarked in February, 1796, and arrived in April at Barbadoes, where four companies of the 28th Foot* were attached to the regiment.

Capture of St. Lucia.

The 14th, with its attached companies, formed part of the expedition against the island of St. Lucia, under command of Lieutenant-General Sir Ralph Abercromby. It sailed from Carlisle Bay, Barbadoes, on 22nd April, 1796, for the rendezvous of the force at Martinique, whence the expedition sailed on the morning of the 26th April for St. Lucia, where the head-quarters landed on 27th April near Pigeon Island, and marched to Choque Bay, to cover the landing of the rest of the troops. They continued in position there a short time, until the batteries against Morne Fortunée were com-

* These companies subsequently rejoined their regiment at Gibraltar.—Colonel Brodigan, *Historical Records 28th Foot*.

pleted, when they were ordered up to take part in the ulterior operations. Prior to the landing, three companies were detached, with a force under Brigadier-General Perryn, on the side of the Grand Cul de Sac, to facilitate the investment of Morne Fortunée, and an attempt was made to drive the enemy from the batteries at the base of the mountain on that side. A battalion under Major Donkin† (44th), composed of three companies of the 14th and the four companies of the 28th, formed part of the force employed on this service. The battalion supported the 44th Regiment, in the column commanded by Colonel Riddell. On advancing to the attack, the battalion was checked at a sudden turn in the winding road cut on the side of a steep hill, by an abattis lined with French troops. Captain James Graves at once sprang up the bank with the aid of a branch, and, being assisted by Captain Henry Cox and Lieutenant Geo. Morris, helped a few soldiers to climb up the hill-side, whence they fired down on the flank of the troops defending the abattis, who at once abandoned it, and the British continued their advance. On arriving at more open ground, the fire of the enemy's batteries was heavy, when Captain James Graves, of the 14th, and Captain John Brown, of the 28th, with a few men of the two regiments, stormed the lower battery, called Chapuis. Captain Brown and Lieutenants W. F. Dalton and John O'Grady, with several soldiers, fell wounded during the advance, but the battery was captured, and was held by Captain James Graves, Lieutenant John Hutchinson, and about forty men. The troops being fired on from a house, it was stormed by Lieutenant Owen with a few men, and the defenders bayonetted. The firing on the right indicating a retrograde movement of the British troops at that point, Sergeant Shaw, of the 14th, was sent to reconnoitre. He returned wounded, and reported the retreat of the British and the advance of a fresh column of the enemy. Under these circumstances the guns were spiked, and the troops retired, fighting their way through a wooded country until they rejoined the column under Brigadier-General Perryn. Owing

† Afterwards Lieutenant-General Sir Rufane Donkin.

to failure of part of the attacking force, the operations were not successful.

The loss in the 14th companies was five men killed and Captain Cox and one sergeant wounded. On sending a flag of truce next day, to enquire for prisoners, the reply was, "The Republicans have made no prisoners."

An attack was afterwards made on the north side of Morne Fortunée; a battery was opened on the works on 16th May; and on 24th the French demanded a suspension of arms, which was followed by the surrender of the island.

After the surrender of St. Lucia, the 14th formed part of the expedition against the island of St. Vincent. A landing was effected on 8th June, 1796, the Caribs having surrendered, and the French troops having retired in a body to the strong fort of La Vigie. The fort was ascertained to be badly off for provisions, and worse as regards water; it was clear it could not hold out many days. Shortly afterwards the Commander-in-Chief received information that the garrison intended to effect an escape at night by descending the course of a deep ravine which led from the town through high and inaccessible rocks. A party of the 14th, consisting of three officers and one hundred men, was sent to occupy the pass. They took up a position in the bed of the stream, behind some large stones, over which the men rested their bayonets. The darkness of the night, and the position between the woods, precluded the possibility of seeing anything, whilst the rushing of the waters obscured every sound. The first intimation the party received of the enemy's approach was the fact of their actually pressing against the points of the bayonets. A desultory fire immediately ensued, which ceased only when the enemy was supposed to have retreated. When day-light broke, a horrible spectacle of killed and wounded presented itself. Such of the garrison as succeeded in returning to La Vigie surrendered next day. Captain Powell, Lieutenants Gibson and Beavan, and the whole party received the thanks of Sir Ralph Abercromby.

These captures having been effected, the 14th returned to Barba-does, where it was stationed during the remainder of the year.

ROLL OF OFFICERS OF THE REGIMENT IN 1796.

(FIRST ROLL AFTER ARRIVAL IN THE WEST INDIES.)

(From the Annual Army List.)

Rank.	Name.	Rank in the Regiment.	Rank in the Army.
Colonel	George Hotham	18th Nov., 1789	*M.G.*, 28th April, 1790
Lieutenant-Colonel	Welbore Ellis Doyle	13th March, 1789	*M.G.*, 26th Feb., 1795
"	*Honble. Alexander Hope	17th Dec., 1794	27th August, 1794
Major	Wiliam Burnett	15th Oct., 1794	1st March, 1794
"	Alex. Macbean	1st Sept., 1795	1st March, 1794
Captain	Fred. Meard	19th Feb , 1794	
"	Charles Griffiths	14th May, 1794	
"	Thomas Green Clapham	2nd July, 1794	
"	Thomas Powell	26th Nov., 1794	
"	†Lewis Tobias Jones	28th Jan., 1795	
"	‡James Watson	11th March, 1795	
"	William Stuart	13th May, 1795	
"	‖Montagu Burrows	2nd Sept , 1795	26th Nov., 1793
"	John Smith	3rd Sept., 1795	20th May, 1794
"	Henry Cox	4th Sept., 1795	24th Sept., 1794
"	James Palmer	5th Sept., 1795	30th April, 1795
Captain - Lieutenant and Captain	James Greaves	1st Sept., 1795	
Lieutenant	John Hutchinson	20th Feb., 1794	
"	Malerius Magrath	18th Sept., 1794	
"	Montagu Hotham	26th Nov., 1794	
"	John Hammell	11th March, 1795	
"	Robert Coleman	5th April, 1795	
"	Thomas Rice	5th April, 1795	
"	John Frankland	13th May, 1795	
"	George Morris	12th August, 1795	
"	Matthew Owen	1st Sept., 1795	
"	—— Tegart	3rd Sept., 1795	
"	Richard Armstrong	4th Sept., 1795	16th May, 1793
"	George Matthew Potts	5th Sept., 1795	14th May, 1793
"	Thomas Kirvan	6th Sept., 1795	28th Feb., 1794
"	John South	7th Sept., 1795	9th May, 1794
"	Archibald Hamilton	8th Sept., 1795	31st Dec., 1794
"	Joshua Gibson	5th Oct., 1795	
"	Lawrence Willis	20th Oct., 1795	
"	Joshua Paul Minchin	21st Oct., 1795	25th June, 1775
"	John Fraser	21st Oct , 1795	1st July, 1795
"	John Collins	21st Oct., 1795	31st July, 1795
"	John Newell	21st Oct., 1795	31st July, 1795
"	Norman Lamont	21st Oct., 1795	19th Sept., 1795
"	Thomas Beavan	21st Oct., 1795	
"	John Grant	9th Dec., 1795	12th Feb., 1794
"	Gabriel Plenderleath	30th Dec., 1795	7th April, 1794
"	John Campbell	31st Dec., 1795	30th May, 1794
Ensign	James Magrath	3rd Sept., 1795	25th March, 1794
"	Jonathan Chetwood	4th Sept., 1795	8th July, 1794
"	John Redcross	26th Sept., 1795	
"	Edward Davis	5th Oct., 1795	
"	James M'Intosh	21st Oct., 1795	22nd August, 1794
"	Peter Ward	21st Oct., 1795	30th April, 1795
"	Thomas Heming	21st Oct., 1795	
"	—— Snow	21st Oct., 1756	
"	Hugh Couston	21st Oct., 1795	
"	— Rumbold	28th Oct., 1795	
"	—— Gregory	4th Nov , 1795	
Chaplain	John Wright	7th June, 1777	
Adjutant	John Hammell	21st Oct., 1794	
Quarter-Master	Peter Caldwell	17th Sept., 1795	
Surgeon	—— Shea	26th Sept., 1795	

* Afterwards General the Honble. Sir Alexander Hope, colonel of the regiment. See *Appendix A*.

† See footnote, page 64.

‡ Afterwards General Sir James Watson, K.C.B., colonel of the regiment.

‖ Afterwards Lieutenant-General Montagu Burrows. See *Appendix Volume*.

Spain having united with France against Great Britain, orders were sent out to attack the Spanish possessions in the West Indies; and early in February, 1797, the 14th proceeded to Cariacou, where an expedition was assembled to attack the island of Trinidad. The fleet sailed on the morning of 15th February, and anchored off Trinidad. Aware of their inability to resist, the Spaniards fired their shipping in the harbour of Port of Spain, and when the troops landed on the 17th the garrison immediately surrendered, delivering up the island. From Trinidad the regiment proceeded to Martinique, where it was stationed for several weeks.

At the beginning of April, 1797, Sir Ralph Abercromby assembled a small force to attack Porto Rico, and the 14th was summoned from Martinique to take part in the enterprise. The fleet entered a narrow channel three leagues to the eastward of the town, and on 18th April the troops landed at Congrejos Point. They encountered a heavy fire of musketry from the Spanish troops lodged in a breast-work on the beach. The 14th were in flat-bottomed boats, manned by Lascars of the Indiamen which had conveyed them. The impetuosity of the troops brooked no delay. Leaping out of the boats and wading ashore, they speedily drove off the Spaniards at the point of the bayonet. Lieutenant-Colonel Burnett was ordered to pursue with all possible speed, to endeavour to get possession of the bridge leading over the river between the town and the beach. So closely were the enemy pursued by the 14th, and especially by the light company, that many threw away their arms and accoutrements, and fairly ran for it. They succeeded in gaining the bridge; but as soon as the 14th reached the bridge-head, the Spaniards blew up the bridge whilst many of their own people were crossing it. The destruction of the bridge caused Sir Ralph Abercromby to alter his plans, which originally had been to take the place by a *coup de main.* Next day, therefore, the army began to throw up batteries against the town; but the enemy kept up such a heavy fire that on the second day after their completion they succeeded in dismounting two guns in one of the batteries and otherwise seriously injuring the works. A strong party was ordered at night to repair the damage. It consisted of three hundred men, under Captain (afterwards Major and Brevet

Lieutenant-Colonel) Powell, of which one hundred and fifty were to act as a working party and the remainder to be placed in advance as a covering party. The same night Major Ronald Hamilton, of the 14th, made an attempt to ford the river, to ascertain if it were practicable for infantry, but, being discovered, was fired upon by an advanced sentry. This created some alarm, and caused an irregular fire to be kept up all night, under cover of which, and the darkness, a body of five hundred Spaniards contrived to cross the river higher up, and, descending along the bank, secreted themselves in the brushwood between the battery and the river. At dawn of day a sergeant and twelve men of the 14th, who had been on picquet in this bush, were called in; and at the selfsame moment, appearing as if by magic, the whole body of Spaniards made a rush at the battery.

Sir Ralph Abercromby, Colonel Hope* (the Adjutant-General), Colonel Maitland† (Military Secretary), and the whole staff of the commander-in-chief had arrived about an hour before to inspect the work, and were at the moment in the battery. The inrush of the Spaniards caused surprise, and the extra number of persons in the battery increased the confusion. The only British with arms were the twelve men of the picquet. The Spaniards were all provided with knives or short swords, evidently intended for the butchery of the working party. For some instants it seemed as though the latter were at the mercy of the enemy, but, recovering themselves, they clubbed their picks and shovels, and fell on the intruders with such vigour that every Spaniard of the party was killed, wounded, or made prisoner before the covering party arrived to the rescue. The working party had five men killed and seventeen wounded. Captain Powell and Lieutenants Gibson and Wrenn were thanked in general orders.‡ The defences of Porto Rico, however, proved too strong

* Afterwards Lieutenant-General Sir John Hope and Earl of Hopetoun, in whose division the 14th was at Corunna. He was brother of the Honble. Alex. Hope, who commanded the 14th at Gueldermalsen.

† Afterwards Lieutenant-General Fred. Maitland. He had been a subaltern in the 14th at the relief of Gibraltar.

‡ Sir Ralph Abercromby entertained a high opinion of the 14th. In the West Indies he always landed with the flank companies, and the regiment furnished a

to hold out any hopes of success, and, after bombarding the town, Sir Ralph Abercromby re-embarked the troops on 30th April, 1797, and returned to Martinique.

1798-1801.

The regiment remained in Martinique during the next three years. Towards the end of the year 1800 it relieved the 70th Foot at Trinidad. At this time Captain Watson was employed recruiting at Bedford, and Lieutenant Carpenter at Woburn.

The Act of Union, taking effect from 1st January, 1801, necessitated an alteration in the blazon of the Union flag, which was forthwith introduced in the colours of the regiments. The red "Saltire" of Ireland was placed on the white St. Andrew's Cross, the white being above the red in two quarters, and the red above the white in the other two. The Shamrock was also entwined with the Rose and Thistle in the Union Wreath.

1802-3.

On 25th March, 1802, a Treaty of Peace was concluded at Amiens, the preliminaries of which had been signed in London 1st October, 1801, according to which Great Britain gave up the captured possessions of France, Holland, and Spain, including among the latter the island of Trinidad. The 14th was thereby released from duty in the West Indies in April, 1803, and returned home, landing at Gosport and marching to Winchester, under Captain Graves, who had performed the duty of commanding officer for over twelve months. After it had been recruited it marched to Silver Hill Barracks, Hastings, and later to Winchelsea.

The Peace of Amiens proved but a brief truce. On 16th May, 1803, France again declared war against Great Britain, and began preparations for an invasion.

---

corporal's guard at his quarters. When he was appointed to the command of the troops in the Mediterranean, with which he proceeded to Egypt, he wrote to Captain Graves: "I regret extremely that I cannot take you with me as I intended, "having found all my staff appointed when I got to London." He afterwards added with pleasantry, "I also greatly regret that the Fourteenth are not on the "expedition, as I do not think any service can go on well without them."

ROLL OF OFFICERS OF THE REGIMENT, 1803.

(ON LEAVING THE WEST INDIES.)

(From the Annual Army List. War Office, 10th January, 1803).

RANK.	NAME.	RANK IN THE REGIMENT.	RANK IN THE ARMY.
Colonel - -	Hon. Geo. Hotham	18th Nov., 1789	*Gen.*, 29th April, 1802
Lieut.-Colonel	William Burnett - -	9th Nov., 1796	
Major - - -	T. Green Clapham -	10th May, 1799	
,, - -	Jas. Watson - - -		
Captain - -	Wm. Stewart - -	13th May, 1795	
,, - -	Montagu Burrows -	2nd Sept., 1795	*Maj.*, 29th April, 1802
,, - -	Montagu Hotham -	11th June, 1796	
,, - -	Nath. Burslem - -	22nd Nov., 1797	4th April, 1795
,, - -	Geo. Miller - - -	15th Jan., 1798	
,, - -	Wm. Wood - - -		
Capt.-Lieut. and Capt.	Jas. Greaves - - -	1st Sept., 1795	
Lieutenant -	Wm. Shea - - -	3rd June, 1795	
,, - -	Digby Thos. Carpenter	22nd July, 1797	
,, - -	Anth. Lewis Wrenn	8th March, 1798	
,, - -	Peter Henry - - -	29th August, 1798	
,, - -	Wm. Griffin - - -	22nd Oct., 1798	10th Nov., 1796
,, - -	Chas. Gosselin - -	6th March, 1799	
,, - -	Jacob Watson - -	30th August, 1799	
,, - -	D. Ogilvie Stewart -	3rd July, 1801	
,, - -	Thos. Barbridge -	29th Oct., 1802	
,, - -	Wm. Moore - - -		
Ensign - -	Moses Clark - - -	23rd Dec., 1800	
,, - -	David K. Fawcett -	7th Nov., 1801	
,, - -	Benj. H. Junck - -	9th July, 1802	
,, - -	Thos. Bailey - - -	29th Oct., 1802	
,, - -	Geo. Morris - - -	29th Oct., 1802	*Adjutant*
,, - -	Walter Rice - - -	5th Nov., 1802	
Paymaster -	John Ready - - -	27th Sept., 1798	
Qr.-Master -			
Adjutant - -	*Geo. Morris - - -	5th June, 1797	*En.*, 29th Oct., 1802
Surgeon - -	Thos. Jackson - -	17th June, 1800	
Asst.-Surgeon	Thos. Draper - -	29th Jan., 1800	

1804-6. – Two Battalions.

In January, 1804, the 14th Regiment was in Winchelsea Barracks, where it remained as part of a force of fourteen thousand foot (line and embodied militia), which, with two thousand eight hundred cavalry and fifty guns, was stationed, under command of Lieutenant-General David Dundas, in Kent and Sussex, while the French armies lay encamped at Boulogne.

A Second Battalion, the first the regiment ever possessed, was added in the summer of 1804. It was formed at Bedford by Lieutenant-Colonel the Honble. William Bligh* of a late second battalion 54th Foot, with Lieutenant Algeo, 14th Foot, as adjutant, and a small draft of officers and men of the regiment, who were distributed as recruiting parties at Leicester, Lichfield, Manchester, Northampton, Norwich, and other places, the new battalion being formed by general recruiting. In the autumn of 1805 the battalion went to Colchester; in the spring of 1806 it was at Chelmsford Barracks; and in the autumn of that year it went to Ireland, leaving a depôt company at Aylesbury.

When the French withdrew from Hanover, which they had seized on the renewal of the war, the British Government decided to send an army, under Lord Cathcart, into the North of Germany, to co-operate with the Prussians and Russians against France from the side of Holland. The advance of the expedition—ten thousand troops under Lieutenant-General George Don—left the Downs early in November, 1805, arrived at Verde at the end of the month, and took up a position on the Lower Weser, with outposts on the Hunt River.

In this force was the 1st battalion 14th, Lieutenant-Colonel William Burnett commanding, mustering nine hundred and sixty-seven of all ranks. It was brigaded with the 1st battalion 4th Foot, 1st battalion 23rd Fusiliers, and five companies 95th Rifles, under Major-General the Honble. Edward Paget. The brigade was quartered in and around Bremen.

---

* Lieutenant-Colonel the Honble. William Bligh, second son of the third Earl of Darnley. He served in Egypt with the 54th, and commanded the 2nd battalion 14th from its formation until 1808, when he left the army after his marriage. He died in 1845, aged 70.

ROLL OF OFFICERS OF THE REGIMENT, 1805.

(FIRST APPEARANCE OF THE REGIMENT WITH TWO BATTALIONS.)

(From the Annual Army List.)

Rank.	Name.	Rank in the Regiment.	Rank in the Army.
Colonel - -	Geo. Hotham - -	18th Nov., 1789	*Gen.*, 29th April, 1802
Lieut.-Colonel	Wm. Burnett - -	9th Oct., 1796	*Col.*, 25th Sept., 1803
,, - -	Honble. Wm. Bligh	16th Aug., 1804	30th June, 1800
Major - - -	Jas. Watson - - -	3rd Dec., 1802	
,, - -	Jas. Greaves - - -	22nd Oct., 1803	
,, - -	Montagu Burrows -	1st August, 1804	29th April, 1805
,, - -	Montagu Hotham -	2nd August, 1804	
Captain - -	Nath. Burslem - -	22nd Nov., 1797	4th April, 1795
,, - -	Geo. Miller - - -	15th Jan., 1798	
,, - -	Wm. Wood - - -	3rd Dec., 1802	
,, - -	Wm. Shea - - -	25th Dec., 1802	
,, - -	Jas. Franc. Costello -	25th May, 1803	1st July, 1795
,, - -	Peter Henry - - -	25th June, 1803	
,, - -	Wm. Griffin - - -	3rd Sept., 1803	
,, - -	Marinus Kennedy -	5th Nov., 1803	
,, - -	Trevor Stannus - -	1st August, 1804	9th July, 1803
,, - -	Peter Johnson - -	2nd August, 1804	9th July, 1803
,, - -	R. G. Hare - - -	3rd August, 1804	9th July, 1803
,, - -	Eyre Coote - - -	3rd August, 1804	5th Nov., 1803
,, - -	Lucas Hunt Jackson	4th August, 1804	
,, - -	Jacob Watson - -	5th August, 1804	
,, - -	John Ready - -	26th August, 1804	20th August, 1803
,, - -	Henry Madden - -	27th August, 1804	
,, - -	Thos. Johnson - -	28th August, 1804	
,, - -	Jas. Reynolds - -	6th Sept., 1804	
,, - -	Wm. Fawcett - -	7th Sept., 1804	
Lieutenant -	D. Ogilvie Stewart -	3rd July, 1801	
,, - -	Wm. Moore - - -	16th May, 1802	
,, - -	Benj. J. Junck - -	24th Dec., 1802	*Adjutant*
,, - -	Harry Green - - -	25th Dec., 1802	24th April, 1800
,, - -	J. H. Aldgeo - - -	25th Dec., 1802	*Adj.*, 3rd July, 1800
,, - -	John Andrews - -	3rd Feb., 1803	4th April, 1800
,, - -	Thos. Wadman - -	26th May, 1803	29th June, 1801
,, - -	David K. Fawcett -	25th June, 1803	

RANK.	RANK.	RANK IN THE	
		REGIMENT.	ARMY.
Lieutenant -	W. L. Brookman -	24th Sept., 1803	20th March, 1801
,, - -	Walter Rice - - -	18th Feb., 1804	
,, - -	Wm. Lupton - -	5th August, 1804	3rd Dec., 1796
,, - -	Jas. Mayo - - -	6th August, 1804	15th Feb., 1800
,, - -	Hippie Edgell - -	7th August, 1804	26th Sept., 1801
,, - -	Michael Nickson -	8th August, 1804	25th August, 1802
,, - -	Franc. O'Connor -	9th August, 1804	27th March, 1801
,, - -	Philip F. Hall - -	10th August, 1804	7th May, 1800
,, - -	Thos. Dunn - - -	11th August, 1804	
,, - -	Thos. Ryder Graves	28th August, 1804	
,, - -	W. J. Newland - -	29th August, 1804	
,, - -	Jas. Barwick - - -	7th Sept., 1804	
,, - -	Alex. Burnett - -	27th Sept., 1804	
,, - -	Fred. W. Hoysted -	22nd Nov., 1804	
,, - -	Thos. E. Clarke - -	23rd Nov., 1804	12th October, 1799
Ensign - -	Thos. Sword - - -	8th Oct., 1803	
,, - -	AmbroseNealSavage	3rd Dec., 1804	
,, - -	Wm. Arthur Hodges	7th April, 1803	
,, - -	Stephen Shute Rowe	6th July 1804	
,, - -	Geo. Rawlins - -	8th August, 1804	9th July, 1803
,, - -	John Parsonage - -	9th August, 1804	9th July, 1803
,, - -	— Goble - - -	10th August, 1804	9th July, 1803
,, - -	W. S. Bertrand - -	11th August, 1804	9th July, 1803
,, - -	John Williams - -	14th August, 1804	
,, - -	Wm. Turnor - - -	15th August, 1804	
,, - -	Franc. Moffett - -	16th August, 1804	
,, - -	John Dyson - - -	29th August, 1804	
,, - -	— Wilkinson - -	30th August, 1804	
,, - -	Chas. Benn - - -	31st August, 1804	
,, - -	Jas. Munro Robinson	1st Sept., 1804	
,, - -	C. Levyns Barnard -	28th Sept., 1804	
Paymaster -	Alex. Foulerton - -	12th Nov., 1803	
Adjutant - -	Benj. H. Junck - -	25th June, 1803	*Lieut.*, 24thDec.,1802
,, - -	J. H. Aldgeo - - -	28th August, 1804	*Lieut.*, 3rd July, 1800
Qr.-Master -	Henry Brown - -	25th June, 1803	
,, - -	Jas. Mercer - - -	28th August, 1804	
Surgeon - -	Theo. Jackson - -	17th June, 1800	
Asst.-Surgeon	Thos. Draper - -	25th Jan., 1800	
,, - -	W. J. Parker - - -	8th Nov., 1804	

The rest of the army, under Lord Cathcart, arrived in the course of December; but the French victory over the Austrians and Russians at Austerlitz, on 2nd December, 1805, established the preponderance of French power on the Continent, and, together with the treaty concluded soon after at Vienna, frustrated the British plans, and decided Lord Cathcart to re-embark the troops and return home. The embarkation began during the last days of January, 1806. The 14th embarked at Bremen early in February, and on landing in England marched to Ashford, Kent. From Ashford the 1st battalion proceeded to Shorncliffe, where it was brigaded with the 9th and 91st regiments, under Major-General Rowland Hill (afterwards General Viscount Hill). The brigade, together with the 43rd Light Infantry, was reviewed at Shorncliffe by H.R.H. the Duke of York, who expressed himself much pleased with the appearance and discipline of the several corps. In December, 1806, the 1st battalion moved to Ireland, and was quartered at Fermoy.

On the decease of General Hotham, Major-General Sir Harry Calvert, Adjutant-General of the Forces, succeeded to the coloncley of the regiment, from the 5th West India Regiment, by commission dated 8th February, 1806.

In a "Statement of the Services of the Fourteenth Regiment of Foot," rendered in conformity with instructions from the Adjutant-General's Department, dated Horse Guards, 6th November, 1822, it is recorded that the whole of the books of the 1st battalion were lost, first, on passage to the Continent in 1805, and again, a second time, on transit to Ireland in December, 1806. No particulars of either circumstance appear to be now procurable.

1807-9.—Two Battalions.

Two changes of some importance to the army were effected about this time. The first was the legalising of enlistments from the regular militia. Although several line battalions of short-service soldiers had been formed of volunteers from the militia at the close of the last century, the practice had been always regarded as unconstitutional,

A representative meeting of lieutenants of counties and militia commanding officers in London in 1799 had declared unanimously that the practice of enlisting men from the embodied militia was "destructive to the force and degrading to all concerned in it;" * and when the militia was re-embodied on the renewal of the war with France in 1803, the act of knowingly enlisting embodied militiamen was made punishable by a fine of £20 for each offence. This was now changed, and militia regiments were permitted and enjoined to furnish a certain quota of volunteers to the regular army every year. The call was speedily responded to, and during the succeeding years of the war the average number of militia-volunteers so furnished was twelve thousand annually. The other change referred to was the introduction of "short service," which the advocates of the measure believed would not only bring more men to the colours, but prove an antidote to desertion, then so rife. Under "Mr. Wyndham's Act," as it was called, men could enlist for *seven years* without pension; re-engage for a second period of seven years, or *fourteen years* in all, with conditional pension; and for a third period of seven years, or *twenty-one years* in all, with claim to life-pension as in the case of enlistments for "unlimited" service.†

The 1st battalion remained at Fermoy during the first five months of 1807, and then returned to England, and sailed for Madras in June, 1807. The battalion, which was commanded by Lieutenant-Colonel James Watson, had an embarked strength of forty-nine officers, sixty-four sergeants, twenty-two corporals, fifty-nine drummers, and one thousand and fifty-five privates, all "life-service" soldiers. It arrived at Madras in November, 1807, and was quartered at Fort St. George, where detachments in the *Wyndham* (Indiaman), under command of Major Burslem, and in the *Wexford* (Indiaman), under Lieutenant-Colonel Montagu Burrows, which had parted company on

* See *Annual Register*, 1799.

† It may be noted that, as stated by the late Chaplain-General, the Rev. A. G. Gleig, the bulk of Wellington's Peninsular Army was composed of "short-service" men, who left at the end of the war in 1814—to which was owing the number of young soldiers in the Waterloo army.

the voyage (the latter detachment putting into the Cape to repair damages), appear to have rejoined.

Denmark having become involved in hostilities with Great Britain, a small expedition, of which the 1st battalion 14th formed part, was despatched from Madras at the beginning of 1808, against the Danish settlement of Tranquebar, situate at one of the mouths of the Cauvery river, in the Carnatic. The place surrendered to the British on 8th February, 1808, and the battalion of 14th returned, and proceeded to Bengal, where, under command of Lieutenant-Colonel Watson, it was stationed at Fort William during the next eighteen months.

Meanwhile the 2nd battalion, stationed at Kinsale, and afterwards at Cashel, had profited by the alteration in the Militia Act. Since its formation it had barely mustered a couple of hundred strong; but a fine draft of one hundred and twenty volunteers from the Royal Bucks Militia and strong drafts of volunteers from the Bedford, Berkshire, Hertford, and Notts militias now brought its numbers up to about seven hundred of all ranks.

In Portugal, where Sir Arthur Wellesley had landed in August, 1808, the victories of Roliça and Vimiero had been followed up by the supersession of Sir Arthur by Sir Harry Burrard and Sir Hew Dalrymple, and the conclusion of the Convention of Cintra. In October, Lieutenant-General Sir John Moore, who had succeeded to the command in Portugal, in accordance with instructions from home marched into Spain to co-operate with the Spanish patriots against the French armies in that country. Reinforcements for Sir John Moore were at the same time despatched from the United Kingdom, under the command of Lieutenant-General Sir David Baird, which were landed at Vigo and Corunna. Among these reinforcements was the 2nd battalion 14th Foot, under command of Lieutenant-Colonel Jasper Nicolls,* which embarked at Cork and landed at Corunna. Two officers of the regiment, Major F. S. Tidy and Captain James Dunlop, were attached to General Baird's staff, the former as deputy-adjutant-general, the latter as deputy-assistant-

---

* See *Appendix Volume.*

adjutant-general. Baird's force, including the 2nd battalion 14th, marched up-country, and effected a junction with Moore's army at Majorga on 20th December, 1808. A redistribution of regiments then took place, and the 2nd battalion 14th, with the 2nd, or Queen's, the 1st battalion 5th, and 1st battalion 32nd, were brigaded together under Major-General Rowland Hill, in Lieutenant-General the Honble. Sir John Hope's division.

Moore, on reaching Salamanca, found, despite the protestations of the Spanish Junta and the representations of Mr. Frere, the British plenipotentiary, that the Spanish patriots were utterly unreliable; the Spanish armies were badly commanded, and everywhere defeated; and the French were advancing in overwhelming force on Madrid. He had decided on retiring from Salamanca into Portugal by way of Ciudad Rodrigo; but, on urgent representations that Madrid could hold out, and in opposition to his own better judgment, he decided on making a diversion in its favour by attacking Soult on the Carrion, with which object the junction with Baird was effected at Majorga. On 21st December, 1808, the British, twenty-nine thousand strong—admirable troops, as Napier describes them, robust, well-disciplined, and needing but the experience of a campaign or two to make them perfect—had their head-quarters at Toro, and the cavalry, under Lord Paget (afterwards Marquis of Anglesey), had a brilliant affair with the French at Sahagun. On 23rd December Moore advanced with his whole force. The infantry were within two hours' march of their opponents, when an intercepted letter brought the news that Napoleon in person had entered Madrid, and that the French armies, which now numbered three hundred thousand men, were in full march on Salamanca and Benavente. A retreat into Portugal was impracticable, and it was decided to retire on Vigo and Corunna. The army was at once put in motion towards the Esla, the greater portion crossing by the bridge at Benavente on 26th December. There the British had a skirmish with some of the Imperial cavalry, and took General Le Fevre and some men prisoners. The retreat continued over two hundred and fifty miles of bare and desolate country, buried in snow or deluged with heavy rains; the men, footsore and weary, losing heart and discipline, hurrying on by toilsome marches, uncertain of food, destitute of fuel; baggage, ammunition, stores, and even specie being destroyed, to prevent their falling into the hands of the enemy.*

Battle of Corunna.

On 5th January, 1809, the army reached Lugo, bivouacing in order of battle, the prospect of an engagement restoring some amount of spirit and order in the ranks; but Soult, who was following close with twenty thousand men, showed no disposition to engage, and the march was resumed on the night of the 9th, the army, with the exception of Crauford's and Charles Alten's brigades, which retired on Vigo, reaching Corunna, by way of Bentanzos, on 11th January, 1809. After resting a couple of days, the troops bivouaced in order of battle outside the town, to await the arrival of the transports. Hill's brigade was posted on the left of the position. On 15th January the transports had all arrived, and the sick and wounded and what remained of the baggage were embarked; the army was preparing to follow. About two o'clock in the afternoon of the following day, 16th January, 1809, the French were seen descending the opposite heights. A sanguinary battle ensued, in which Moore received his mortal wound. The French attack was made in three columns, the brunt falling on Lord William Bentinck's brigade, on the British right, where a French column captured the village of Elvinas and, dividing, endeavoured to turn Baird's division. Another column attacked the British centre; a third menaced the left, where Hill's brigade was posted, having in its front the village of Palavia Abaxo, through which the high road to Madrid passes. The French at one time got possession of the village, but were driven out by a very gallant charge of some companies of the 14th, led by Lieutenant-Colonel Jasper Nicolls. In the end the French were repulsed at all points; by 5 o'clock the engagement was confined to a distant cannonade; by 6 o'clock the French were everywhere retreating in disorder;

---

* Bags containing £25,000 were thrown over the side of a mountain with this object.

King's and Regimental Colour, original 2nd Battalion.
Made circa, 1812.

the success of the army, achieved under the most disadvantageous conditions over a very superior force, being dimmed only by the loss of its illustrious chief.

Sir John Hope, who succeeded to the command of the army, in his despatch, dated on board H.M.S. *Audacious*, 18th January, 1809, wrote:

"The enemy not having rendered the attack on the left a serious "one, did not afford the troops stationed in that quarter the oppor-"tunity of displaying that gallantry, which must have made him "repent the attempt. The Picquets and Advanced Posts, however, "of the brigades under Major-Generals Hill and Leith and Colonel "Catline Crawford conducted themselves with determined resolution, "and were ably supported by the Officers Commanding Brigades, "and by the Regiments of which they were composed.

"It is peculiarly incumbent on the Lieutenant-General to notice "the vigorous attack made by the 2nd battalion 14th Foot, under "Lieutenant-Colonel Jasper Nicolls, which drove the enemy out of "the village on the left, of which he had for a moment possessed "himself."

After the battle, the 14th, with the rest of Hill's brigade, took up a position nearer the ramparts, to cover the embarkation, leaving the picquets to keep up the bivouac fires. The embarkation of the troops commenced the same night, closing with Beresford's and Hill's brigades next day. There was a good deal of hurry and confusion at the last, but, with the aid of the men-of-war, all were got on board, and at 4 o'clock on the afternoon of the 17th the fleet and convoy made sail, going right before the wind, and, after a boisterous passage through the Bay of Biscay, reached Plymouth on 23rd January.*

Colour-Sergeant Thomas Garrett, of the regiment, is mentioned in the regimental records as having distinguished himself at the battle of Corunna; but further particulars, unfortunately, are want-

---

* Lieutenant C. M. Brannon and some men were taken prisoners at Lugo, 8th-9th January, 1809.

ing. Lieutenant-Colonel Jasper Nicolls received a gold medal; and long after, the word "Corunna" was inscribed on the colours by royal authority.

The battalion was landed at Plymouth and Portsmouth as the transports came in, and, when the men could be moved, was collected at Buckingham and Aylesbury. No complete returns of the losses of the army were ever published, if ever completed; but the battalion pay-list for the quarter ending 24th March, 1809, shows the greater part of the men sick in hospital,* and accounts for the deaths of sixty-six non-commissioned officers and men with the battalion between 25th December, 1808, and 24th March, 1809, without distinguishing those killed in action.

Whilst the battalion was stationed at Buckingham and Aylesbury, the county title of the regiment was changed from "Bedfordshire" to "Buckinghamshire."

The battalion received drafts of three hundred and seventy volunteers from the Royal Bucks, Royal Westminster, 2nd Royal Tower Hamlets, Hertford, and Warwick militias, the larger number engaging for life service.

Thus recruited, the battalion was not long before it was again on active service. The British Government had formed a plan for destroying Antwerp, where Napoleon had formed an extensive arsenal, which was a standing menace to our eastern coasts. The movement was intended also to create a diversion in favour of Austria, whither Napoleon had hurried from Spain. Antwerp was at the time badly garrisoned, most of the troops having been withdrawn to reinforce the Grand Army at Vienna. A fine army of twenty-one thousand men, in four divisions, was got ready, the command of which was given to the Earl of Chatham, brother of the deceased Prime Minister, William Pitt. The first division sailed on 28th July, the others following on 29th-30th July. A landing was effected on the island of Walcheren, in the German Ocean, near the mouth of the Scheldt; but instead of pushing on Antwerp, Lord Chatham invested Flushing, the chief

---

* Ship-typhus was very prevalent among the troops returning from Corunna, who carried infection with them wherever they went.

port of the island. The 2nd battalion 14th, Lieutenant-Colonel Jasper Nicolls in command, marched from Buckingham to Portsmouth, and, landing at Walcheren early in August, took part in the siege. The regimental returns show that at this time it had present with head-quarters thirty-one officers, five staff, thirty-seven sergeants, twenty drummers, and seven hundred and sixty-four rank and file. Captain John Marshall was a deputy assistant-adjutant-general with the army. On the evening of 12th August the flank companies, supported by the rest of the battalion, were directed to storm one of the Dutch entrenchments in front of the position occupied by the troops under Major-General Thomas Graham (afterwards Lord Lynedoch). A detachment of the King's German Legion co-operated in the attack. Lieutenant-Colonel Nicolls and the 14th led the assault with great gallantry, the entrenchment being carried in a few minutes, a gun and thirteen prisoners captured, and a lodgment established within musket-shot of the walls of the town. On the following day the line-of-battle ships bombarded the town, which was soon in flames, and presented an awful scene of destruction. The same evening one of the principal batteries was stormed by the 36th, 71st, and a light battalion of the King's German Legion, and on the morning of the 15th the garrison surrendered, marching out with the honours of war on the 18th. The 14th Regiment was thanked in General Orders for its distinguished conduct.

Embarking at Flushing, the battalion was ready to sail up the Scheldt for an attack on Antwerp, but the delays which took place enabled the enemy to prepare additional means of defence. Louis Buonaparte had arrived with large reinforcements; Bernadotte had assumed command at Antwerp, and was placing the city in an active state of defence. After referring the question to the seven lieutenant-generals of the army, it was decided that an attack on Antwerp had become impracticable, and that no good could result from minor operations. Malarial fever was by this time rife in the British camps. It was therefore resolved to re-embark the troops and return home, leaving only a small force on the island of Walcheren, to prevent the French fleet leaving the Scheldt. The 2nd battalion 14th was among the returning regiments, and before the end of September was in Steyning barracks, near Brighton, where it was quartered when the Jubilee of King George III. was celebrated, on 25th October, 1809. Towards the end of the year it was decided to bring off the small force, under General Don, remaining at Walcheren, some of the regiments of which had not a single man fit for duty. A brigade under Major-General George Townshend Walker, in which was the effective portion of the 2nd battalion 14th, under Major James Wood, was despatched for that purpose. The basin and naval defences of Flushing were blown up on 10th December, 1809, and the troops embarked and returned to England, the battalion of the 14th being again collected at Steyning barracks.

1810.—First Battalion.

In the autumn of 1810 the 1st battalion 14th Regiment, which, under command of Lieutenant-Colonel James Watson, had remained at Fort William since its arrival there, was withdrawn from Bengal, to take part in an expedition against the Isle of France, or Mauritius, the chosen haunt of the French cruisers and privateers that harried the India and China trade. The battalion sailed for Isle Rodriguez, the rendezvous of the expedition, and on 28th November the fleet, with the troops on board, sighted the Isle of France. A landing was effected in the Bay of Mapou, whence the troops advanced through a thick wood, skirmishing at times with the French sharpshooters, and, on reaching the open country, moved direct on Port Louis. As the soldiers suffered greatly from want of water, a halt was made at the stream by the powder-mills, about five miles out of the town. Next day the advance was resumed, and there was some sharp fighting, in which the 1st battalion 14th had one man killed and two wounded. The troops took up a position opposite the lines covering the town on that side, and on the following day the French Governor, General de Caen, agreed to surrender the island to Major-General Abercromby. Lieutenant Giles Dove, of the regiment, died of fatigue during the operations. After the capture, the battalion, which was thanked in General Orders for its conduct, embarked for Madras, leaving a large number of sick behind in Mauritius, in charge of Lieutenant Bernard. The rest of the year it spent in camp near Fort St. George, Madras.

1810.—Second Battalion.

Meantime the 2nd battalion, Lieutenant-Colonel Montagu Burrows commanding, had embarked in March of the same year for Malta. On reaching Gibraltar, it was ordered to land, and two companies, under Captains Everard and Ramsay, were detached to Tarifa for the defence of that town. These companies, which were attached to the 47th at Tarifa, returned to Gibraltar in June, and the battalion then resumed its voyage to Malta, where it arrived on 23rd June, 1810.

On the next page is the roll of officers of the regiment, as it appears in the *Annual Army List* for 1810. Some notes have been added, showing the names of those present in the Corunna campaign.

The monthly returns record the deaths of the undermentioned officers of the regiment during the years 1809-10. It will be seen that their names all appear in the annexed roll.

Lieutenant	J. M. Robinson	-	lost at sea (date not stated).
„	J. Gray	-	died 9th November, 1809.
„	R. Heathcote	-	„ 22nd September, 1810.
„	Giles Dove	-	„ 29th November, 1810, in Isle of France.

ROLL OF OFFICERS OF THE REGIMENT, 1810.

(From the Annual Army List. With additional notes)

Rank.	Name.	Rank in the Regiment.	Rank in the Army.
Colonel	Sir Henry Calvert	8th Feb., 1806	*M.G.*, 25th Sept., 1803
Lieut.-Colonel	[1] James Watson	15th May, 1806	
,,	[2] Montagu Burrows	14th May, 1807	
,,	[3] Jasper Nicolls (c)	31st March, 1808	29th October, 1807
Major	[4] Nathaniel Burslem	7th March, 1805	
,,	George Miller	15th May, 1806	
,,	[5] William Wood (c)	14th May, 1807	
,,	[6] F. Skelly Tidy (c)	10th Sept., 1807	12th April, 1807
Captain	William Shea	25th Dec., 1802	
,,	William Griffin	3rd Sept., 1803	
,,	Marinus Kennedy	5th Nov., 1803	
,,	Trevor Stannus	1st August, 1804	9th July, 1803
,,	Peter Johnstone	2nd August, 1804	9th July, 1803
,,	Eyre Coote	3rd August, 1803	5th November, 1802
,,	Jacob Watson	5th August, 1804	
,,	William Fawcett	7th Sept., 1804	
,,	William Moore	29th May, 1806	
,,	D. K. Fawcett (c)	14th May, 1807	
,,	James Dunlop (c)	15th May, 1807	29th November, 1806
,,	[7] Matthias Everard (c)	21st May, 1807	23rd April, 1807
,,	W. H. Hall	29th May, 1807	6th June, 1805
,,	Robert Ramsay (c)	30th May, 1807	30th April, 1807
,,	[8] John Marshall (c)	21st April, 1808	28th November, 1806
,,	J. M. Waller (c)	12th May, 1808	12th November, 1807
,,	[9] Henry Nooth (c)	2nd June, 1808	22nd August, 1804
,,	[10] Cecil Bisshopp (c)	23rd June, 1808	
,,	T. H Light	18th Aug., 1808	
,,	Oct. Temple	15th Sept., 1808	15th December, 1804
,,	Thomas Dunn	22nd Sept., 1808	
,,	W, A. Hodges	22nd Dec., 1808	
Lieutenant	George Rawlins	13th May, 1805	30th January, 1805
,,	W. H. Bertrand	16th May, 1805	
,,	[11] William Turnor (c)	23rd May, 1805	*Adjutant*
,,	John Dyson	4th August, 1805	
,,	M C. Wilkinson	5th Augnst, 1805	
,,	Charles Benn	6th August, 1805	
,,	James Munro Robinson	7th August, 1805	
,,	James Roche	18th Sept., 1805	
,,	John Gray	19th Sept., 1805	
,,	Gerald Rochfort	25th Sept., 1805	
,,	A. P Browne	26th Sept., 1805	
,,	Giles Dove	17th Oct., 1805	
,,	— Knollis	14th Nov., 1805	
,,	Adam Hunter	28th Nov., 1805	
,,	Lewis Algeo	25th Sept., 1806	
,,	Thomas Savage	26th Sept., 1806	
,,	W. H. Coghlan (c)	2nd Oct., 1806	
,,	Ralph Heathcote	9th Oct., 1806	

[1] Afterwards Gen. Sir John Watson, K.C.B. See Succession of Colonels, *Appendix A*.
[2] Afterwards Lieut.-Gen. Montagu Burrows. See *Appendix Volume*.
[3] Afterwards Lieut.-Gen. Sir Jasper Nicolls, K.C.B. See *Appendix Volume*.
[4] Afterwards Col. Nath. Burslem, K.H. Died, 1854.
[5] Afterwards Gen Sir William Wood, K C.B. See Succession of Colonels, *Appendix A*.
[6] Afterwards Col. F S. Tidy, C B. Died, 1834. See *Appendix Volume*.
[7] Afterwards Major-Gen. Mathias Everard, C.B., K.H. See *Appendix Volume*.
[8] Afterwards Lieut.-Col. John Marshall, K.H.
[9] Afterwards Colonel H. Nooth, unattached.
[10] Afterwards Lieut.-Gen. Cecil Bisshopp, C.B. Died, 1858.
[11] Afterwards Major-Gen. William Turnor. Died, 1858.

N.B.—The officers marked (c)—together with the undermentioned, who were not in the regiment at the date of this roll—made the Corunna campaign: Lieutenants John Armstrong and W. J. Rae; Ensigns Roger Swiny, Wm. Deighton, H. Pierard, and John Carr; and Assistant-Surgeon Loinsworth.

RANK.	NAME.	RANK IN THE REGIMENT.	RANK IN THE ARMY.
Lieutenant	— Maxwell	6th Nov., 1806	
,,	William Betts	21st Jan., 1807	*Adjutant*
,,	Hector McLean	2nd April, 1807	
,,	Henry Hill	9th April, 1807	
,,	George Bolton (c)	21st May, 1807	11th December, 1860
,,	Hercules John Heyland	21st May, 1807	
,,	Edward Waller (c)	25th June, 1807	9th April, 1807
,,	Thomas Hewitt Baylie	25th June, 1807	
,,	John Casimir Harold (c)	16th July, 1807	28th May, 1807
,,	William Akenside (c)	6th August, 1807	2nd January, 1807
,,	Thomas Hall	29th Aug., 1807	
,,	Charles Myler Brannon (c)	3rd Dec., 1807	29th January, 1807
,,	Thomas Caulfield	4th August, 1808	
,,	James Bancroft Ainsworth	18th Aug., 1808	
,,	Kenneth M'Kenzie	22nd Sept., 1808	
,,	George White (c)	23rd Sept., 1808	
,,	H. Bruere Armstrong	27th Oct., 1808	23rd January, 1808
,,	Henry Gamble	22nd Dec., 1808	
,,	Thomas Kirkman	9th March, 1809	
,,	John Grant	25th May, 1809	1st April, 1795
,,	Thomas Jenour	25th May, 1809	
,,	Charles Tothill	27th May, 1809	
,,	Thomas Way (c)	27th May, 1809	
,,	George Garnon	28th May, 1809	
,,	William Fellowes	29th May, 1809	
,,	Henry Johnson (c)	30th May, 1809	
,,	Dominic C. Lynch	31st May, 1809	
,,	[1]Rich England	1st June, 1809	
,,	John Ball	15th June, 1809	12th March, 1801
,,	H. J. Henley (c)	15th June, 1809	
,,	George Mackenzie	29th June, 1809	4th May, 1809
,,	H. L. Franklin	19th Oct., 1809	8th May, 1805
Ensign	Geo. Thurles Finucane	22nd Sept., 1808	
,,	E. L'Estrange	10th Nov., 1808	
,,	Charles Gray	5th Jan., 1809	Vol. with the Regiment
,,	Henry Gillman	23rd Jan., 1809	Vol. with the Regiment
,,	Richard Stack	24th Jan., 1089	Vol. with the Regiment
,,	Samuel Parker	9th Feb., 1809	
,,	— Nickisson	9th March, 1809	
,,	William Fowler	10th April, 1809	from Lt. Bedford Mil.
,,	James Gordon	11th April, 1809	from Lt. R. Westminster Mil.
,,	Simon Kent	12th April, 1809	from Lt. 2nd Tower Ham. Mil.
,,	Thomas Powell	13th April, 1809	from Lt. R. West Militia
,,	— Howe	19th April, 1809	from Lt. R. Bucks Militia
,,	— Bryan	20th April, 1809	from Lt. R. Bucks Militia
,,	D. Hazlewood	10th May, 1809	from Lt. R. Herts Militia
,,	— Ewer	14th Sept., 1809	from Lt. R. Bucks Militia
,,	Henry Marshall	27th Sept., 1809	
,,	Chas. Rayner Newman	28th Sept., 1809	
,,	George Fitzherbert	5th Oct., 1809	
Paymaster	Thomas Smart	23rd July, 1807	
,,	William Wilkinson (c)	3rd March, 1808	
Adjutant	William Betts	30th May, 1805	*Lieutenant*
,,	William Turnor (c)	6th March, 1806	*Lieutenant*
Quarter-Master	Henry Brown	25th May, 1803	
,, ,,	James Mercer (c)	28th Aug., 1804	
Surgeon	Thomas Jackson	17th June, 1800	
,,	James Safe (c)	20th March, 1806	
Assistant-Surgeon	W. J. Parker	8th Nov., 1804	
,, ,,	John Smith	30th Jan., 1806	
,, ,,	John Cassidy (c)	21st April, 1808	
,, ,,	John Trigge	29th June, 1809	

[1] Afterwards Gen. Sir Rich. England, G.C.B., K.H., Commanding 3rd division in the Crimea. Died, 1883. He served with the battalion at Walcheren, and was on the staff in Sicily when promoted out of the regiment.

1811.—First Battalion.

Conquest of Java.

After the capture of the Isle of France, the British Government resolved to complete its dominion in the East by the conquest of the Island of Java, of which the Dutch had held undisputed possession for more than one hundred years. The size of the island—six hundred and forty miles long by a hundred broad—the luxuriance and fertility of the soil (the mountain districts yielding the vegetables and grain of Europe, the plains the delicious fruits and other valuable products of the tropics, without need of much tillage, and in such abundance that it was sometimes called the Granary of the East)—rendered the island a valuable possession to the United Provinces, and its capital, Batavia, was the metropolis of the Dutch settlements in the Eastern Seas. Holland had fallen under the sway of Napoleon, and it was deemed necessary to deprive her of her colonial possessions. A considerable force of troops was placed under command of Lieutenant-General Sir Samuel Auchmuty for that purpose, of which the 1st battalion 14th formed part.

The greater portion of the regiment embarked in March, 1811, on board the following ships of war ordered to cruise off the island, in which service they had many opportunities of distinguishing themselves, in destroying the enemy's gunboats, and in other enterprises on the coast: H.M.S. *Caroline*, *Cornelia*, *Doris*, *Sir Francis Drake*, and *Minden*. On 23rd May, 1811, a party of the 14th was on board H.M. light-armed frigate *Sir Francis Drake*, 36 guns, on passage from Rembang to Sourabaya, when a flotilla of fourteen Dutch gunboats was discovered inshore. The gunboats were chased by the frigate's boats, and nine of them captured without the loss of a single man. Lieutenant Henry Gillman and twelve privates of the 14th were in the boats.* Soon after Lieutenant Henry Gillman was killed in another encounter with the enemy, on 4th June, 1811. Lieutenant H. J. Heyland, 14th, and Ensign Oliver Brush, 89th, with

---

* *James's Naval History*, vol. v., p. **269**, *et seq.*

forty soldiers, were engaged with the boats' crews of H.M. brig-sloop *Procris*, Captain Maunsell, in a very sharp engagement with six Dutch gunboats protecting a convoy of forty to fifty prows. Five of the gunboats were captured, and the sixth blew up. Eleven men were wounded in the boats, including two soldiers.* In his despatch, Captain Maunsell highly commended the conduct of Lieutenant H. J. Heyland and the detachments under his orders. Detachments of the 14th and 89th regiments, Royal Marines, and blue-jackets were landed from H.M.S. *Minden*, 74 guns, Captain Hoare, near Bantam, on the coast of Java, and in two contests defeated five hundred of the enemy's chosen troops, which had been sent from Batavia to attack them. Captain John Watson, Lieutenants Rochfort, McLean, and L'Estrange, and Ensign Jennings, of the 14th, and Lieutenant Dunscombe, of the 89th, particularly distinguished themselves on these occasions.

The head-quarters of the 14th, under Lieutenant-Colonel James Watson, left Madras on 18th April, 1811, and on 4th August landed at the village of Chillingching, twelve miles from Batavia, towards which city the army directed its march. The French and Dutch troops set fire to the magazines, and abandoned the city, which was taken possession of by the British.

Advancing from Batavia on 10th August, the British found three thousand Franco-Batavian troops in a strong position, defended by abattis, behind Weltfreden. The post was carried at the point of the bayonet, many of the enemy being killed, and the remainder retreating to an entrenched position at Cornelis, between the great river Jacatra and the aqueduct of Slaken. The conduct of Captain Stannus, commanding the light company of the 14th, and of Lieutenant Coghlan, commanding the rifle company,† was highly commended in Colonel Robert Rollo Gillespie's report of the action. The regiment had Ensign Nickisson and three men wounded.

In the strong position at Fort Cornelis, defended by seven strong

---

* *James's Naval History*, vol. v. A painting of this action, presented by King William IV, is in the Painted Hall, Greenwich Hospital.

† The rifle company was formed of the best shots in the several companies, who were armed with rifles supplied by the Madras Government.

redoubts and numerous batteries, were assembled ten thousand Gallo-Batavian troops, with a large and powerful force of artillery. The weather was too hot and the British too few to allow of an attack in form, and it was therefore decided to attempt the place by assault. The position was stormed by the British on 26th August, 1811, when some two thousand of the enemy were killed, and five thousand more, including three general officers, made prisoners. The rest dispersed, excepting a few men who accompanied the Dutch commander, General Jansens, in his flight. The 14th riflemen and flank companies were in the principal attack, under Colonel Gillespie, the light company being in the right flank battalion, under command of Major George Miller, 14th regiment, and the grenadier company with the supporting brigade under Colonel Gibbs, 59th regiment. The grenadiers had their captain, a sergeant, and two men killed, and over sixty wounded, by a terrific explosion of a powder magazine in a redoubt they had just carried. The battalion companies were engaged in an attack, under Colonel William McLean, 69th regiment, on the opposite side of the position. They distinguished themselves on this as on other occasions, and their commanding officer, Lieutenant-Colonel Watson, was commended by Colonel Gillespie in his despatch. Colour-Sergeant William Hanniford is mentioned in the regimental records as having highly distinguished himself on the occasion.

The victory of Cornelis terminated the Dutch sovereignty of Java. General Jansens was pursued up-country, and on 16th September the 14th was engaged in storming the fortified position of Jattoo, when the remaining Franco-Batavian troops were routed. General Jansens was afterwards forced to surrender, and the rich island of Java became a possession of the British Crown. It was restored to Holland by the Treaty of Vienna in 1814.

The loss of the 14th Foot at the storming of Fort Cornelis was Captain Marinus Kennedy, two sergeants, and nine rank and file killed; Major George Miller, Captain Trevor Stannus, Lieutenants W. H. Coghlan and Kenneth McKenzie, seven sergeants, and eighty-three rank and file wounded, and one rank and file missing.

In his despatch, Lieutenant-General Sir Samuel Auchmuty stated: "The superior discipline and invincible courage, which have so "highly distinguished the British Army, were never more fully dis- "played, and I have the heartfelt pleasure to add, that they have "not been clouded by any acts of insubordination."

The strength of the battalion at the capture of Java was forty-eight officers and one thousand one hundred and forty-five non-commissioned officers and men. The commanding officer, Lieutenant-Colonel James Watson, received a gold medal, and years afterwards the word "Java" was inscribed on the colours of the regiment, to commemorate its distinguished services in connection with the most splendid acquisition made by British arms in the year 1811. Thirty-seven years after the event the survivors received the British and East India Company war medals with "Java" clasp.

1811.—Second Battalion.

The 2nd battalion remained at Malta this year, with a detachment at Gozo. Captain John Marshall,* who had served on the staff at Walcheren, was on the Adjutant-General's staff in Malta, and Lieutenant Richard England† in Sicily.

1812.—First Battalion.

The 1st battalion, which remained in Java after the capture of the island, had its head-quarters at Samarang at the beginning of the year.

The Sultan of Mataran, a potentate who governed a portion of the interior, relying on his power, and the strength of his fortified palace of D'Joejocarta, meditated a plan for the expulsion of all Europeans from the island, and commenced a series of aggressions which it became necessary to stop without delay.

To effect this, a force, consisting of H.M. 14th and 78th regi-

---

* Afterwards Lieutenant-Colonel J. Marshall, K.H.

† Afterwards General Sir Richard England. He was promoted out of the regiment this year.

ments and some native troops, under Colonel Robert Rollo Gillespie, was sent to attack the "Cratton," or fortified palace of the Sultan. The enterprise was a difficult and perilous one, and most gallantly carried out. The Cratton, a fortified enclosure three miles round, with a lofty bastioned rampart mounting one hundred guns, and a very broad wet-ditch, crossed by a few drawbridges only, was defended by seventeen thousand well-equipped troops, whilst a population of one hundred thousand armed Javanese swarmed around it. The attacking force mustered one thousand five hundred men only.

The assault was made on 20th June, 1812, in two columns, whereof the main column, commanded by Lieutenant-Colonel Watson, 14th regiment, escaladed, led by the grenadiers of the regiment, under Captain Peter Johnstone, in their usual gallant style. The garrison held their ground stubbornly and well; but after much hard fighting, the second column having in the meantime effected an entrance over one of the drawbridges, the place was carried, and the capture of the Sultan rendered the victory complete.*

Colonel Gillespie recorded in orders that

"To Lieutenant-Colonel Watson, who commanded the leading "column, the Commander of the Forces cannot convey the sense he "entertains of his distinguished bravery, and of the quickness and "celerity with which he conceived and executed the attack.

"The animated style in which Captain Johnstone and Lieutenant "Hunter crossed the ditch at the head of the 14th grenadiers, and "escaladed the ramparts under the fire of the east bastion, could only "be equalled by the order and zeal of their followers."

The conduct of Lieutenant Hill, who, with a Bengal havildar, reconnoitred the approaches beforehand, and of Lieutenant McLean, commanding the rifles of the regiment, was also commended. The total loss was one hundred killed and wounded, of which the 14th had eight men killed (Lieutenant McLean died of his wounds) and thirty men wounded.

Thorn, the historian of the Java expedition, writes:

---

* "The gallant 14th proceeded to scour the ramparts, and the capture of the Sultan rendered the victory complete."—*London Gazette*, 1812.

Dawsons Ph. Sc.

General Sir James Watson, K.C.B.
Served 75 Years in the 14th Regt
Died August 1862.

"The forbearance of the troops when flushed with success was the " more honourable, as they had a fresh recollection in their minds of " what their comrades had suffered in being mangled and tortured to " death.* The females in the inner apartments of the palace were all " respected, and the property protected. So strict, indeed, was the dis- " cipline observed, that not a single person was molested, nor did any " outrage take place."

1812.—Second Battalion.

In January, this year, Lieutenant-Colonel Montagu Burrows, with the flank companies of the battalion under Captains R. Ramsay and T. H. Light, proceeded from Malta to Sicily, to serve under Lord William Bentinck. They appear to have returned to Malta some weeks later.† Detachments were stationed at Gozo and the Boschetta Isles.

1813.—First Battalion.

In this year a small force, consisting of a detachment of the 1st battalion 14th and some native troops, under command of Lieutenant-Colonel Watson, 14th regiment, was despatched from Batavia against the piratical state of Sambas, on the north-west coast of Borneo. For years the armed prows of the Sultan of Sambas had been accustomed to lurk in the mouths of the Sambas and Borneo rivers, ready to dart out on wind-bound merchant ships, the crews of which were treated with abominable cruelty. On 3rd July, 1813, after some sharp fighting, in which Captain John Watson and Lieutenant Jennings, 14th regiment, were wounded, the defences were captured and destroyed. The thanks of the Governor-General in Council, on the completion of this service, were notified in a General Order, dated Fort William, 4th September, 1813 :—

"The Official report received from the Government of Java of the ' successful result of the operations of the Detachment which proceeded

---

* Six men of H.M. 22nd Light Dragoons had been cut off when scouting near the Cratton, and put to death with cruel torture.

† See Monthly Returns 2nd battalion 14th Regiment, 1812.

"from Java under the command of Lieutenant-Colonel James Watson,
"H.M. 14th Regiment, against the piratical State of Sambas, on the
"West Coast of the Island of Borneo, which terminated in the capture
"of all the Batteries, fortified posts and defences of the Sultan, and
"in the complete discomfiture of Pangerang Anom and his adherents,
"having been published under the authority of Government on 26th
"ult.:

"The Rt. Honble. the Governor General in Council has great satis-
"faction in recording his testimony of public approbation and unbounded
"applause of the judgment, coolness, and intrepidity displayed by
"Lieutenant-Colonel Watson, Captains Watson and Brooks, and the
"other officers and men, both European and Native, employed on that
"important service."

The Battalion proceeded from Java to Bengal in October, this year, and, after a sojourn at Fort William, moved into cantonments at Berhampore.

During its service in Java the battalion became reduced to five hundred and fifty of all ranks, inclusive of detached details. It lost the under-named officers:

In 1811, *Captains* Shea (died in I. of France), M. Kennedy (killed in action, 2nd August); *Lieutenants* T. Caulfield (died, 18th April), H. Gillman (killed in action, 4th June), W. H. Coghlan (died of wounds, 30th August), C. Benn (died, 23rd September), M. C. Wilkinson (died, 13th October).

In 1812-13, *Major* G. Miller (died, 12th July); *Lieutenants* H. McLean (died of wounds, 23rd June), L. Algeo (died, 4th August); *Lieutenant* A. P. Brown (died, 9th April, 1813).

Brevet Lieutenant-Colonel Burslem remained behind as deputy barrackmaster-general at Batavia, and Lieutenant Heyland as assistant to the resident at Cadoe.

1813.—Second Battalion.

The 2nd battalion remained in Malta during this year, with a detachment under Lieutenant Gordon* at Lampedusa, a rocky islet between Malta and Tunis. An outbreak of the plague occurred in Malta this year, which carried off four thousand five hundred

* Lieutenant James Gordon died in 1817.

persons. Rigid measures of police were adopted on the first appearance of the pestilence in May, framed on the assumption that the disease was communicable by contagion only, and not by infection. Guards of troops were placed over infected houses, cordons were drawn round the pest hospitals, and various precautions enjoined; but, although a native Maltese was shot in the streets of Valetta for concealment of the disease soon after the outbreak, it was not until after the arrival of the new Governor, Sir Thomas Maitland, in October, when the pestilence was already on the turn, that the regulations were uniformly enforced with a stringency that speedily effected the desired results. The 4th December, 1813, was proclaimed a "day of purification" for the whole island; and on 13th December the village of Casal Curmi was "put out of the King's peace," also by proclamation. In this miserable, low-lying "casal" the plague appeared to have become endemic, owing, it was alleged, to systematic thieving of infected property by the inhabitants. A cordon of troops was placed round the village, within which the inhabitants were locked in their houses day and night, the keys being in charge of the troops, who supplied their prisoners with food day by day. A commission, consisting of the adjutant-general and two regimental officers, was appointed to administer martial law within the cordon while the proclamation continued in force. A medical inspection took place daily till the cases ceased. At the New Year, Malta was declared free of the plague.

Cases of plague occurred among the troops, consisting of the 2nd battalion 14th Regiment, 3rd Garrison battalion, De Roll's Regiment, the Sicilian Regiment, and a detachment of Royal Artillery, many of which recovered. There are no detailed medical reports of the garrison extant; but the Monthly Returns of the 14th Regiment show only five deaths, from all causes, in the battalion during the year, and ninety-three men invalided home at its close.* The return of the battalion for December shows a

* Additional particulars will be found in medical works on the plague, particularly that of Staff-Surgeon J. D. Tully (*History of the Plague in Malta, Corfu, etc.* London, 1821). The printed returns of the Malta Board of Health

strength of one thousand one hundred and seventy-six non-commissioned officers and men (all English, except forty-two), five hundred and forty-one of whom were enlisted for life service.

1813.—A Third Battalion Formed.

A third battalion was added to the 14th Regiment this year. It was formed at Weedon Barracks, by Lieutenant-Colonel the Honble. James Stewart,* of volunteers from the militia. Since the legalisation of volunteering from the embodied militia, it had been the rule to give ensigncies in the line to militia subalterns bringing a sufficient number of volunteers. Many militia officers had thus joined the regiment during the years 1808-12. A further alteration in the Militia Act now authorized the granting of rank in the line, *equal to that held by them in the militia*, to militia captains and subalterns bringing certain quotas of militia volunteers, such rank to be temporary for the first nine months, and then to be made permanent if the war continued.

The following militia officers were thus appointed to the new battalion:

Captains.—John Robertson, Royal Westminster Militia; Henry Nicolls, Bedfordshire Militia; Henry Coxwell, Hertfordshire Militia; J. G. Doran, Berkshire Militia; Gray E. Boulter, Notts Militia; and Thomas Gould, Hertfordshire Militia.

Lieutenants.—Charles Pycroft, Hertfordshire Militia; R. P. Condonne, Bedfordshire Militia; — Stephens, Berkshire Militia; Cham. Hall, Bedfordshire Militia; William Booth, Berkshire Militia; Charles

---

for the period are in the British Museum, but do not account for the Military Pest Hospital. The name of Colonel Burrows appears as one of the leading members of the Society for the Relief of the Poor, which, in a place like Malta, had a most important part in dealing with the calamity.

* Lieutenant-Colonel the Honble. James Henry Keith Stewart, C.B., a son of the seventh Earl of Galloway, and brother of Lieutenant-General the Honble. Sir William Stewart, G.C.B., with whom he had served as captain in the 95th Rifles. Lieutenant-Colonel Stewart left the 14th for the 3rd Guards (Scots Guards), with which he was at Waterloo. He retired from the service soon after, and died in 1836, aged fifty-three.

Atkinson, Berkshire Militia; Francis Beardsley, Notts Militia; and J. J. Moore, Royal South Lincoln Militia.

The following received ensigncies:—Lieutenant Alfred Cooper, West Essex Militia; Lieutenant W. Reed, South Devon Militia; Ensign J. Nickelson, Hertfordshire Militia; Ensign Channel, Berkshire Militia; and Ensign Westcote, Notts Militia.

Lieutenant Akenside,* 14th Regiment, who had served with the 2nd battalion at Corunna, was appointed adjutant, and, at the special request of Colonel Stewart, two veteran riflemen of the 95th—Sergeants A. Ross† and William Graham‡—were appointed respectively quarter-master and sergeant-major of the battalion.

1814.—First and Second Battalions.

The 1st battalion, under Lieutenant-Colonel Watson, C.B., remained in cantonments at Berhampore, Bengal, during this year.

Early in the year an Anglo-Sicilian force, under command of Lieutenant-General Lord William Bentinck, sailed from Sicily for the long-projected descent on the coast of Italy. After operations at Leghorn, Spezzia, Pisa, etc., these troops took and occupied Genoa on 18th April, 1814. Just a week later the 2nd battalion 14th Regiment, under command of Lieutenant-Colonel Montagu Burrows, arrived from Malta, as a reinforcement, having taken possession of several places on the Italian coast on their way.§

After thirty days' quarantine, the battalion landed, and went into garrison at Genoa, where a draft from England, under Major W. Fawcett, which had been doing duty at Messina, and the detachment from the island of Lampedusa joined head-quarters. In July, 1814, in consequence of reported disorders at Sarzana, Major R. S. Tidy, 14th regiment, was despatched thither with three hundred men of the 2nd battalion 14th Regiment and two guns. Major

---

* Adjutant at Waterloo.

† Quarter-Master at Waterloo.

‡ Sergeant-Major at Waterloo and Bhurtpore. Afterwards captain half-pay, unattached.

§ On the authority of Captain Burrows, R.N., Chichele Professor of Modern History, University of Oxford.

Tidy found, on arrival, that the alleged resistance to authority had no foundation of fact, and that exactions were being illegally levied on the inhabitants on the plea of supplying the British troops. He returned to Genoa, leaving a company at Sarzana—a measure approved by Major-General Honstedt, King's German Legion, commanding the British troops in Genoa.*

1814.—Third Battalion.

Early in the year the 3rd battalion 14th Regiment, which was still at Weedon, was ordered to hold itself in readiness for foreign service, and had already begun its march to the coast, when the entry of the Allies into Paris in April caused the order to be countermanded.

On 30th May, 1814, a General Peace was concluded between France and the Allied Powers. Reductions in the army were speedily announced, and the militia officers of the battalion, after the completion of their nine months, were retired from 24th September, 1814, receiving permanent half-pay as line officers in return for their services.† Their places were filled chiefly from the half-pay list.

In the autumn the battalion, of which Major and Brevet Lieutenant-Colonel Tidy was now in command, was ordered to Plymouth, to embark for America; but the Treaty of Peace between Great Britain and the United States, concluded at Ghent on 24th December, 1814, although not known on the other side of the Atlantic in time to avert the unfortunate attempt on New Orleans, prevented the despatch of fresh troops from home. The 3rd battalion 14th Regiment accordingly remained at Plymouth, awaiting disbandment, which was ordered to take place on 24th March, 1815, the last day of the current financial year.

1815.—Third Battalion. Waterloo Campaign.

Whilst the Congress of Vienna was reconstructing the map of

---

* Mrs. Ward's *Recollections of Colonel Tidy.*

† Of the 14th Regiment, Lieutenants Pycroft, Hall, Booth, Atkinson, and Moore were still drawing half-pay at the time of the Crimean War.

Europe and the nations were looking forward to a period of peace, Napoleon reappeared on the scene. Leaving Elba suddenly, he landed at Cannes on 1st March, 1815, and was everywhere welcomed by his old troops. On 20th March he entered Paris. In the interim, Great Britain, Austria, Russia, and Prussia had declared war against the usurper, and hurried preparations for a conflict began once more. The order for disbanding the 3rd battalion 14th Regiment was forthwith rescinded, and on 21st March, 1815 (within three days of the date that had been fixed for its disbandment), it received orders for Belgium, to join the forces there collecting under the Prince of Orange. Under command of Major and Brevet Lieutenant-Colonel R. S. Tidy, the battalion landed at Ostend on 31st March, and was sent to the neighbourhood of Grammont. Other regiments of horse and foot were sent over in quick succession; the Duke of Wellington assumed the chief command; and the army was organised in brigades and divisions. The 3rd battalion 14th Regiment, with the 23rd Royal Welsh Fusiliers and the 51st Light Infantry, were formed into the 4th Brigade, commanded by Brevet Colonel H. H. Mitchell,* 51st regiment, which constituted part of the Fourth, or Lieutenant-General the Honble. Sir Charles Colville's, division.

On 13th June, 1815, the campaign opened with Napoleon's attack on the Prussians at Wavre. On the days of the fighting at Quatre Bras and Ligny, Mitchell's brigade was at Enghein and Nivelles; on the evening of the latter (17th June) it was, with the rest of its division, in the vicinity of the Duke's position in front of Waterloo. At sunrise on the memorable 18th June an order was given for Colville's division to fall back on Hal, eight miles distant, where it passed the day out of sight and sound of the mighty conflict raging near.† Mitchell's brigade was excepted from the order, and remained on the field with Lord Hill's corps, on the extreme right

---

* Lieutenant-Colonel and Brevet Colonel H. H. Mitchell, C.B., 51st Regiment, a Peninsular officer, died in 1817.

† It is a singular fact that although the reverberations of the firing were distinctly noticeable at Dover and Hythe on that Sunday afternoon, the brigades of Colville's division at Hal were ignorant of a battle having been fought.

of the allied position, being attached for the day to Sir Henry Clinton's division.

When the battle commenced, soon after 11 a.m., the disposition of the brigade appears to have been as follows: An avenue of fine trees (which succumbed to the injuries they received, and have long since disappeared) then led from the Nivelles road to the Chateau of Houguomont on its left. In the part of this avenue abutting on the road was posted the light company of the 23rd Fusiliers. Hard by, an abattis had been thrown across the Nivelles road, and close to it was posted a company of the 51st Regiment, under Captain, afterwards Major, Ross. (The 51st, it may be remarked, still wore their old sea-green facings.) On the right of the Nivelles road, and in prolongation of the line of avenue from the Chateau, ran a hollow lane towards Braine l'Alliance. Along this hollow way were extended four companies of the 51st Regiment and the light company of the 14th Regiment, the latter forming the right of the line of skirmishers, which thus faced the extreme left of the French position. About two hundred yards in rear of the line of skirmishers was the rest of the 51st, in support; and further to the rear, in column, stood the 14th Regiment in reserve. The remaining companies of the 23rd Fusiliers were away to the left of the Nivelles road, in rear of the 2nd, or Byng's, Brigade of Guards. In these relative positions the regiments of Mitchell's brigade remained during the furious attacks on Houguomont, and subsequently on the left and left centre of the Duke's position, which occupied the earlier part of the day. In the afternoon fresh attempts were again made on Houguomont. Major Ross, 51st regiment, in a letter written some twenty years after the battle, says that a body of French cuirassiers broke through the skirmishers of Mitchell's brigade about 2 p.m., and surrendered in rear of them. Some of them afterwards attempted to break away from the weak escort of dragoons in charge of them, and were shot at the abattis and in the hollow way when attempting to gallop back to their own side.* When Byng's Guards were taken

* Siborne's *Waterloo Letters*, p. 315. Probably, although the time assigned to it differs, this is the incident related in Dalton's *Waterloo Roll*, on the authority

to reinforce Houguomont, their place was occupied by the 23rd Regiment and some Brunswickers. The other regiments of Mitchell's brigade were also advanced, bringing them more under the fire of the French guns on the opposite ridge. Eventually, according to the statement of Lord Albemarle, the 14th took up a position "about "100 yards from the Nivelles road, with the left of their right "company resting on the abattis across the road." The late General Cavalié Mercer's horse-artillery troop was at this period of the battle a little to the right front of the 14th, and he has left a vivid picture of the situation, than which a more trying one to young troops can hardly be conceived. He says :

"Meantime the roar of cannon and musketry in the main position "never slackened ; it was incessant, as was the smoke arising from it. "Amidst the fire from time to time were to be seen still more dense "columns of smoke rising into the air, like a great pillar, then spreading "out into a mushroom head. These arose from the explosions of ammu-"nition-waggons, which were constantly taking place, although the noise "which filled the whole atmosphere was too overpowering to allow them "to be heard.

"Amongst the multitudes of French cavalry continually pouring over "the front ridge, one corps came sweeping down the slope entire, and "was directing its course straight to us, when suddenly a regiment of "light dragoons (I believe of the German Legion)* came up the ravine at

---

of a friend of the late Captain Samuel Goddard, 14th regiment. That officer, who was afterwards quarter-master of the regiment at the capture of Bhurtpore, was then a sergeant in the 3rd battalion. "He was with an advanced party of "skirmishers of the 14th, and about 4 p.m. the reflex wave of some French "cuirassiers passed through them. They were of course fired at by the 14th "skirmishers, and several of them bit the dust. One poor wounded Frenchman "was thrown from his horse, and a comrade nobly rode back and offered him the "help of his stirrup. An active light company man of the 14th, named Whitney, "who had shot one cuirassier, having reloaded, was about to fire on the French-"man, when Sergeant Goddard interposed, saying, 'No, Whitney ; don't fire. "Let him off. He is a noble fellow !' "—Dalton, *Waterloo Roll*, p. 225.

* This was the 2nd Light Dragoons, King's German Legion, the same corps which, as a heavy regiment, had so greatly distinguished itself, under Baron Bock, at Salamanca. It was now light cavalry, and wore scarlet uniforms with black velvet facings. It had been detached from its brigade (Sir William Dornberg's) "before 4 p.m." to look after some French cavalry menacing Braine l'Alleud, and "rejoined its brigade at 6.30 p.m., when the French cavalry had disappeared "from Braine l'Alleud." (See Beamish, *History of the German Legion*, vol. ii. pp. 206-282.)

"a brisk trot on their flank. The French had barely time to wheel up to "their left and push their horses into a gallop, when the two bodies came "into collision. They were at a very short distance from us, so we saw "the charge perfectly. There was no check; no hesitation on either "side; both seemed to dash on in a most reckless manner, and we fully "expected to have seen a horrid crash.—No such thing. Each, as if by "mutual consent, opened their files on coming near, and passed rapidly "through each other, cutting and pointing, much as one might pass the "fingers of the right hand through those of the left. We saw but few "fall. The two corps re-formed afterwards, and in a twinkling both dis- "appeared—I know not how or where. It might have been about 2 p.m. "that Colonel Gould, R.A., came to me, or perhaps a little later.* Be "that as it may, we were conversing on the subject of our situation, which "appeared to have become rather desperate. He remarked that in the "event of a retreat there would be but one road, which, no doubt, would "immediately be blocked up. . . . . We were still talking on this subject "when suddenly a dark mass of cavalry appeared for an instant on the "main ridge, and then came sweeping down the slope in swarms, remind- "ing me of an enormous surf breaking over the prostrate hull of some "stranded vessel, and then running hissing and foaming up the beach. "The hollow space between was in an instant covered with horsemen, "crossing, turning, riding about in all directions, apparently without any "object. Sometimes they would come pretty near us; then they would "retire a little. There were lancers among them, hussars, and dragoons. "It was a complete *melée.* On the main ridge no squares were to be "seen; the only objects were a few guns standing in a confused position, "with their muzzles in the air, and no artillerymen near. After caracoling "about for a few minutes, the crowd began to separate and draw together "in small bodies, which continually increased, and now we really appre- "hended being overwhelmed as the first line had been.

"For a moment an awful silence pervaded that part of the position, to "which we anxiously turned our eyes. 'I fear all is over,' said Colonel "Gould, who still remained by me; and this time I could not withhold "my assent to his remark, for it did indeed appear to be so. Meantime "the 14th, springing from the earth, had formed their square, whilst we, "throwing back the guns of our right and left divisions, stood waiting in "momentary expectation of being enveloped and attacked. Still they "lingered in the hollow, when suddenly loud and repeated shouts (not "English hurras) drew our attention to the other side.

"There we saw two dense columns of infantry, pushing forward at a "brisk pace towards us, crossing the fields as though they had come from "Merke Braine. Everyone, both of the 14th and ourselves, pronounced "them to be French; yet still we delayed opening fire upon them.

---

* Written several days after the battle.

" Shouting, yelling, and singing as they came right for us, and now not
" being more than 800 or 1,000 yards from us, it seemed folly to allow
" them to come nearer unmolested. The commanding officer of the 14th,
" to end our doubts, rode forward to ascertain what they really were, but
" soon returned, assuring us they were French. The order was already
" given to fire, when luckily Colonel Gould recognized them as Belgians.
" While our attention was occupied by these people, the cavalry had
" vanished, nobody could say how or where. We breathed again." *

Mercer's guns were galloped away to the left soon after, and he saw no more of Mitchell's brigade ; but from the narrative of Lord Albemarle it appears that the 14th and adjacent squares† received and repulsed at least one formidable charge of a magnificent body of Imperial cuirassiers late in the afternoon. When the shades of evening were drawing over the field, and the general advance took place, large bodies of French lancers, which had reappeared on the opposite ridge, as well as the knots of sharpshooters in the foreground, melted away. The 14th bivouaced for the night close to the gates of Houguomont.

The official returns show that the 3rd battalion 14th Regiment, Major and Brevet Lieutenant-Colonel R. S. Tidy commanding, went into action on 18th June with thirty-eight officers, thirty-three sergeants, eleven drummers, and five hundred and forty-eight privates present and fit for duty, and that it had seven men killed, and one officer, four sergeants, and sixteen men wounded.

The battalion pay-list, now in the Public Record Office, accounts for eight men dead, viz.: Privates Thomas Dorman, Thomas Gibbards, James Hunt, Robert Jolly, George Postlethwaite, Jacob Parr, Thomas Williams, and William Williams. The only officer wounded was Ensign Alfred Cooper. Among the non-commissioned officers wounded were Sergeant-Major William Graham and Drum-Major Sunderland.

The following were the officers present with the battalion: ‡

---

* Mercer's *Journal of the Waterloo Campaign*, vol. i. p. 305, *et seq.*

† Lord Albemarle speaks of a square of Brunswickers. Major Ross, 51st regiment, believes that there were no Brunswick troops just about. Some stray men of the Brunswick infantry fell in with his company of the 51st, having known him as aide-de-camp to Major-General Berniewicz, of the Brunswick troops, in the Peninsula. (See Siborne's *Waterloo Letters.*)

‡ The roll given in Cannon's *Historical Records* is incomplete, as it only gives the names of Waterloo officers still with the regiment when the medals were issued.

Field Officers :—Major and Brevet Lieutenant-Colonel R. S. Tidy,[1] commanding; Major J. Keightley,[2] Captain and Brevet Major George Marlay, C.B.[3] (acting).

Captains Thomas Ramsay, William Turnor,[4] William Ross, Richard Adams, Christopher Wilson, J. Loraine White,[5] and William Hewett.[6]

Lieutenants W. Akenside, C. M. Brannan, Samuel Beachcroft, George Baldwin, J. Nickelson, L. Westwood, H. Boldero, and J. C. Hartley.

Ensigns W. Reed, George M'Kenzie, C. Fraser, A. A. Adamson, W. Keowan, J. M. Wood, A. Ormsby,[7] J. R. Smith, A. Cooper, J. Bowlby, J. P. Mathews, R. J. Stackpoole, R. Holmes, R. B. Newenham, and Honble. G. T. Keppel.[8]

Paymaster P. Melton. *Adjutant* Lieutenant George Buckley. *Quarter-Master* Alexander Ross. *Assistant-Surgeons* A. Shannon and H. Terry.

Volunteer Montagu Burrows.[9]

The following extracts are taken from *Fifty Years of My Life*, by the late Earl of Albemarle. They contain the most circumstantial account of the proceedings of the regiment during and just before the battle that has yet appeared in print.*

LORD ALBEMARLE'S NARRATIVE.

"The 3rd battalion of the 14th Foot, which I now joined (April, "1815), was one which in ordinary times would not have been con-"sidered fit to be sent on foreign service at all, much less against an "enemy in the field. Fourteen of the officers and three hundred of the "men were under twenty years of age.† These last, consisting princi-"pally of Buckinghamshire lads fresh from the plough, were called at

[1] Colonel R. S. Tidy, C.B. (See *Appendix*.) [2] Colonel J. Keightley, C.B., afterwards commanding 11th and 35th regiments. Died 1852. [3] Lieutenant-Colonel George Marlay. Served as a D.A.G. in Peninsula (C.B. and gold medal). Died 1830. [4] Major-General W. Turnor, C.B. Died 1860. [5] Captain Loraine White, Military Knight of Windsor. Died 1875. [6] Colonel Hewett. (See *Appendix*.) [7] Afterwards captain, unattached. The last Waterloo officer to leave the regiment. [8] Earl of Albemarle. (See *Appendix*.) [9] A connection of Colonel Montagu Burrows. Received a commission a few days after the battle. Died young, in the 41st Foot.

* See also the letters of Major-General, then Captain, W. Turnor, and Ensign Keowan, in *Appendix Volume*.

† The last monthly return rendered before the battle of Waterloo showed that the battalion had eighteen sergeants, eight drummers, and three hundred and thirty rank and file enlisted for life service, and nineteen sergeants, two drummers, and two hundred and forty-two rank and file for limited service. Of the non-commissioned officers and men all but seven were English.

"home 'The Bucks,' but their un-buckish appearance procured for them "the nickname of the 'Peasants.'

"Our colonel, Lieutenant-General Sir Harry Calvert, bore the same "name as a celebrated brewer; and as the Fourteenth was one of the "few regiments in the service with three battalions, we obtained the "additional nickname of 'Calvert's Entire.'*

"In my commanding officer, Lieutenant-Colonel Francis Skelly Tidy, "I found a good-looking man, above the middle height, of soldier- "like appearance, of a spare but athletic figure, of elastic step, and "of frank, cheerful, and agreeable manners. He had been present "at the reduction of all the French islands in the West Indies, had "served under Baird and Wellesley in Spain, in 1808, and in the "Walcheren expedition the following year. When I reported myself, 'Tidy was in high spirits at having procured for his regiment a "prospective share in the honour of the forthcoming campaign. The "Battalion had been drawn up in the square at Brussels the day "before, to be inspected by an old General of the name of Mackenzie,† "who no sooner set eyes on the corps than he called out: 'Well, I "never saw such a set of boys, both officers and men.' Tidy asked the "General to modify the expression. 'I called you boys,' said the veteran; "'and so you are; but I should have added, I never saw so fine a set "of boys, both officers and men.'

"Still the General could not reconcile it to his conscience to declare "the raw striplings fit for active service, and ordered the Colonel to "march them off the ground and to join a brigade then about to "proceed to garrison Antwerp. Tidy would not budge a step. Lord "Hill happening to pass by, our Colonel called out: 'My lord, were "you satisfied with the behaviour of the "Fourteenth" at Corunna?' "'Of course I was; but why ask the question?' 'Because I am sure "your lordship will save this fine regiment from the disgrace of garrison "duty.'

"Lord Hill went to the Duke, who had arrived that same day at "Brussels, and brought him to the window. The regiment was inspected "by His Grace, and their sentence reversed.

"In the meanwhile a priggish Staff Officer, who knew nothing of the "countermand, said to Tidy in mincing tones: 'Sir, your brigade is "waiting for you. Be pleased to march off your men.' 'Ay, ay, Sir,'

---

* The 2nd battalion was known as Calvert's "All-Butt;" the 3rd as Calvert's "Entire."

† Major-General Kenneth Mackenzie, commanding at Antwerp, afterwards Lieutenant-General Sir Kenneth Douglas, Bart. Died 1836. Lord Albemarle appears not to have known that he was the Mackenzie who, as an old subaltern of fifteen years' standing, had signalized himself in command of the 14th light company at Dunkirk in 1793.

"was the rough reply; and with a look of defiance my Colonel gave the "significant word of command: 'Fourteenth, TO THE FRONT! Quick "march.'

"From henceforth our regiment formed part of Lord Hill's Corps. . . .

"For four days in a week, from daylight to nine in the morning, we "were generally engaged in regimental drill. The other two days were "devoted to exercise in brigade movements.

"Our brigade, under the command of Brigadier Mitchell, was com-"posed of the 14th, 23rd, and 51st regiments. . . . .

"Time hung somewhat heavily on the hands of us, officers, in the "Acren cantonment; a swim across to Dender, or a stroll into Gram-"mont, where we made acquaintances with the 23rd, 51st, and 52nd "regiments, formed our principal recreations. Our men were more "agreeably and usefully employed; they were quite at home with the "'Peasants' upon whom they were billeted, and clubbed their rations of "bread, meat, and schnapps with the vegetables, cheese, butter, and beer "of their hosts.

"Whenever not on duty they were to be seen assisting the Boers and "Boerrinen in their various labours.

"Before they left the cantonment they had weeded the flax and the "corn, and the potato crop of that year was entirely of their planting.

"Races on a grand scale came off at Grammont on the 13th June.

"*June* 15. I was this afternoon, about sunset, one of a group of officers "assembled near the principal inn at Acren, when a Belgian, dressed in "a blouse, told us the French had crossed the frontier. I well remember "the utter incredulity with which his statement was received by us all; "but it proved to be perfectly correct. At daybreak that morning "Napoleon opened the campaign by attacking the first corps of the "Prussian army, commanded by Count Ziethen, in the neighbourhood of "Charleroi.

"The following morning, as I was proceeding to fall in with my com-"pany as usual, I found the regiment in heavy marching order, and all "ready for a start. They had received 'the route' to Enghien. . . . .

"Hurrying back to my billet, I swallowed hastily a few mouthfuls of "food, and with the assistance of my weeping hostess, packed up my "baggage. It was consigned to the care of the baggage guard, and when "I entered on the Waterloo campaign, all my worldly goods consisted "of the clothes on my back.

"As we passed through the village, our drums and fifes playing '*The* "*girl I left behind me*,' we were greeted with the cheers of the men and "the wailing of the women.

"At Enghien we received a fresh route for Braine-le-Comte. During "this afternoon we could hear the booming of the artillery at Quatre "Bras. We entered Braine-le-Comte after dark, and left it before the "break of day.

"*June* 17. We were now ordered to Nivelles. As we approached the
" town, we met several spring carriages of the Waggon Train, full of
" the men wounded at Quatre Bras. . . . .

" We were detained two hours at Nivelles, to allow some Belgian cavalry
" to pass through our ranks. We resumed our march at three in the
" afternoon. Before we reached our ground, the rain came down in tor-
" rents, and in a few moments wetted us to the skin.

"Ascending the rising ground on which Mont St. Jean is situated,
" the Colonel pointed to a spire in the distance. 'That,' said he, 'is
" Waterloo.' . . . .

" Prior to taking up our position for the night, the regiment filed past a
" large tub full of gin. Every officer and man was, in turn, presented
" with a little tin-pot full. As soon as each man was served, the remains
" in the tub were tilted over on to the ground. We soon after halted and
" piled arms.

" Looking to the south, that is to say in the direction of the ground we
" had lately traversed, we heard heavy firing to our left.* This proceeded
" from La Haye Sainte, where Picton had ordered two brigades of artil-
" lery to play upon the French infantry, which was pressing upon the
" Anglo-Allied Forces in retreat upon Waterloo from Quatre-Bras.

" For about an hour before sunset, the rain that had so persecuted us
" on our march relieved us for a time from its unwelcome presence, but
" as night closed in, it came down again with increased violence, and
" accompanied by thunder and lightning. For a time I abode as I best
" could the pitiless pelting of the storm. At last my exhausted frame
" enabled me to bid defiance to the elements. Wearied with two days of
" incessant marching, I threw myself on the slope of the hill on which I
" had been standing. It was like lying in a mountain torrent. I never-
" theless slept soundly till two in the morning, when I was awoke by my
" soldier servant, Bill Moles. . . . .

"*June* 18. During the first hour after sunrise on the morning of the
" 18th, our regiment, like the rest of the troops, were occupied in cleaning
" and drying their arms (a very necessary business after such a night as
" we had passed through). That done, we had a rigid inspection of every
" musket and ammunition pouch. We then piled arms, and fell out till
" the bugle recalled us to the ranks. . . . .

" There was, I should suppose hardly any British soldier in the field
" that morning who did not understand that we were there, not to give, but
" to receive battle ; and who was not surprised that hour after hour should
" pass away without any indication from the enemy that he intended to
" pay us a visit. . . . . We had been under arms for six hours, when a
" numerous cavalcade appeared on the crest of the opposite hill (evidently
" some great man and his suite) ; they were so near that a small body of

---

* See also Mercer's *Waterloo*.

" Volunteer Riflemen of the present day could easily have emptied every
" saddle.

" My comrades and I made sure that we had seen Napoleon himself.
" We were wrong ; it was Jerome Buonaparte, whose division was posted
" on the extreme left of the French line, facing Houguomont; he had just
" received his Imperial brother's order to give the signal of battle. Almost
" the moment he disappeared from view, a single cannon-shot was fired ;
" a pause of two seconds was distinctly perceptible, and then arose a roar
" of artillery, which did not cease for the next eight hours. For some
" time after the firing had begun, Mrs. Ross, our quartermaster's wife,
" remained with the regiment. She was no stranger to the battlefield,
" and had received a severe wound in Whitelock's disastrous retreat from
" Buenos Ayres, 1807. . . . . At length she was prevailed upon to leave
" the field, and passed the rest of Sunday in the belfry of a neighbouring
" church, where she enjoyed a better view of the battle than could have
" been obtained by the commander of either army.

" From the spot we then occupied we could see neither friend nor foe.
" Our arms were piled, and we were waiting for orders to fall in. I was
" one of a group assembled round our sergeant-major, James (William)
" Graham, who was fighting some of his Peninsular battles 'o'er again.'
" Suddenly the spokesman fell to the ground ; a chance musket ball had
" struck him on the neck. Although in great pain, nothing would induce
" him to leave the field."

Albemarle describes how as junior ensign he proceeded to his usual place with the colours, and was shortly after relieved by another ensign, and that a colour-sergeant, thinking it a good opportunity for instructing the two "young officers" as to what they were to expect, said : " 'Now,
" you see, the enemy always makes a point of aiming at the colours; so if
" anything should happen to either of you young gentlemen, I up with
" the colour and defends it with my life.' One of the first casualties of
" the day happened to this man, Colour-Sergeant Moore. . . . . Colville's
" Division, the one to which we belonged, being at Hal, eight miles dis-
" tant, we were attached to the second infantry division, under Lieutenant-
" General Sir W. Clinton. . . . .

" Along a portion of a road, principally consisting of a hollow way,
" Siborne tells us, were posted in advance some light troops of the Anglo-
" Allied Army.

" They formed a part of the fourth brigade of the fourth division (under
" Colonel Mitchell) attached to the 2nd Corps, commanded by Lieutenant-
" General Lord Hill. The brigade consisted of the 3rd battalion 14th
" Regiment (Lieutenant-Colonel Tidy) ; 23rd Fusiliers (Colonel Sir H.
" Ellis) ; and the 51st Light Infantry (Lieutenant-Colonel Rice), in the
" following manner :—Along that portion of the Houguomont avenue
" which is nearest to the Nivelles road was extended the light company
" of the 23rd Regiment. On its right was an abattis which had been
" thrown across the great road ; and close upon the right of this artificial

"obstacle a company of the 51st was posted. Four more companies of "this regiment, and the light company of the 14th, were extended along "the hollow way alluded to as stretching across the ridge on the extreme "left of the French position. The remainder of the 51st stood in column "of support about 200 yards in rear of the hollow way. The 23rd Regi- "ment was stationed on the left of the Nivelles road, on the reverse slope, "and immediately under the crest of the main ridge, in rear of the 2nd "Brigade of Guards. The 14th Regiment was posted in column on the "southern descent of the plateau, on which was assembled the 2nd British "Division.

"To arrive at this position, we descended the plateau we had hitherto "occupied, and entered upon a narrow ravine covered with brushwood. "Shot and shell came occasionally into the ravine; but as we were out of "sight of the French artillery, they did us no harm. How long we "remained in this place I have no idea. . . . .

"About 3 in the afternoon Napoleon, who in the early part of the "action had directed his principal attack on our left and left centre, sent "strong reinforcements to his troops attacking Houguomont—that part of "the field in which the 14th were posted. In consequence of this, General "Byng's Brigade advanced towards the chateau, and an open space was "left between Halkett's and Kemp's Brigades. Sir J. Shaw Kennedy "pointed out the chasm to the Duke, who said to him: 'I shall order the "Brunswick Troops to the spot, and *other troops besides;* go you and get "all the German Troops of the Division to the spot where you are, and all "the guns that you can find.' . . . .

"I presume that our regiment formed a portion of the *other troops*, "whom the Commander-in-Chief sent to fill up the hiatus; for it must "have been about this time that Captain Bridgman, one of Lord Hill's "aides-de-camp, brought us the order to advance. We marched in "column of companies. The hill in our front was fringed with the "enemy's cannon, and we advanced to our new position in a shower of "shot and shell. Turnor, the captain of my company, writing home on "the 19th June from the field of battle, says: 'The whole day we were "exposed to the fire of several batteries of artillery, and particularly of "two guns brought to bear on us.' I can well remember the interest I "took in those two guns—an interest heightened by the consideration "that I formed part of the living target against which their practice was "pointed.

"Fifteen years after the battle, I was present in Paris at the Grands "Couverts, the annual dinner which the older Bourbon princes were in "the habit of eating in public. A French officer on duty entered upon a "subject of his own choosing, but one generally avoided by his country- "men—Waterloo. He told me he was an artillery officer, posted in that "action on the extreme left of the French line, and that his orders were "to fire on three British regiments, the colours of which were buff, blue,

" and green, thus proving beyond all doubt that it was against our brigade
" that his fire was directed.

" We halted and formed square in the middle of the plain. As we were
" performing this movement, a bugler of the 51st who had been out with
" skirmishers and had mistaken our square for his own, exclaimed, 'Here
" I am again, safe enough.' The words were scarcely out of his mouth
" when a round shot took off his head and spattered the whole battalion
" with his brains, the Colours and ensigns in charge coming in for an
" extra share. A second shot carried off six of the men's bayonets ; a
" third broke the breast-bone of a lance-sergeant (Robinson) whose piteous
" cries were anything but encouraging to his youthful comrades. The
" soldier's belief that 'every bullet has its billet' was strengthened by
" another shot striking Ensign Cooper, the shortest man in the Regiment,
" and in the very centre of the square. We were now ordered to lie down.
" Our men lay packed like herrings in a barrel. I seated myself on a
" drum. Behind me was the Colonel's charger, which, with its head
" pressed against mine was mumbling my epaulette, while I patted his
" cheek. Suddenly my drum capsized, and I was thrown prostrate, with
" the feelings of a blow on the right cheek. I put my hand to my head,
" thinking half my face was shot away, but the skin was not even abraded.
" A piece of shell had struck the horse on the nose, exactly between my
" hand and my head, and killed him instantly.

" We soon received an order to seek the shelter of a neighbouring hill.

" Our new position was further in advance, but less exposed to the
" enemy's fire. We were now about 100 yards from the Nivelles chaussée,
" near to the abbatis, on which the left wing of our right company rested.
" In our front were some riflemen in grey uniforms faced with green, their
" hats looped up on one side, and bearing the Hanoverian badge of the
" white horse. They lined the road, and were engaged with some French
" skirmishers in the corn-fields on the opposite side. . . . .

" Captain Mercer's battery of Horse Artillery was on our right flank,
" and a little in advance. In his book on Waterloo he refers to the posi-
" tion of the regiment, and also to the fright in which all that portion of
" the field was when the approach of strangers was first noticed from the
" rear." The battle at this time was generally regarded as lost.

Both Albemarle and Mercer describe the awful effect of the enemy's artillery fire, and both record the sufferings of some artillery horses, which were horribly mutilated by the bursting of the enemy's shells.

" The steadiness of our peasant lads, which had already been tolerably
" tried, was about to be subjected to another test. On our right flank
" there appeared an armed force some thousands strong, who advanced
" towards us singing and cheering. They wore the dress which the
" prints of the day described as belonging to the French Army. Charles
" Brennan, an Irish lieutenant, who had served all through the Peninsular
" war, called out, 'Och, then, them's French safe enough!' 'Hold your
" tongue, Pat!' thundered out our Colonel ; 'what do you mean by

"frightening my boys?' but the expression of his countenance showed "that he shared Pat's apprehensions. . . . . The new-comers were General "Chassé's Dutch-Belgian division. They had been posted in the early "part of the day at Braine l'Alleud, and were now ordered to the front. "They had so recently formed a part of Napoleon's army that the slight "change in their uniforms escaped the notice of a casual observer.

"Towards evening the 14th was the right-hand Infantry Regiment of the "British line. . . . . Our instructions were to keep a good look out upon a "strong body of the Cavalry of the Imperial Guard. We now occupied "the crest of a gentle eminence, and looked down upon what, from a few "blades still standing, was shown to have been in the morning a field of "rye, ripe for the sickle. It had now, from the action of horse, foot, and "artillery, been beaten down to the appearance and consistency of an "Indian mat.

"From the reverse of the hill in front of us there now appeared the "enemy our Colonel had taught us to expect. They were a magnificent "body of horsemen, and wore black helmets, and, if my memory does not "deceive me, black cuirasses.

"As soon as they reached the ascent of our hill, they advanced towards "us at the pas de charge. For a moment they left us in doubt which "square they intended to honour, but gave the preference to our left-"hand neighbour, a regiment of Brunswickers, which was at wheeling "distance from ours. After one or two vain attempts to pierce the square, "they went some fifty paces to our rear.

"Their presence amongst us procured us a momentary respite from "the fire of the enemy's artillery. They now repassed between the "two battalions. As soon as they were clear of our battalion, two faces "of the attacked square opened fire. At the same time the British gun-"ners on our right who, at the approach of the Cuirassiers, had thrown "themselves at the feet of our front rank men, returned to their guns and "poured a murderous fire of grape into the flying enemy. . . . . When "the smoke cleared away, the Imperial horsemen were seen flying in "disorder. The matted hill was strewed with dead and dying, horses "galloping away without riders, and dismounted Cuirassiers running out "of the fire as fast as their heavy armour would allow them.

"This is the last incident I remember of that eventful Sunday."

According to Albemarle, the regiment bivouaced that night in the vicinity of Houguomont.

On the morrow of the battle, the 23rd Fusiliers, which had followed the wreck of the French army as far as La Belle Alliance, marched back and rejoined its brigade, which then advanced towards Nivelles; Colville's division now forming the advance of the army. The 14th entered Nivelles playing *Ca Ira*, and thence marched to

Mons, in which neighbourhood they bivouaced. From Mons they marched to Valenciennes, and next day to Le Cateau Cambresis.

There, in a Divisional Order, Lieutenant-General Sir Charles Colville congratulated the 4th Brigade* on the share they had in achieving the glorious victory of Waterloo, adding :

"The 23rd and 51st fully maintained their former high character, whilst "the very young 3rd battalion 14th, in this its first trial, displayed a steadi- "ness and gallantry becoming of veteran troops."

On entering France, the Duke of Wellington invited Louis XVIII. to repair to Le Cateau Cambresis, and, to preclude risk to the king's person, ordered the fortress of Cambray to be summoned. The garrison refused to surrender, and drove off the troops which approached on 23rd June. An attack by escalade was ordered for next day. Two brigades which had not been engaged on the 18th, were ordered to make the attack, whilst Mitchell's brigade made a feint on the Paris Gate. Chance allowed the latter to be turned into a real attack, and Mitchell's brigade was in possession of the town before the others could effect an entrance.†

The battalion resumed its march with the army to Paris, which it entered on 4th July, 1815, encamping with its brigade on the Bois de Boulogne. Brevet Lieutenant-Colonels Tidy and Keightley were made C.B., and each officer and man present at Waterloo received a grant of two years' service for the day, and subsequently a silver medal. In common with the other corps actually engaged, the 3rd battalion 14th Regiment received permission, conveyed in a Horse Guards Letter, dated 23rd July, 1815, to inscribe the word "WATERLOO" on its colours. The notification of the royal commands was not published in the *London Gazette* until 24th December,

---

* The other brigades of Colville's division received the medal as having been constructively present, but did not inscribe "Waterloo" on their colours.

† Lord Albemarle writes: "Having got close to the wall with a few haystack "ladders tied together, we resolved to try our luck in a real attack. My position "happened to be on the bridge with part of the 51st and all my own men, who "were getting over the top of the gate, which being tedious, we knocked it "in; an inhabitant actually let down the drawbridge, and we walked in and "marched in subdivisions to the Grand Square, in most regular order, in column "of battalion."—*Fifty Years of My Life.*

1815. The battalion remained in camp in the Bois de Boulogne, taking part in the magnificent reviews of the allied troops before the Emperors of Austria and Russia and the Kings of France and Prussia, until November, when it marched into cantonments in the village of Le Massy, about nine miles north of Paris. When the Army of Occupation was formed, it returned home, landing at Dover at Christmas, 1815.*

1815.—Second Battalion.

Whilst the British and Prussians were yet on the march from Belgium towards Paris, and the armies of other nations were crowding into France, something like anarchy was prevailing in the south of the country. The excitable temperament of the people, the widespread distress, the feebleness of the royalist executive, the revival of old feuds, political and religious, all contributed to a reign of disorder, resulting in dangerous disturbances at Marseilles, Nismes, and elsewhere. At Marseilles, on 25th-26th June, 1815, over a hundred lives were lost under circumstances which, it has been said, "proved the Satanic wisdom of the chiefs of the Gironde when they "sent for and awaited 'Les Fedérés de Marseilles' to head the insur-"rection of the 18th August, 1792."† In the midst of these scenes,

---

* Lord Albemarle thus describes the landing: "Public feeling had undergone "a great revulsion in regard to us soldiers. The country was saturated with "glory, and was brooding over the bill that it had to pay for the article. "Waterloo and Waterloo men were at a discount. We were made painfully "sensible of the change. If we had been convicts disembarking from a hulk we "could not have met with less consideration. 'It s us pays they chaps' was the "remark of a country bumpkin as we came on shore. It was a bitter cold day "when we landed; no cheers, like those which greeted the Crimean army, "welcomed us home. The only persons who took any notice of us were the "custom-house officers, and they kept us under arms for hours in the cold, while "they subjected us to a rigid search. Those functionaries were unusually on the "alert because a day or two before a brigade of artillery, with guns loaded to the "muzzles with French lace, had slipped through their fingers. Our treatment was "all of a piece. Towards dark we were ordered to Dover Castle, part of which "building served as a prison. Our barracks were strictly in keeping with such "a locality—cold, dark, gloomy, dungeon-like. No food was to be got but our "rations, no furniture but what the barrack store afforded."—*Fifty Years of My Life.*

† Alison, *History of Europe, 1815-52*, vol. i.

yet remembered as the "White Terror," the 2nd battalion 14th Regiment arrived in France.

The battalion, under command of Lieutenant-Colonel Burrows, was still in garrison at Genoa (where the British troops appear to have been extremely popular) when Sir Hudson Lowe arrived from Heidelberg and assumed command of the troops, on 19th June, 1815. A week later couriers arrived with at first most conflicting accounts of the battle in Belgium. Lord Exmouth also arrived at Genoa with a naval squadron, bringing tidings of the state of affairs in the south of France. On 3rd July, 1815, Sir Hudson Lowe embarked with three thousand British and foreign troops, including the 2nd battalion 14th Regiment, in Lord Exmouth's squadron, which arrived off Toulon, where the tricolour was still flying, on 9th July, and landed the troops at Marseilles without opposition on the 12th. Three days later Napoleon sought refuge on board H.M.S. *Bellerophon*, Captain Maitland, in Aix Roads. On 1st August Sir Hudson Lowe* received intimation that he had been chosen as the fallen emperor's custodian, and started for Paris, leaving Lieutenant-Colonel Burrows in command at Marseilles. There appears to have been a proposal to march on Toulon and Nismes, in conjunction with a force of Austro-Hungarians under Count Nugent. The annexed letter from the Duke of Wellington was received in reply to Colonel Burrows' application for instructions on the subject:

"Paris, 20th August, 1815.

"Sir,—

"I had the Honour of receiving your letter of 15th inst. You are not "to proceed to the attack and blockade of Toulon, and are not to quit "Marseilles for the purpose of acting hostilely against the King of France "or his troops or for any purpose whatever, except in your own defence, "without receiving further orders from me. I beg you to communicate "these instructions to Admiral Lord Exmouth and Lieut.-General Count "Nugent.

"I am, &c.,

"WELLINGTON.

"To Col. Sir M. Burrows." †

---

* At his departure, Sir Hudson Lowe received the thanks of the city for having saved it from pillage at the hands of the mob by the timely arrival of his troops. He was presented with a silver urn, inscribed "Marseilles—Ever Grateful."

† Wellington, *Supplementary Desp.*, vol. xii., p. 612. It will be observed that Colonel Burrows was addressed by a title to which he had no claim.

Colonel Burrows appears to have been still in command at Marseilles at the time of Murat's flight from Corsica in October.* Major-General Phillips afterwards assumed the command, and when the troops were withdrawn, in December, 1815, issued the following order, dated on board the *Intrepid* transport at sea :

"DIVISION ORDER.

"6th Dec., 1815.

"The Major General Commanding has great pleasure in communi- " cating to the British and Piedmontese Troops under his command, " the letter he has received from the Marquis de Montgrand, Mayor of " Marseilles. The great tribute of praise it contains must be highly " gratifying to everyone concerned, but to none more so than the Major- " General himself, who so fully knows how well it has been merited by " the whole of the Division he has had the honr to command.

"Marseilles, 5th Dec., 1815.

"General,—I received with the most unfeigned sensibility the letter " you did me the Honour to write to me expressing at the moment of " your departure with the Troops under your command your satisfaction " at the reception they have met with in this Town. The British Troops " have a right to the affection and esteem of the Marsellians. These are " the sentiments that dictated their conduct towards them. I feel person- " ally happy in having contributed towards the comfort of this Division of " His Britannic Majesty's Army, and thereby obeying the intention of the " King, my Master, and the wishes of my fellow citizens.

"I feel also great pleasure in acknowledging, General, that yourself and " your predecessors in the command, have infinitely lightened the charge " which inevitably resulted from the residence of the Troops amongst us " by the kind discretion you have used every demand that was made from " them, and by the facility always given to assist such demands. I must " likewise render due homage to the solicitude you have constantly shewn " to preserve the most exact discipline among the troops under your " authority, and in severely punishing the slightest misbehaviour. The " organ of all the constituted authorities of this Town and of the univer- " sality of my fellow Citizens, I request, General, that you will receive and " offer to the whole of the Division under your command in Marseilles, " the sincere wishes which we all form for yourself and this estimable " portion of the British Army. The personal relations I have had the " Honr to maintain with yourself and with all the Officers of the Troops " you command have been a subject of habitual satisfaction to me. I " shall ever preserve the most pleasing remembrances of it.

---

* See Wellington, *Supplementary Desp.*, vol. ix., p. 189. Letter dated Marseilles, 9th October, 1815.

"Accept, General, the assurances of the most distinguished considera-
"tion with which I have the Honour to be

"Your most obedt. and humble servant,
"(Signed) THE MARQUIS DE MONTGRAND,
"Mayor."

The battalion landed and went into quarters at Malta at the beginning of the New Year.

1815.—First Battalion.

Meanwhile the 1st battalion in India had not been inactive. The arrival at Calcutta, early in October, 1814, of the new Governor-General, Lord Moira, afterwards Marquis of Hastings, inaugurated a more spirited and high-handed policy than had been pursued by his predecessor, Lord Minto. This was speedily manifest on the Nepaul frontier. Lord Minto had represented in strong terms the need of curbing Ghoorkha aggression; but he had made great concessions, in the hope of avoiding a war. He demanded that the Ghoorkhas should withdraw from districts they had wrongfully appropriated; but his demands were treated with scorn, and the Ghoorkhas prepared to fight. Lord Moira took more vigorous measures. Within two months after his arrival a force had been collected on the frontier, in four divisions, to make simultaneous attacks on as many points in the long line of Ghoorkha conquests so soon as the rains had subsided sufficiently for the purpose. Lord Moira appears to have relied chiefly on the fourth division, consisting of H.M. 24th Regiment and six thousand native troops, which was to march straight on the Ghoorkha capital, Khatmandu. It started from Dinapur, under command of Major-General Bennet Marlay, H.E.I.C.S., on 26th November, 1814.

After wasting time in "mischievous indecision," * Major-General Marlay, influenced probably by Lord Moira's unqualified disapproval of his proceedings, left his command suddenly and without notification early one morning in February, 1815. He was succeeded at

* Mill, *History of India*, edited by Professor Wilson, vol. viii., p. 50.

the end of the month by Major-General George Wood. During the month, the division had been reinforced by the arrival of the 1st battalion 14th Regiment, under command of Lieutenant-Colonel James Watson, from Berhampore, and other corps.

But Major-General George Wood, like his predecessors, was possessed by a spirit of caution and procrastination, and like them entertained a belief of the insuperable difficulties presented to an invader by the then unexplored fastnesses of the lower Himalaya. He pleaded the advanced state of the season as an excuse for confining his operations to the plains. A march to Janakpur, on the Tirhoot frontier, showed that the Ghoorkhas had entirely abandoned the lowlands. Some harassing marches and countermarches were then made by the division, and several Ghoorkha stockades were destroyed. Meanwhile Colonel Gardner, the well-known leader of Gardner's irregular horse, had obtained some advantages over the enemy in the Nepaulese province of Kamaon, and on 8th April, 1815, Colonel Jasper Nicolls, H.M. 14th Foot, quartermaster-general of the King's troops, assumed command of a native brigade, consisting of battalions of the 5th and 14th Bengal Native Infantry, a native flank battalion, six guns, and two mortars, in front of Almora. A spirited action, fought by part of this force on 22nd April, 1815, was followed up by so vigorous an attack on the Ghoorkha position on the hill of Setauli, in front of Almora, on the 25th, that the Ghoorkha garrison sent in a flag of truce, and, after a brief negotiation, surrendered. This was the first real blow struck at the Ghoorkha power. It was followed by the permanent annexation of the provinces of Kamaon and Gorhwal, and the opening of negotiations between the Nepaulese and British governments, during which the greater part of the troops were placed in temporary cantonments along the border line from the Gandak to the Kusi. Thus ended what is sometimes called the First Nepaul War. At this time the 14th was sent to Dinapur, a military cantonment on the south bank of the Ganges, where it remained until October, 1815, when it descended the river in boats to Cawnpore, and there spent the rest of the year. Thus it had no part in the so-called Second Nepaul War—the successful operations, under Sir

David Ochterlony, which followed the renewal of hostilities in November, 1815, and ended with the conclusion of a satisfactory peace with the Government of Nepaul in March, 1816.

The campaigns against Nepaul were included in those for which the East India Company's war medal, 1799-1826, was granted to surviving officers and men thirty-five years afterwards. The medal, a handsome silver one, struck in 1850, was designed by William Wyon, R.A., and bore upon one side the Queen's head, and on the other a seated figure of Victory, "in quiet pose most fitted "to record services long past," having in her left hand a laurel, in her right an olive branch. A lotus flower is beside her, and in the background a palm-tree and an oriental trophy. The medal was worn with a sky-blue ribbon.

1816.—Third Battalion Disbanded.

After some hurried changes of quarters between Dover, Hythe, Deal, and Ramsgate, each move at a moment's notice, the 3rd battalion, Major and Lieutenant-Colonel Tidy, C.B., commanding, was disbanded at Deal on 17th February, 1816. Some of the officers and all the effective non-commissioned officers and men were transferred either to the 1st or the 2nd battalion of the regiment.

1816.—Second Battalion.

Soon after the arrival of the 2nd battalion 14th Regiment at Malta in January, this year, a Horse Guards Letter, dated 2nd January, 1816, was received, directing that the word "CORUNNA" be inscribed on the colours and appointments of the 2nd battalion, in consequence of the distinguished conduct of the battalion in the battle there on 16th January, 1809.*

Another Horse Guards Letter, of the same date, directed that, as the men of the 3rd battalion had been transferred to the 1st and 2nd battalions, the word "WATERLOO" should be inscribed on the colours and appointments of the surviving battalions, in commemoration of the distinguished gallantry displayed by the 3rd battalion on 18th June, 1815.†

* See *Appendix*. † *Ib.*

Colours of 3rd Battalion of the 14th—or Buckinghamshire Regiment of Foot,—disbanded 1816. Carried at Waterloo.

ROLL OF OFFICERS OF THE REGIMENT, 1815.

(From the Annual Army List.)

Rank.	Name.	Rank in the Regiment.	Rank in the Army.
Colonel	Harry Calvert	8th Feb., 1806	*L.G.*, 25th July, 1810
Lieut.-Colonel	James Watson	15th May, 1806	*Col.*, 4th June, 1814
,, ,,	Montagu Burrows	14th May, 1807	*Col.*, 4th June, 1814
,, ,,	Jasper Nicolls	31st March, 1808	*Col.*, 4th June, 1814
,, ,,	Nathaniel Burslem	13th Oct., 1814	1st March, 1811
Major	Fras. Skelly Tidy	10th Sept., 1807	*L.C.*, 4th June, 1813
,,	William Fawcett	22nd April, 1813	
,,	Peter Johnstone	1st Dec., 1813	
,,	William Percival	12th Jan., 1814	21st June, 1813
,,	John Keightley	13th Jan., 1814	
,,	Eyre Coote	13th Oct., 1814	4th June, 1814
Captain	Jacob Watson	5th August, 1804	*Major*, 4th Jan., 1814
,,	William Moore	29th May, 1806	
,,	Matthias Everard	21st May, 1807	23rd April, 1807
,,	Robert Ramsay	30th July, 1807	30th April, 1807
,,	John Marshall	21st April, 1808	28th Nov., 1806
,,	Henry Nooth	2nd June, 1808	*Major*, 4th Jan., 1814
,,	Cecil Bisshopp	23rd June, 1808	
,,	T. H. Light	18th August, 1808	
,,	Octavius Temple	15th Sept., 1808	
,,	Thomas Dunn	22nd Sept., 1808	
,,	George Marlay	14th June, 1810	*M.*, 21st June, 1813
,,	George Rawlins	5th July, 1810	
,,	Thomas Ramsay	18th Oct., 1810	17th May, 1810
,,	W. S. Bertrand	7th March, 1811	
,,	William Turnor	15th August, 1811	
,,	John Dyson	28th Nov., 1811	
,,	James Roche	5th Nov., 1812	
,,	Thomas Savage	29th April, 1813	
,,	Thomas Hall	16th Sept., 1813	
,,	Gerald Rochfort	1st Dec., 1813	
,,	Charles Knolles	3rd Dec., 1813	
,,	William Ross	24th Dec., 1813	16th Dec., 1813
,,	Harcourt Morton	12th Jan., 1814	
,,	Richard Adams	13th Jan., 1814	
,,	John Maxwell	7th July, 1814	
,,	William Betts	1st Nov., 1814	
,,	Henry Hill	2nd Nov., 1814	
,,	Charles Stanhope	3rd Nov., 1814	
,,	Christopher Wilson	4th Nov., 1814	6th October, 1814
,,	J. L. White	5th Nov., 1814	11th Dec., 1806
Lieutenant	George Bolton	21st May, 1807	
,,	Hercules J. Heyland	21st May, 1807	
,,	Thomas H. Baylie	25th June, 1807	

Rank.	Name.	Rank in the Regiment.	Rank in the Army.
Lieutenant .	John C. Harold . .	16th July, 1807	23rd May, 1807
,, . .	William Akenside .	5th August, 1807	2nd Jan., 1807, *Adjt.*
,, . .	*Charles Myler Brannan	3rd Dec., 1807	29th January, 1807
,, . .	James B. Ainsworth .	18th August, 1808	
,, . .	Kenneth M'Kenzie .	22nd Sept., 1808	
,, . .	H. B. Armstrong . .	27th Oct., 1808	23rd Jan., 1806, *Adjt.*
,, . .	Henry Gamble . .	22nd Dec., 1808	
,, . .	Thomas Kirkman . .	9th March, 1809	
,, . .	Thomas Jenour . .	25th May, 1809	
,, . .	Thomas Way . .	27th May, 1809	
,, . .	Henry Johnson . .	30th May, 1809	
,, . .	Dominick C. Lynch .	31st May, 1809	
,, . .	H. J. Henley . .	15th June, 1809	
,, . .	George Mackenzie .	29th June, 1809	4th May, 1809
,, . .	H. L. Franklin . .	19th Oct., 1809	8th May, 1808
,, . .	George T. Finucane .	15th March, 1810	
,, . .	Edward L'Estrange .	19th April, 1810	
,, . .	D. Hazlewood . .	13th Dec., 1810	
,, . .	Richard Stack . .	3rd March, 1811	
,, . .	Samuel Park . .	4th March, 1811	
,, . .	William Fowler . .	6th March, 1811	
,, . .	James Gordon . .	7th March, 1811	
,, . .	Simin Kent . .	28th March, 1811	
,, . .	George Fitzherbert .	31st July, 1811	
,, . .	Thomas Powell . .	1st August, 1811	
,, . .	Henry Mansell .	14th August, 1811	
,, . .	C. R. Newman . .	15th August, 1811	
,, . .	C. Fras. Jennings .	27th Nov., 1811	
,, . .	Samuel Beachcroft .	28th Nov., 1811	
,, . .	Edward Pender . .	27th August, 1812	
,, . .	Jacob Meek . .	22nd Oct., 1812	
,, . .	James McDermott .	2nd Nov., 1812	
,, . .	William Buckle . .	3rd Nov., 1812	
,, . .	†Charles Chichester .	4th Nov., 1812	
,, . .	Visey Temple . .	5th Nov., 1812	
,, . .	William Jappie . .	29th April, 1813	
,, . .	John Campbell . .	1st Dec., 1813	
,, . .	Thomos Hemans . .	2nd Dec., 1813	
,, . .	E. H. Cosens . .	3rd Dec., 1813	
,, . .	J. E. MacArthur . .	4th Dec., 1813	
,, . .	Robb James . .	5th Dec., 1813	
,, . .	James Sidney . .	26th Jan., 1814	
,, . .	James H. Patterson .	27th Jan., 1814	
,, . .	Grenville Pigott . .	26th May, 1814	
,, . .	Charles Grove . .	13th Oct., 1814	
,, . .	Christopher F. Holmes	1st Nov., 1814	
,, . .	Edward Gyfford . .	2nd Nov., 1814	

* This officer's names are spelt differently in successive Army Lists.

† Afterwards Colonel Sir Charles Chichester, K.S.F. Greatly distinguished himself as second in command of the British Legion during the Carlist war in Spain, 1835-37. Died lieutenant-colonel commanding 81st Regiment in Canada, in 1848. Sir James Alexander describes him as one of the best commanding officers in the British army.

RANK.	NAME.	RANK IN THE REGIMENT.	RANK IN THE ARMY.
Lieutenant	Patrick McKie	3rd Nov., 1814	
,,	James Grant	6th Nov., 1814	
,,	J. C. O'Hara Dickeens	7th Nov., 1814	
,,	J. C. Lambie	8th Nov., 1814	
,,	George Baldwin	9th Nov., 1814	
,,	*Orlando Felix	10th Nov., 1814	
,,	Thomas Woodford	7th Dec., 1814	
,,	W. Harrison Hill	8th Dec., 1814	
Ensign	James Healy	1st Nov., 1813	
,,	George J. Bower	2nd Nov., 1813	
,,	Robert Jones	2nd Dec., 1813	
,,	John Nickelson	25th Dec., 1813	
,,	Lyttleton Westwood	25th Dec., 1813	
,,	William Reed	13th Jan., 1814	
,,	George Mackenzie	22nd Jan., 1814	
,,	Walter N. Williams	23rd Jan., 1814	
,,	J. Campbell Hartley	24th Jan., 1814	
,,	F. R. Tane	25th Jan., 1814	
,,	Gilbert Pasley	26th Jan., 1814	
,,	R. B. Newenham	27th Jan., 1814	
,,	C. Fraser	10th Feb., 1814	
,,	A. F. F. Adamson	3rd March, 1814	
,,	William Keowen	21st April, 1814	
,,	J. Manly Wood	19th May, 1814	
,,	Arthur Ormsby	2nd June, 1814	
,,	J. R. Smith	13th Oct., 1814	
,,	A. Cooper	1st Nov., 1814	
,,	James Bowlby	2nd Nov., 1814	
,,	J. Powell Mathews	3rd Nov., 1814	
,,	R. J. Stackpoole	8th Nov., 1814	
,,	E. F. Morden	9th Nov., 1814	
,,	R. Birt Holmes	10th Nov., 1814	
Paymaster	William Wilkinson	3rd March, 1808	
,,	J. H. Matthews	21st Oct., 1813	
,,	Robert Mitton	7th Oct., 1814	
Adjutant	H. Bruin Armstrong	21st March, 1811	*L.*, 23rd Jan., 1806
,,	William Akenside	27th Jan., 1814	*L.*, 2nd Jan., 1807
Qr.-Master	William Harris	25th Jan., 1810	
,, ,,	Luke Lambert	21st Nov., 1811	24th January, 1807
,, ,,	Alexander Ross	20th Jan., 1814	
Surgeon	Thomas Jackson	17th June, 1800	23rd January, 1800
,,	George Adams	3rd June, 1813	
,,	W. P. O'Reilly	27th Jan., 1814	
Asst.-Surgeon	W. J. B. Parker	8th Nov., 1804	
,, ,,	John Smith	30th Jan., 1806	
,, ,,	James Trigge	29th June, 1809	
,, ,,	Alexander Shannon	27th Jan., 1814	
,, ,,	Henry Terry	21st March, 1814	

* This officer was transferred to the 95th Rifles in May, 1815, and was wounded with that corps at Quatre Bras. In the 14th Regiment pay-list he is shewn as drawing rations with the 14th through the campaign. He died a major-general in 1860. In later years he was a distinguished Egyptologist, and was first to decipher the names and titles of the Pharaohs.

On 26th April, 1816, the 2nd battalion, under command of Colonel Montagu Burrows, proceeded from Malta to the Ionian Islands. Head-quarters were at first stationed at Corfu, with detachments in Cephalonia, Zante, St. Maura, etc.* When Lieutenant-Colonel Tidy joined after the disbanding of the 3rd battalion, he was appointed by the Lord High Commissioner, Sir Thomas Maitland, to the post of Capo de Governo, or Resident, at Cephalonia.

Previous to the arrival of the battalion in the Greek Islands, the plague had made its appearance at Corfu, and afterwards in Cephalonia and Zante, and detachments of troops were employed, as before in Malta, in forming cordons round infected places. A garrison order, dated Zante, 20th August, 1816, records the thanks of Lieutenant-Colonel Count Rivarola,† senior officer in the islands of Zante, Cephalonia, and Ithaca, to Lieutenant Jacob Meek, 14th regiment, for his conduct when in command of a detachment of the regiment employed on this duty at the village of Comitato, in Cephalonia :

"Lieutenant Meek,‡ of the 14th Regiment, who has commanded in "Comitato, deserves great praise for the zeal by him evinced. In short, "all the officers and men employed in this harassing duty supported the "exertions of their superiors, and the whole have displayed proper judg-"ment and energy in the performance of a most arduous task."

---

* Lord Albemarle thus writes of St. Maura, where he was on detachment : "We were ill off for society at Santa Maura as at Zante. . . . . Shooting was "our principal amusement, and of this we had abundance. The lagoon swarmed "with water-fowl, and on the island there was no lack of partridge. The contents "of our sportsmen's bags helped greatly to lighten our monthly bills. We had a "Scotch brother-officer for caterer, one Lieutenant Mackenzie, and he managed "admirably. A cow fed on the line wall supplied us with milk and butter. We "had a pound of meat each for our ration. Fish, wine, and fruit were at nominal "prices. *Our money contributions to the mess rarely exceeded 5½d. a day.*— *Fifty Years of My Life*, revised edition, p. 186.

† Sir Francesco Count Rivarola, afterwards colonel Royal Malta Fencible Regiment, the officer who was inspector of police during the plague in Malta in 1813.

‡ Afterwards Captain Jacob Meek. He obtained his company in the 14th in 1825, and retired as captain, half-pay, unattached, in 1834. He died in 18[illegible]4. Staff-Surgeon J. D. Tully, in his *History of the Plague in Malta, Corfu, etc.*, speaking of Lieutenant Meek's services, refers in the highest terms to the "unceasing exertions" of this officer, who was in command at Comitato, and who volunteered to superintend the work of expurgation carried out by a corps of expurgators, all of whom had had the plague. There was a very fatal outbreak of

1816.—First Battalion.

The 1st battalion 14th Regiment remained in cantonments at Cawnpore throughout this year.

1817.—First Battalion.

Owing to various causes, chiefly the remissness of past administrations, the talookdars or zemindars* of the Dooab—the country betwixt the Jumna and Ganges, annexed in the days of Lord Lake—although not intended to be exempt from British rule, had been allowed to exercise supreme authority within their estates, to maintain large bodies of military retainers, and to convert their residences into feudal strongholds, where they defied interference. Attempts to enforce civil or criminal justice within these talooks were baffled or resisted; criminals were sheltered or aided to escape; gangs of robbers were harboured for a share of their spoil. Lord Moira determined to take advantage of the concentration of troops in the western provinces to put an end to this state of things. Dyaram, rajah of Hatrass, in the province of Agra, was one of the most powerful and contumacious of these zemindars, and against him measures were first directed. His fort of Hatrass was of the usual construction, of mud or sunburnt brick, with towers at the angles. It mounted many guns, was girt with a particularly wide and deep ditch, and was reputed all but impregnable. Dyaram maintained a force of eight thousand men, three thousand of them horsemen clad in chain-mail.

A strong division, consisting of H.M. 8th and 24th Light Dragoons, the 14th and 87th regiments, several regiments of native and irregular cavalry, seven battalions of native infantry, a force of artillery and

---

malarial fever at Argostoli at the time the plague was raging, which carried off a good many 14th Regiment men, chiefly old soldiers who had had Walcheren fever and young soldiers fresh out from home.

* The word "zemindar" receives different meanings. (See Mill, *History of India*, edited by Wilson, vol. vii.) Here it implies a hereditary landholder (not necessarily the proprietor of the soil), receiving the taxes of a particular group of villages.

pioneers, with a train of seventy-one mortars and Howitzers and twenty-four battering-guns, took the field, under command of Major-General Sir Dyson Marshall, and appeared before Hatrass on 12th February, 1817. Overtures of peace were made to the rajah, but as he would not submit to the disarmament of his fort, the siege commenced with a pitiless storm of shells and Congreve rockets, directed against the town and fort. On the arrival of the troops before the place, some enquiry had arisen as to the depth of the ditch, which a soldier of the 14th, whose name, unfortunately, has not been put on record, volunteered to ascertain. Fastening a stone to a long cord, he proceeded alone after dark and procured the required information—an act of cool intrepidity which evoked much admiration at the time. On the third day of the siege orders were given for the assault; and two columns formed for the purpose at 2 p.m., the right column led by the 14th,* under Brevet Major Everard, the left by the 87th; but at the request of the commanding-engineer the attempt was put off till next day. At night the garrison and inhabitants retired into the fort, so that when Brigadier Watson, with the 14th, took possession of the town on the following day—they had to use scaling-ladders to get in, as the gates were blocked—they found but a few aged people and many dead, mute testimony to the severity of the fire poured into the place. The besiegers' efforts were then directed against the fort, which returned the fire with vigour. General Marshall humanely offered a safeguard for their property if the garrison would send out the women and children, to be escorted out of the rajah's territory; but they were deaf to all entreaty. They were prepared for the worst; and the worst soon came in unexpected fashion. A chance shell from a British mortar penetrated the great powder-magazine of the fort, exploding three hundred thousand pounds of powder, the accumulation of years, and causing an explosion that was felt at Meerut, one hundred and fifty miles away. The interior of the fort, some five

---

* Lieutenant John Shipp, of the 87th, who was there, speaks of the 14th as "a beautiful corps." (See *Extraordinary Military Career of John Shipp.* London: new edition, 1890.)

hundred persons (mostly women and children sheltering in the stone chambers beneath), and much cattle were destroyed, but the defences suffered lightly, and when the mountain of dense black smoke, vomiting tongues of fire, lifted a little from the ruin below, the batteries reopened, to which the garrison pluckily responded with even more vigour than before. This was a ruse to cover the evacuation of the fort, which took place the same night. The British troops met the outgoing garrison at the gates, and some were cut down; but many escaped, among them Dyaram * himself. The fort and town were dismantled; the neighbouring zemindars submitted; and such anomalies as feudal forts and fortified villages in the midst of a peaceful population in one of the richest regions of Bengal, became things of the past. The service ended, the 14th returned to Cawnpore, where they spent some months.

Military operations of greater magnitude were at hand—the execution of a plan for the extirpation of the great Pindaree confederacy, which formed a prominent feature in the governor-general's policy.

The Pindarees originally were bodies of irregular cavalry, serving without pay for the sake of pillage, in the armies of the Mohammedan rulers of India, of which they formed a recognised element. Later, they transferred themselves to the Mahratta States. They lived by plunder, making sudden and distant raids, and dividing the spoil with the native States that gave them shelter. For long, the Pindarees abstained from entering the Company's territories; but a successful incursion through Bundelkund and Rewa into the neighbourhood of Mirzapur, in 1812, had given them confidence, and their murderous forays were becoming so frequent as to call for prompt and effective repression. As they were known to have the support of the great Mahratta States, preparations were made on a proportionately large scale. Two army corps were formed for the purpose—the Army of the Deccan and the Grand Army.

The Army of the Deccan, composed of King's and native troops from the Madras Presidency, together with some brigades and detachments from the other presidencies, was formed on the Madras

* Dyaram returned two years afterwards, and was allowed a pension by the Company.

frontier, under the command of Lieutenant-General Sir Thomas Hislop. The Grand Army--twenty-nine thousand foot, fourteen thousand horse, and one hundred and forty guns in all—was composed of troops from the Bengal Presidency, and was formed in Bengal in four divisions, under the personal command of the governor-general, now the Marquis of Hastings. Of this, the right or second field division was organised at Agra, under the command of Major-General (afterwards Sir Rufane) Donkin, and consisted of H.M. 11th Light Dragoons and 14th Foot, with three regiments of native cavalry and some irregular horse, eight battalions of native infantry, and eighteen guns. The infantry was divided into three brigades, whereof the first brigade consisted of H.M. 1st battalion 14th Regiment, from Cawnpore, and the 23rd and 63rd Bengal Native Infantry. As Colonel Jasper Nicolls, C.B., was quartermaster-general of the army, and Colonel James Watson, C.B., had been appointed to command a native brigade in the left division of the army, which was formed under Major-General Sir Dyson Marshall at Kalinjar, in Bundelkund, the 14th Regiment took the field under command of Major Peter Johnstone.

The preliminary movements of the armies gradually drove back the Pindarees from their haunts on the Nerbudda towards the north and west. During this period Asiatic cholera made its appearance in Lord Hastings' army, from which, between the outbreak in May-June and the sudden disappearance of the pestilence on 23rd November, it carried off seven hundred and sixty-four fighting men and eight thousand camp followers. On 9th November, 1817, Donkin moved his division from Agra to a position at Dholpore, on the Chumbal, where he threatened alike the Maharajah Sindia and Ameer Khan, whose good faith towards the Government was more than doubtful. At the beginning of December, leaving a detachment to guard the fort at Dholpore, he moved forward, skilfully cutting off all communication between the bodies of Pindarees flying before Marshall's division and Ameer Khan in Mewar. The neutrality of Sindia and Ameer Khan being secured, Donkin's division recrossed the Chumbal on 13th December, and when the Pindarees, disappointed of shelter at Gwalior and at Mewar, turned towards Jawab, the division marched to the westward in pursuit, halting at Janapur early in the New Year.

1817.—Second Battalion Disbanded.

The 2nd battalion, which had remained in the Ionian Islands since its arrival there in 1816, embarked at Cephalonia in the autumn of this year, and proceeded to Malta, where it remained a few days. The peace of Europe being now apparently established on a firm basis, further reductions of the army took place, and the 2nd battalion was ordered home to be disbanded. It landed at Portsmouth on 24th-25th November, and was disbanded on 23rd December, 1817, at Chichester, by Major and Lieutenant-Colonel Tidy, C.B. Four hundred and twenty rank and file were transferred to the 1st battalion.*

1818.

A War Office Letter, dated 29th November, 1817, directed the new establishment of the regiment, after the disbanding of the 2nd battalion to be as follows :

Ten companies : one colonel, two lieutenant-colonels, two majors, ten captains, twenty-two lieutenants, eight ensigns, one paymaster, one adjutant, one quartermaster, one surgeon, two assistant-surgeons, one sergeant-major, one quartermaster-sergeant, one paymaster-sergeant, one armourer-sergeant, one schoolmaster-sergeant, fifty sergeants (including ten colour-sergeants), fifty corporals, one drum-major, twenty-one drummers and fifers, nine hundred and fifty privates; total, one thousand one hundred and twenty-eight. One recruiting company : one captain, two lieutenants, eight sergeants, eight corporals, nineteen privates. Total of the regiment, one thousand one hundred and sixty-six.

---

* The regimental books of the battalion were directed to be deposited with the agents on disbandment. Messrs. Cox and Co., the agents in question, state that they have not and never had these books in their possession. The monthly returns and duplicate pay-lists of the two disbanded battalions of the regiment are preserved complete in the Public Record Office.

After a halt of a few days at Sonapur, General Donkin called in his detachments, which had been watching the neighbouring passes, and shifted his head-quarters to Shahpura, subsequently occupying various hill-forts in Mewar, which had been seized by the fugitive leader Jesnunt Rao, among them Komalner, reputed to be one of the strongest in India. By the end of January, 1818, the principal objects of the campaign had been accomplished. The Pindarees were broken up; most of their leaders had surrendered themselves; the power of the Mahratta States had been broken by the overthrow of the Peishwa at Mahidpore on 21st December, 1817, and the other successes of the Army of the Deccan. The Grand Army was accordingly broken up, and the remaining operations were entrusted to detached columns. The 14th Regiment remained in the field until April, 1818, when it went into cantonments at Meerut.

1818-24.

At Meerut, in pleasant quarters in the midst of a wide grassy plain, the 14th spent seven years, maintaining a high reputation for discipline and good conduct in cantonments. The following were the chief official events of this time:

A Horse Guards Letter, dated 24th June, 1818, directed that the word "JAVA" be inscribed on the colours and appointments of the regiment, in commemoration of the gallantry displayed by it at the conquest of the island of Java in August, 1811.*

In a General Order of 5th October, 1819, the Most Noble the Governor General in Council communicated to the regiment, in common with the rest of the army, the thanks of both Houses of Parliament, voted to the officers and troops engaged in the late Pindarree and Mahratta campaigns.

New colours were presented to the regiment by Major and Brevet Lieutenant-Colonel Tidy, C.B., commanding, whose speech on the occasion, delivered impromptu, is recorded in the following words: †

---

* See *Appendix*.

† Mrs. Ward's *Recollections of Colonel Tidy*.

"Soldiers,—I cannot let so good an opportunity as the one which now "presents itself go by without saying a few words on the sacred connec-"tion that ought to exist between the soldier and the standard under "which he serves. When once a man has entered the service of any "Power, he engages with his life to defend the flag of the country which "pays him, and there are instances where mercenary troops under a "foreign banner have shown the most devoted bravery in its defence. "What then ought we to feel when our ensigns are unfurled, which con-"tain the emblems of the three kingdoms from whence we spring? The "united crosses of St. George, St. Patrick, and St. Andrew our squadrons "and battalions have carried from one extremity of the universe to the "other; and after an unparalleled struggle of more than twenty-five years, "the minister of the King whom we honour and obey was enabled to "stand up in the midst of assembled Europe and dictate the will and "pleasure of his master to them all! It matters not in what part of the "globe your exertions have been made—whether on the heights of Cor-"nelis or on the plains of the European continent, they were alike "honourable, and tended equally to the subjugation of our most deter-"mined and gallant, but most inveterate enemy. The colours from "which you have just parted have been long and honourably preserved "by the regiment, so long, indeed, that the very circumstances render "it necessary that they should be replaced. The compliments of to-day "belong exclusively to them: henceforward the colours which have just "been unfurled claim, next to God and the King, your most devoted "attachment. I shall say no more than that I trust some few of us, "perhaps many, may live, and even serve, till we bid farewell to the "colours now before us in as ragged, and therefore as honourable, a "condition as the old friends from whom you have just parted."

On 29th January, 1820, King George III. died, and the Prince Regent ascended the throne as George IV.

In accordance with a circular letter from the Adjutant-General of the Forces, dated Horse Guards, 6th November, 1822, a "Statement of Services of the Regiment," such as it was possible to compile in the absence of any early records, was prepared and sent in, with summaries of the services of the disbanded battalions. On this "Statement" was based the work of Mr. Cannon, which has been corrected and expanded in the present volume.

ROLL OF OFFICERS OF THE REGIMENT, 1821.

(From the Annual Army List.)

Rank.	Name.	Rank in the Regiment.	Rank in the Army.
Colonel	Sir Harry Calvert, G.C.B.	8th Feb., 1806	*Lt.-Gen.*, 25th July 1810
Lieut.-Colonel	Jas. Watson, C.B.	15th May, 1806	*Col.*, 4th January 1814
,, ,,	John McCombe	24th Aug., 1820	*Col.*, 12th August, 1819
Major	Francis Skelly Tidy, C.B.	10th Sept., 1807	*Lt.-Col.*, 4th June, 1813
,,	Peter Johnstone	1st July, 1812	
Captain	Jacob Watson	5th August. 1804	*Major*, 4th June, 1814
,,	Matthias Everard	21st May. 1807	23rd April 1807
,,	John Marshall	21st April, 1808	*Major*, 12th Aug., 1819
,,	Cecil Bisshopp	23rd June, 1808	
,,	Geo. Rawlins	5th July, 1810	
,,	W. S. Bertrand	7th March 1811	
,,	Wm. Turnor	26th March, 1818	15th August, 1811
,,	Gerald Rochfort	20th May, 1818	3rd March, 1813
,,	Thos. Hall	21st May. 1818	16th Sept., 1813
,,	Edward Raynsford	28th Oct. 1818	17th Feb. 1814
,,	Charles Knolles	1st April, 1819	3rd March, 1813
Lieutenant	Wm. Akenside	6th August. 1807	2nd January, 1807
,,	Chas. Moylan Brannan	3rd Dec., 1807	23rd January, 1807
,,	Jas. B. Ainsworth	18th Aug., 1808	
,,	Kenneth M'Kenzie	22nd Sept., 1808	
,,	H. Bruin Armstrong	27th Oct., 1808	*Adjt.*, 23rd Jan., 1806
,,	Thos. Kirkman	9th March, 1809	
,,	Thos. Jenour	25th May, 1809	
,,	Thos. Way	27th May, 1809	
,,	Hy. Johnson	30th May, 1809	
,,	Martin Crean Lynch	31st May, 1809	
,,	Geo. Mackenzie	29th June, 1809	4th May, 1809
,,	Hugh Lloyd Franklin	19th Oct., 1809	8th May, 1805
,,	Geo. Thurles Finnucane	15th Mar., 1810	
,,	Ed. L'Estrange	19th April, 1810	
,,	D. Hazlewood	13th Dec., 1810	
,,	Rich. Stack	3rd March, 1811	
,,	Wm. Fowler	6th March, 1811	
,,	Simin Kent	28th March, 1811	
,,	Hy. Mansell	14th Aug., 1811	
,,	Chas. Rayner Newman	15th Aug., 1811	
,,	Chas. Fras. Jennings	27th Nov., 1811	
,,	Ed. Pender	27th Aug., 1812	
,,	Jacob Meek	23rd April, 1818	22nd October, 1812
,,	Jas. Grant	14th Oct., 1818	6th November, 1814
,,	Jas. McDermott	9th Dec., 1819	2nd November, 1812
Ensign	Gilbert Pasley	26th Jan., 1814	
,,	R. B. Newenham	27th Jan., 1814	
,,	Wm. Keowan	21st April, 1814	
,,	J. Manly Wood	19th May, 1814	
,,	A. Ormsby	2nd June, 1814	
,,	J. R. Smith	13th Oct., 1814	
,,	A. Cooper	1st Nov., 1814	
,,	J. Bowlby	2nd Nov., 1814	
Paymaster	J. H. Matthews	21st Oct., 1813	
Adjutant	H. B. Armstrong	21st March, 1811	*Lieut.*, 23rd Jan., 1806
Quarter-Master	W. Harris	25th Jan., 1810	
Surgeon	Thos. Jackson	17th June, 1800	25th January, 1800
Asst Surgeon	Jas. Trigge	29th June, 1809	
,, ,,	Thos. F. Colton	17th Aug., 1815	

1825-26.

Capture of Bhurtpore.

Events occurred at this time which brought the regiment again into the field. The Rajah of Bhurtpore (a state in Rajpootana), Baldeo Singh, had become attached to the British Government, with which he had formed an alliance, offensive and defensive, and had procured a guarantee for the succession of his child, Bhulwant Singh, to the throne; but among his subjects were many who were hostile to British interests, particularly in the army. Baldeo Singh died in August, 1824. From that time up to March, 1825, nothing occurred calling for the interference of the British Government. Information then reached Sir David Ochterlony, the British commissioner, that a revolution had been effected; that Doorjun Sal, the nephew of the late ruler, had attacked the regent mother and uncle of the boy rajah, and seized the latter after a sacrifice of many lives. The revolution had been effected with skill, secrecy, and despatch; but there was no popular manifestation of the Jat population at large against the rajah. It was, however, considered expedient to draw together an army capable of enforcing the claims of Bhulwant Singh, should such a course become necessary. Troops were collected for the purpose under Major-General Reynell, which Sir David Ochterlony proposed to augment by as many corps as he could spare from his own division. Soon after Sir David Ochterlony died. Later in the year the Government felt itself called upon to take active measures for the protection of the rightful heir, and a proclamation was issued on 25th November, 1825, in which Sir Charles Metcalfe—Ochterlony's successor—called on the usurper, Doorjun Sal, to surrender the principality to the lawful rajah, and retire into the British domains on a suitable provision being made by the British Government for all his rights, present and future. In the event of further opposition, it was significantly added, the British Government must perform its duty.

The troops, which had been recalled to their quarters, were accordingly directed to reassemble, and when Lord Combermere, who had succeeded the Honble. Sir Edward Paget in the chief command in India, arrived at Agra on 1st December, 1825, he found two divisions, each consisting of one brigade of cavalry and three brigades of infantry, with horse and foot artillery attached, assembled—one at Agra, under command of Major-General Jasper Nicolls, C.B., the other at Muttra, under command of Major-General Thomas Reynell, C.B.—and ready for the field.

The 14th Regiment, nine hundred strong, had joined the first, or Muttra division, from Meerut, during November. As Colonel McCombe and Lieutenant-Colonel W. T. Edwardes were appointed to command brigades, and Major and Brevet Lieutenant-Colonel Tidy was absent as deputy-adjutant-general with the army in Burmah, the command of the regiment again devolved on Brevet Major Matthias Everard. The 14th, with the 23rd and 63rd regiments, Bengal Native Infantry, formed the first, or Brigadier-General McCombe's, brigade of the first, or Major-General Reynell's, division of the army.

As the proclamation remained unheeded, the divisions were set in motion on 9th December, and on 10th December a force of the Muttra division, including two companies of the 14th Regiment, was detached to seize a strong post of the enemy on the Jheel Bund, north of the fortress, which commanded the water supply of the ditches. The enemy was speedily dislodged, the water cut off, and the embankment of the Bund turned into an entrenched post, which was held by a company of the 14th Regiment and some sepoys.

The place was then fully invested. On 12th-13th December a powerful battering-train of one hundred pieces arrived in camp, under escort of Brigadier-General W. T. Edwardes' brigade of the Agra division. On the 24th December two breaching batteries opened; other batteries were thrown up and opened in succession; and the siege and the defence were pushed on with vigour.

On 5th January, 1826, dispositions were made for the assault, which was postponed at the desire of the commanding engineer, as the breaches were considered not to be easy enough. The defences of the fortress were peculiarly difficult to breach, as not only were they held together by logs of a very tough description of timber, built into them, which resisted shot with remarkable pertinacity, but the sun-dried mud or brick of which they were composed, when it fell, formed vast impassable mounds of fine dust. Mining operations had to be resorted to on a large scale, and at last, after many delays, orders were given for the assault to be made at 2 a.m. on 18th January, the signal to be the springing of a great mine, charged with ten thousand pounds of powder, right under the Cavalier bastion.

There were two principal breaches, besides several minor ones. The assault of the right main breach, or Cavalier breach, was assigned to Major-General Reynell, with the undermentioned brigades formed in double column :

Brigadier Paton's.—Four companies H.M. 14th Regiment, under Major Matthias Everard, five companies 41st Native Infantry, and 6th Native Infantry.

Brigadier-General McCombe's.—Four companies H.M. 14th Regiment, under Major Cecil Bisshopp, the 23rd Native Infantry, and 63rd Native Infantry.

A column, consisting of two companies Bengal Europeans, with some native infantry and Ghoorkhas, under Colonel J. Delamaine, was to assault the Jungeenah breach, further to the right, and a reserve column, under Brigadier Whitehead, consisting of the remainder of H.M. 14th Regiment and the 18th and 32nd Native Infantry, was left at the head of the trenches.

A simultaneous assault was to be made on the left main breach by a brigade of Major-General Nicoll's division.

It was 8.30 a.m. before the great mine was sprung, by which it was supposed about five hundred of the enemy were killed. Brigadiers McCombe and Paton, Lieutenant Irvine, of the Engineers, and Lieutenant Daly, of the 14th Foot, were severely injured, the latter having his leg amputated on the spot.* There was a momentary check, but on General Reynell giving the word "Forward" the men rushed up the breach, Majors Everard and Bisshopp leading with the 14th companies. On learning that the brigadiers were wounded, Colonel Nation, of the 23rd Native Infantry, went to the front and took command, but was shot down on the ramparts in front of the 14th. At this moment the right column carried the Jungeenah breach. The brigades of the right attack, with the 14th at their head, then pushed along the ramparts to the right, clearing them, and taking bastion after bastion, the enemy's artillerymen only yielding up their guns with their lives. There was no check until reaching the Fateh Burj—the so-called "Bastion of Victory," where Lord Lake had made his last unsuccessful attack in 1805. There the column met H.M. 59th heading General Nicolls' column, which had made a simultaneous assault on the left main breach, and carried it after desperate fighting, in which Brigadier W. T. Edwardes, 14th Foot, and other officers of the column were killed. The 14th light company, which with the grenadiers had formed part of General Reynell's column, penetrated into the town with part of Nicolls' column, pressing the retreating garrison so closely up to the citadel that three or four hundred of the latter were shut out, with whom a furious fight ensued, in which every man of the enemy was bayonetted. At 4 p.m. the citadel surrendered, and on the following morning Lord Combermere entered it at the head of the 14th Foot, which he placed in garrison there as a compliment to the regiment for its distinguished gallantry. Brigadier-General McCombe, 14th regiment, was appointed governor. The usurper, Doorjun Sal, was captured whilst attempting to escape, and was sent a prisoner to Allahabad, and on 20th January the young rajah was installed on the throne of the country by Lord Combermere and Sir Charles Metcalfe in the presence of the 14th Regiment, acting as a guard of honour.

---

* Afterwards Major Robert Daly, for very many years a captain of cadets at the Royal Military College, Sandhurst. Cadets of the pre-Crimean epoch will remember "Old Peg," as he was called by irreverent youth on account of his wooden leg. He died in 1855.

Inscription on Picture.

View of the assault on the Fortress of Bhurtpore by the British troops, under the personal command of His Excellency General The Right Honorable Lord Combermere, on the 18th January, 1826.

(Drawn on the spot by Captain G. B. P. Field.)

Printed and Published at the Asiatic Lithographic Press, Park Street, Chowringhee, Calcutta, 1827.

Inscription on Frame.

ASSAULT OF BHURTPORE, 18-1-1826.

Presented by the Quarter Master General in India to The Prince of Wales's Own West Yorkshire Regiment, late 14th Foot, this being the Regiment shown in the drawing—18-7-1887.

Assault on the Fortress of Bhurtpore, 18th January, 1826.

Lord Combermere, in his despatch to the Governor-General, wrote :

"I have the pleasure, your lordship; that the conduct of everyone
" engaged was marked by a degree of zeal which calls for my unqualified
" approbation ; but I must particularly remark the conduct of H.M. 14th
" Regiment, commanded by Major Everard, and H.M. 59th, commanded
" by Major Fuller. These corps having led the columns of assault, by
" their steadiness and determination decided the fate of the day."

In Divisional Orders the following appeared :

"Major-General Reynell congratulates the troops of his division, Euro-
" pean and native, engaged in the storming of Bhurtpore this morning,
" upon the brilliant success, which attended their gallant exertions. It is
" impossible for him to convey half what he feels in appreciating the
" conduct of H.M. 14th Regiment, that led the principal storming column.
" It has impressed his mind with stronger notions of what a British regi-
" ment is capable of when led by such men as Major Everard,* Major
" Bisshop,† and Captain Mackenzie, than he ever before possessed. The
" Major-General requests Major Everard will assure the officers and
" soldiers of the 14th Regiment that they more than realised his expecta-
" tions."

Lieutenant-Colonel W. T. Edwardes,‡ 14th Regiment, an officer of high promise, fell at the head of the second brigade in Major-General Jasper Nicolls' division, pierced by many wounds. The regiment had Captain H. B. Armstrong mortally wounded ; two sergeants, twenty-nine men, and three Lascars killed ; Lieutenant-Colonel J. McCombe, Lieutenants Richard Stack, Robert Daly, Edward C. Lynch, Volunteer Tulloh, two sergeants, ninety-eight men, and three Lascars wounded.

Colonel J. McCombe and Majors Matthias Everard and Cecil

* Afterwards Major-General Matthias Everard, C.B., K.H. Died 1851.

† Afterwards Major-General Cecil Bisshopp, C.B. Colonel 56th Foot. Died 1858.

‡ Lieutenant-Colonel Wilbraham Tollemache Edwardes served some time in the 73rd Regiment. He retired from it on half-pay at the peace, and had recently joined the 14th as junior lieutenant-colonel. He was killed at the head of the second brigade in General Jasper Nicolls' Agra division of the Bhurtpore Army.

Bisshopp were made C.B.; and royal authority was afterwards given to inscribe the victory on the regimental colours.

So fell the great Jat fortress, which had long been regarded as a standing menace to British rule. After reviewing the divisions, Lord Combermere departed for the presidency. Leaving a sufficient garrison in Bhurtpore under Major-General, now Sir Jasper, Nicolls, K.C.B., the army broke up for its cantonments, and the 14th returned to its old quarters at Meerut.

The following is from a narrative written a few years ago by one of the last survivors of Bhurtpore—the late Captain W. L. O'Halloran, formerly of the 38th regiment, who was present at the siege as an ensign in the 14th regiment:

"The night was cold and gloomy, as the anxious sentinels of an out-" lying picquet paced the skirts of a deep wood. One by one they were " suddenly called in, and the picquet was noiselessly withdrawn. By the " same dim light, a British corps assembled in mysterious silence. Neither " drum nor bugle sounded as hasty groups emerged from their tents and " simultaneously formed their respective companies. A strong column " soon closed up, the men resting on their arms in quiet expectation.

"An occasional clatter of horses' hoofs, as staff officers rode past, " betokened unusual preparations. A group drew rein on the flank of " the column, now in close formation. A familiar voice gave hasty orders, " and in few, but impressive, words addressed the men, informing them " that the fortress was about to be assaulted, and that the honour of lead-" ing one of the principal attacks was entrusted to their valour.

"The corps now threaded a dark, thick wood, and emerged on a line of " trenches and batteries which it occupied, opposite a huge rampart and " cavalier. Shot, shell, and salvoes of artillery broke the stillness of " the night. As day dawned, the peculiar chill of that hour caused a " vigorous stirring of strong limbs, while the return of the cheerful light " gradually unloosed the tongues of expectant groups that, with laugh " and jest and banter, gaily talked over the impending conflict with " careless determination, and high hopes of distinction and prize " money.

"Presently a few brief words passed along the trenches. In an instant " each flew to his station, and all was stern discipline. Every eye turned " to the lofty 'cavalier,' that frowned down secure in its massive strength " and seeming inaccessibility.

"Turbaned heads occasionally peered for a moment over the parapet, " followed by a succession of shots from either side. For five weeks past " the curtain had been plied with artillery, but no impression was visible " in its solid front, nor were there any means apparent by which armed

" men could gain its summit. The round swift ball entered the earthen
" mound, and, burying itself deeply, left no trace of its path.

" But hark ! a rushing sound ! and a meteor suddenly springs upwards
" to meet the stars. The very ground itself trembles beneath the feet.
" That lofty mound actually moves, slowly it heaves and expands, then
" plunges bodily upwards, carrying with it the bodies of eight hundred
" human souls. Those blackened corpses are hurled into the air mingled
" with huge fragments of earth, and buried in the ruins. A signal rocket
" had sprung the largest mine that had ever been exploded in war.* Solid
" masses of terrific size dot the air, and are borne bodily in the direction
" of the trenches, which are speedily enveloped in an intense darkness.
" The stormers recoil, and are violently borne back some paces. That
" moment of awful gloom and suspense appears an age.

" But ere that fearful darkness passed away, a forward impulse was
" bearing the stormers along the trenches. Swiftly they rushed over the
" mangled bodies of comrades who were crushed by those huge clods, and
" rapidly dashed across an intervening space into the ditch beyond. But
" ere the breach was gained, the enemy had recovered from the unex-
" pected shock of the explosion, and gallantly opposed further progress.

" The immediate leader † of the column had been struck down by the
" force of that terriffic explosion. His place was quickly filled, but scarcely
" had his successor ‡ in the post of honour reached his station than gallant
" blood flowed once more, and a third,§ though not unscathed, was called
" to lead. A severe conflict ensued, for fierce was the struggle of those
" sable warriors.|| Driven from point to point, every successive battery
" being carried by the bayonet, the enemy slowly retired, pouring in their
" fire in front and flank. For one instant during the engagement the
" flushed cheek blanched, the parted lip trembled, the advancing foot of
" the victor was arrested, and he stood transfixed in horrible expectation
" as the cry of 'A mine!' was raised. The besieged had, in their turn,
" fired a train, which, though partial in its effects, consigned a few of our
" men to the same fate that so many of their own had just experienced.

" Several columns of assault swept the whole line of the ramparts, while
" some penetrated into the heart of the town, and made a dash at the
" citadel itself. Sufficient blood had now been spilt¶ to atone for reverses
" in bygone days, which, it was boastfully asserted, had furnished Euro-
" pean bones sufficient to line as a trophy the face of the Fateh-Burj, or

---

* A mine charged with 10,000lbs. of powder, immediately under the cavalier.

† Colonel McCombe, H.M. 14th Foot.

‡ Lieutenant-Colonel Nation, H.C.S. 24th Native Infantry.

§ Major Everard, H.M. 14th Foot.

|| Upwards of seven thousand were slain.

¶ Thirty-nine officers and five hundred and twenty-nine men of the storming parties were killed and wounded on 26th January, 1826.

" Bastion of Victory ; for, some twenty years before, successive assaults
" had four times been repulsed near the spot where the contest ended.*

" The survivors of the well-fought day now retraced their steps over the
" ground which they had deluged with blood, and witnessed the mute
" writhing of wounded men, and maimed bodies in heaps being slowly
" consumed, for matches had set fire to their clothing, and they were
" unable even to crawl away. Yet not a groan escaped them, and nought
" but stern defiance could be traced in their dying features. The mother,
" and the infant too, were there, having sought their protectors when the
" din of war was over ; and sat round in groups, mute, passive, and
" petrified with horror at seeing them bleed and burn to death. Intense
" terror rooted them to the spot, powerless and heartbroken.

" On the conquerors passed in triumph ; the stirring music and floating
" banners and the glad sun filled them with elation, and effaced any
" momentary impression which the sad sights of the day had left. They
" counted the treasure bought with blood, and were eager to imbrue their
" hands once more.

" Next day the citadel surrendered. Possession was taken of the town,
" but not one out of a dense population remained. The home had been
" abandoned, the loved hearth deserted, merciless shells had spread de-
" struction around, and strangers plundered the accumulated earnings of
" many years.

" Famine soon raged among the self-banished race. Hunger drove
" them back into the midst of those they feared, and it was piteous to see
" those miserable wretches furiously struggling for a handful of grain. In
" mercy, bayonets were pointed to keep back the starving crowd, lest they
" should trample each other to death while the scanty mouthful was being
" distributed. Old men and young children fought with desperation for
" each grain that dropped ; neither age, nor sex, nor human tie was recog-
" nised by that famishing throng."

1826-31.

On 3rd September, 1826, the much respected colonel of the regiment, Lieutenant-General Sir Harry Calvert, Bart., G.C.B., died, and the regiment was given to General Lord Lynedoch.

The regiment marched from Meerut on 20th October, 1826, and reached Cawnpore on 15th November. On 19th November it embarked in boats on the Ganges, and after a tedious passage arrived at Fort William, there to be stationed, on 15th January, 1827.

---

* Under Lord Lake, in 1805, the English were repulsed with a loss of one hundred and forty-two officers and two thousand eight hundred and forty-one men killed and wounded.

ROLL OF OFFICERS OF THE REGIMENT, 1826.

(From the Annual Army List.)

RANK.	NAME.	RANK IN THE REGIMENT.	RANK IN THE ARMY.
Colonel	Sir Harry Calvert, Bart., G.C.B., G.C.H.	8th Feb., 1806	*Gen.*, 19th July, 1821
Lieut.-Colonel	John McCombe	24th Aug., 1820	*Col.*, 12th August, 1819
,, ,,	Wilbraham Tollemache Edwardes	4th Nov., 1822	*Col.*, 12th August, 1819
Major	Franc. Skelly Tidy, C.B.	10th Sept., 1807	*Lt.-Col.*, 4th June, 1813
,,	John Campbell	6th May, 1824	*Lt.-Col.*, 19th Jan., 1821
Captain	Jacob Watson	5th August, 1804	*Major*, 4th June, 1814
,,	Matthias Everard	21st May, 1807	*Major*, 19th July, 1821
,,	John Marshall	21st April, 1808	*Major*, 12th Aug., 1819
,,	Cecil Bisshopp	23rd June, 1808	*Major*, 27th June, 1825
,,	W. S. Bertrand	7th March, 1811	
,,	William Turnor	26th March, 1818	15th August, 1811
,,	Gerald Rochfort	20th May, 1818	1st July, 1812
,,	Thomas Hall	21st May, 1818	16th September, 1813
,,	William Akenside	6th Sept., 1821	
,,	James Bancroft Ainsworth	25th Dec., 1822	
,,	Kenneth M'Kenzie	27th Jan., 1823	
Lieutenant	Henry Bruin Armstrong	27th Oct., 1808	*Adjt.*, 23rd Jan., 1806
,,	Henry Johnson	30th May, 1809	
,,	Martin Crean Lynch	31st May, 1809	
,,	Geo. Mackenzie	29th June, 1809	4th May, 1809
,,	Geo. Thurles Finnucane	15th March, 1810	
,,	Ed. L'Estrange	19th April, 1810	
,,	D. Hazlewood	13th Dec., 1810	
,,	Rich. Stack	3rd March, 1811	
,,	Hy. Mansell	14th Aug., 1811	
,,	Chas. Rayner Newman	15th Aug., 1811	
,,	Ed. Pender	27th Aug., 1812	
,,	Jas. Grant	14th Oct., 1818	6th November, 1814
,,	Geo. Jas. Bower	1st March, 1819	3rd April, 1815
,,	Jas. McDermott	9th Dec., 1819	2nd November, 1812
,,	Gilb. Pasley	10th June, 1820	
,,	Wm. Cain	6th April, 1822	12th June, 1819
,,	Wm. Maxwell	11th April, 1822	7th December, 1815
,,	Jas. Watson	25th Dec., 1822	
,,	Arthur Ormsby	27th Jan., 1823	
,,	Jas. Ramsay Smith	20th March, 1824	
,,	Robt. Nayler	21st March, 1824	
,,	Allan Stewart	3rd March, 1825	10th July, 1815
,,	Michael Chas. Horner	5th May, 1825	29th April, 1812
,,	Wm. Cockell	23rd June, 1825	
Ensign	Robert Daly	8th August, 1822	
,,	Brownlow Villiers Layard	24th May, 1823	
,,	Ed. C. Lynch	11th Dec., 1823	
,,	W. Littlejohn O'Halloran	11th Jan., 1824	
,,	Chas. J. Otter	20th Dec., 1824	
,,	Ed. Capudon	9th March, 1825	
,,	Ralph Budd	10th March, 1825	
,,	Thos. Holmes Tidy	14th April, 1825	
Paymaster	John Hy. Matthews	21st Oct., 1813	
Adjutant	Hy. Bruin Armstrong	21st March, 1811	*Lieut.*, 23rd Jan., 1806
Quarter-Master	Saml. Goddard	20th March, 1823	
Surgeon	Thos. Jackson	17th June, 1800	25th January, 1800
Asst. Surgeon	Thos. F. Colton	17th Aug., 1815	
,, ,,	Hugh Lindsay Stuart	15th Dec., 1825	

At Fort William a Horse Guards letter was received, dated 18th May, 1827, directing the word "BHURTPORE" to be inscribed on the colours.

A letter, of which the following is an extract, was addressed to Lord Combermere by Sir Herbert Taylor, under date 29th October, 1826:

"H.R.H. sees no objection to the Regiments of H.M. Army who were "employed at Ava, including those employed in Arracan, bearing on their "colours the word 'Ava,' and he will submit to His Majesty that such "distinction should be conferred upon them.

"He will also submit to His Majesty that the regiments employed in "the reduction of Bhurtpore should bear on their colours the word 'Bhurt-"pore.'

"But H.R.H. cannot, without departing from a principle which has "been adopted after due consideration, and which was observed in respect "of the War in the Deccan, recommend to His Majesty that Medals "should be bestowed upon and distributed to the Officers and Troops "employed upon those occasions." *

The regiment was stationed at Fort William for twelve months. On 25th January, 1828, it left for Berhampore, where it arrived on 21st February. A complimentary order was issued by the Governor-General, Lord William Bentinck, on its departure from the seat of Government, expressing his lordship's "high approbation of the very "orderly conduct of the regiment during the period it has been in "this garrison, evincing a state of discipline highly creditable to the "officers, as well as to the men they command."

Colonel McCombe,* C.B., commanding the regiment, died on leave at Calcutta on 12th October, 1828. He was succeeded by Brevet Lieutenant-Colonel George Thornhill, C.B., from the 13th Light Infantry.

---

* The East India Company's war-medal, 1799-1826, was given to the survivors of Bhurtpore in 1850.

* Colonel John McCombe entered the army in 1793 as ensign 52nd Foot. He was some time lieutenant and adjutant of that regiment. In 1801 he got his company in an old 2nd battalion 52nd Regiment, which became the 96th Foot (disbanded 1818). In 1805 he was appointed major in the Royal Corsican Rangers, and commanded a detached wing of that corps at the battle of Maida (gold medal). He served as major and lieutenant-colonel of the Corsican Rangers, under Sir Hudson Lowe, until the corps was disbanded in 1817, and was then appointed

On 26th June, 1830, King George IV. died, and King William IV. ascended the Throne.

The 14th Regiment remained in cantonments at Berhampore until November, 1830, when it returned to Fort William, and was thrown open to volunteering preparatory to leaving India. The men who elected to continue serving in that country having been transferred to other corps, the head-quarters of the regiment, after a service of over twenty-three years in India and the Indian seas, embarked for Europe in the Honble. Company's chartered ship *Recovery* on 27th December, 1830, arrived at Gravesend on 13th May, 1831, and marched to Chatham. The remaining details of the regiment left Calcutta in January, 1831, and arrived at Gravesend in July. On 12th July, 1831, Brevet Lieutenant-Colonel Matthias Everard, C.B., K.H., became lieutenant-colonel. The regiment remained at Chatham until 26th September, when the head-quarters embarked in the *Cygnet*, transport, for the Isle of Wight, and were quartered at the Albany Barracks, Parkhurst, until 6th December, when the regiment crossed over to Gosport, and occupied quarters in Haslar Barracks and Fort Monckton.

Many alterations were carried out in the army at and about the period of the return of the regiment from India.

The practice of enlisting men at option for "limited" (seven years) service, or "unlimited" service, which had been in force during the Peninsular War and afterwards, was abolished by a Horse Guards Order, dated 18th April, 1829. From that time until the introduction of "short service" (ten years in the infantry) in 1848, all recruits were engaged for unlimited service, with claim to pension after twenty-one years. Rates of army pensions were somewhat improved. Pikes, which for years had been carried by the sergeants in place of the old-fashioned "halberds" of fifty years before, were discontinued, the pike being replaced by a short, smooth-bore musket, called a "fusil," of the same bore as the muskets carried by the men.

---

lieutenant-colonel of the 64th Foot at Gibraltar. Soon after he exchanged to the 14th with Colonel Montagu Burrows. He commanded a brigade at Bhurtpore (wounded; mentioned in despatches; C.B.), and died at Calcutta 12th October, 1828.

This continued in use until the introduction of the short Enfield rifle for sergeants during the Crimean War.* The use of regimental-pattern lace on the clothing was discontinued in the case of the rank and file, and restricted to the drummers. The band was dressed in white coats, instead of buff, faced with red. The wearing of silver lace by officers was ordered to be confined to militia and yeomanry corps. The 14th Regiment officers accordingly substituted gold lace for the silver so long worn in the regiment. The officers' dress-coats were shorn of much of the rich lace worn on them since the days of the Regency, and many minor changes were introduced in details of the dress and accoutrements of all ranks.

1832-35.

On 29th February, 1832, the regiment passed over from Gosport to Portsmouth, and occupied the Colewort and Fareham barracks for the next five months. On 16th July, 1832, it embarked for Ireland in H.M. troopship *Jupiter*, arriving at Cove on 22nd. The right wing, under Colonel Everard, proceeded to Cork; the left, under Major Turner, to Spike Island and the forts round Cork Harbour. In August the regiment arrived by wings at Buttevant. In March-April, 1833, the regiment, in three divisions—the first under Major Turner, the second under Major Barlow, the third (head-quarters) under Colonel Everard—moved from Buttevant to Athlone.

General Lord Lyndoch, G.C.B., having been appointed to the colonelcy of the 1st Royal Regiment of Foot, was succeeded in the 14th by Lieutenant-General the Honble. Sir Charles Colville, G.C.B., G.C.H., on 12th December, 1834. Sir Charles Colville was transferred to the 5th Fusiliers in March, 1835, and was replaced by Lieutenant-General the Honble. Sir Alexander Hope, G.C.B., a former lieutenant-colonel of the regiment.

---

* On parade the fusil was carried "advanced" in the right hand, but with the hammer to the front (as flank company officers carried their fusils fifty years before), instead of the trigger-guard as in the case of the short Enfield. A very neat and simple manual exercise was devised for sergeants, which was also adopted on board some of H.M. ships at the time for the exercise of the small-arm men of ships' companies.

ROLL OF OFFICERS OF THE REGIMENT, 1831.

(From the Annual Army List.)

RANK.	NAME.	RANK IN THE REGIMENT.	RANK IN THE ARMY.
Colonel - -	Thos. Lord Lynedock G.C.B. - - - -	6th Sept., 1826	*Gen.*, 9th July, 1821
Lieut.-Colonel	Geo. Thornhill, C.B.	1st May, 1828	
Major - -	Wm. Turnor - -	19th Dec., 1826	
,, - -	Maurice Barlow - -	25th June, 1830	12th June, 1828
Captain - -	Cecil Bisshopp, C.B.	23rd June, 1808	*Lt.-Col.*, 19th Jan., 1826
,, - -	W. S Bertrand - -	7th March, 1811	*Maj.*, 22nd July, 1834
,, - -	Kenneth M'Kenzie -	27th Jan., 1823	
,, - -	Hy. Johnson - -	21st Jan., 1826	
,, - -	Martin Cuan Lynch	22nd June, 1826	
,, - -	Benjamin Whitney -	16th Sept., 1826	
,, - -	George Mackenzie -	19th Dec., 1826	
,, - -	W. John Pym Gore	13th Sept., 1827	27th April, 1827
,, - -	C. Rayner Newman	1st May, 1828	
,, - -	Ed. L'Estrange - -	9th Nov., 1830	
Lieutenant -	D. Mark Hazlewood	13th Dec., 1810	
,, - -	Hy. Mansell - - -	14th August, 1811	
,, - -	Jas. Grant - - -	14th Oct., 1818	6th November, 1814
,, - -	Jas. M'Dermott - -	9th Dec., 1819	2nd November, 1812
,, - -	Wm. Maxwell - -	11th April, 1822	7th December, 1815
,, - -	Jas. Watson - - -	25th Dec., 1822	
,, - -	Arthur Ormsby - -	27th Jan., 1823	
,, - -	Robt. Daly - - -	6th Jan., 1826	
,, - -	Brownlow V. Layard	20th Jan., 1826	
,, - -	Ed. C. Lynch - -	21st Jan., 1826	
,, - -	Ralph Budd - - -	16th March, 1826	
,, - -	John Higginbothum	3rd August, 1826	26th March, 1818
,, - -	Thos. Holmes Tidy	28th Sept., 1826	
,, - -	Chas. Dormer - -	18th Oct., 1826	2nd December, 1819
,, - -	G.C.M.L. Johnstone	14th April, 1827	2nd June, 1825
,, - -	Alex. Enerson - -	2nd August, 1827	15th February, 1827
,, - -	John Du Vernet - -	20th Oct., 1827	8th April, 1826
,, - -	Chas. J. Otter - -	12th Nov., 1827	
,, - -	Spencer Greaves -	3rd June, 1828	
Ensign - -	John Barry Maxwell	26th Jan., 1826	
,, - -	John Kyffin Lloyd -	23rd May, 1827	
,, - -	Matthew C. Wilder -	24th May, 1827	
,, - -	Chas. Campbell -	9th August, 1827	
,, - -	Wm. Graham - -	21st Sept., 1827	*Adjutant*
,, - -	Ough Barring - -	31st July, 1828	
,, - -	H. M. F. Stirke - -	31st July, 1828	
,, - -	Hy. Thos. Hutchins	27th April, 1829	8th November, 1826
,, - -	Ed. Senior - - -	7th Jan., 1830	
Paymaster -	James Johns - - -	29th Nov., 1827	*Lieut.*, 29th Jan., 1811
Adjutant -	Wm. Graham - -	21st Sept., 1827	*Ensign*, 21st Sept., 1827
Qr.-Master -	Saml. Goddard - -	20th March, 1823	
Surgeon -	J. M'Andrew, M.D.	30th April, 1829	
Asst.-Surgeon	Thos. F. Colton - -	17th August, 1815	

In October, 1834, the head-quarters, under Major Turnor, moved from Athlone to Mullingar, detaching a company to Killican, and another to Tullamore. In April, 1835, the regiment moved by divisions from Mullingar into Dublin.

In the Royal Barracks, Dublin, on Waterloo Day (18th June), 1835, new colours were presented to the regiment by Major-General Sir Edward Blakeney, commanding the troops, on a garrison parade consisting of the right flank companies of the 1st Grenadier Guards, the 81st, 85th, and 90th regiments.

A return of the regiment when in Dublin at this time, shews the nationalities of the non-commissioned officers and men as follows:—

English .	31 sergeants,	27 corporals,	12 drummers,	453 privates.
Scotch .	1 ,,	... ,,	... ,,	7 ,,
Irish .	10 ,,	9 ,,	2 ,,	133 ,,

The larger portion of the men were between 20 and 25 years of age. Their heights were :—6ft. and over, 13 ; 5ft. 11in., 28 ; 5ft. 10in., 46 ; 5ft. 9in., 61 ; 5ft. 8in., 121 ; 5ft. 7in., 204 ; 5ft. 6in., 139 ; 5ft. 5in., 24 ; under 5ft. 5in., 11.

Five years had not elapsed since the return of the regiment from India when it received orders to prepare for embarkation for the West Indies. It was inspected by the General Commanding on 9th September, 1835. It was broken up into six service and four depôt companies, and, leaving details in Dublin, proceeded by sea to Cork, whence it marched on 29th October to Fermoy. The service companies returned by the same route to Cork in November.

1836-38.

On 9th and 11th February, 1836, the service companies embarked, in two divisions, for the West Indies. The first division, consisting of Nos. 6, 9, and 10 (light) companies, with all the women and children, embarked in the *Stakesby*. The remainder, Nos. 1 (grenadiers), 3, and 7, left in the *Catherine Stewart Forbes*. The companies landed at Barbadoes on 14th March.

Authority to inscribe the word "TOURNAY"* on the regimental colours and appointments was granted in March, 1836.

On 26th April, 1836, the head-quarters and four companies

---

* See letter in *Appendix Volume*. The *Naval and Military Gazette*, 16th February, 1839, stated that the only officers present at Tournay at that time surviving were Lieutenant-General William Burnett and Colonel Richard Goodall Elrington.

King's and Regimental Colour, XIV. Regiment. Presented, 1835; retired circa, 1853.
Old Drum Major's Staff, dated 1786, and Sash.

embarked for St. Kitts, where they landed on the 29th. The light company was detached to Basseterre, and another company to Tortola, which rejoined on 14th July. Detachments were also found for Nives and Monserrat in May and July this year.

On 28th January, 1837, the head-quarters and three companies embarked for Antigua, where they arrived on 3rd February, and occupied the barracks on Blockhouse Hill and Shirley Heights vacated by the 74th Regiment.

On 19th May, 1837, Lieutenant-General Sir James Watson, K.C.B., who had commanded the regiment in India from 1807 to 1821, was appointed colonel, in the place of General the Honble. Sir Alexander Hope, G.C.B., lieutenant-governor of Chelsea Hospital, deceased.

In June the depôt companies, which had remained in Ireland, proceeded from Waterford to Bristol, and thence to Brecon, South Wales. A large draft from the depôt joined at Antigua in April this year.

On 20th June, 1837, King William IV. died, and Her Majesty Queen Victoria succeeded to the Throne.

The service companies, under Lieutenant-Colonel M. Everard, C.B., K.H., remained at Antigua until the end of the year 1838.

The distinguished services of the 14th Regiment in India from 1807 to 1831 having been brought before Her Majesty by the General Commanding-in-Chief, Lord Hill, at the special request of Sir James Watson, a Horse Guards letter, dated 8th November, 1838, signified the Queen's pleasure that the badge of a ROYAL TIGER, superscribed "INDIA," should be borne on the colours and appointments, to commemorate the services of the regiment in that part of Her Majesty's dominions.*

On Christmas Day, 1838, the head-quarters and one company of the regiment embarked at Antigua in the *Duke of York* for Barbadoes. On the following morning the vessel went on the rocks at St. Rose, island of Guadaloupe. Happily no lives were lost, but there was great destruction of regimental property. The orderly-room documents, and, indeed, everything with the head-quarters, except what the men carried, was lost. The French authorities provided means for the conveyance of the troops to Barbadoes.

---

* See letter in *Appendix Volume.*

ROLL OF OFFICERS OF THE REGIMENT, 1841.

(From the Annual Army List.)

RANK.	NAME.	RANK IN THE	
		REGIMENT.	ARMY.
Colonel - -	Sir J. Watson, K.C.B.	21st May, 1837	*L.-G.*, 10th Jan., 1837
Lieut.-Colonel	M. Everard, C.B., K.H.	12th July, 1831	19th January, 1826
Major - - -	Maurice Barlow -	25th June, 1830	12th June, 1828
,, - -	James Watson - -	10th Jan., 1840	
Captain - -	Martin Crean Lynch	22nd Jan., 1826	
,, - -	C. Rayner Newman -	1st May, 1826	
,, - -	D. M. Hazlewood -	10th July, 1831	
,, - -	George Douglas - -	17th July, 1835	26th June, 1835
,, - -	Thos. Holmes Tidy -	2nd Oct., 1835	
,, - -	Joseph Smith - - -	19th August, 1836	13th March, 1827
,, - -	John Watson - - -	25th August, 1837	11th August, 1837
,, - -	G. A. Wilson - - -	26th April, 1839	
,, - -	John Manley Wood -	28th August, 1840	*Maj.*, 28th June, 1838
,,	Sir J. E. Alexander -	11th Sept., 1840	18th June, 1830
Lieutenant -	Ralph Budd - - -	16th March, 1826	
,, - -	H. F. M. Strike - -	6th Sept., 1824	
,, - -	John Dwyer - - -	25th Sept., 1835	21st November, 1834
,, - -	Collett Leventhorpe -	2nd Oct., 1835	
,, - -	A. W. Campbell - -	25th Dec., 1835	
,, - -	Edward Archdall -	25th Nov., 1836	
,, - -	Henry Pigott - - -	1st Sept., 1837	14th June, 1839
,, - -	William Douglas - -	1st August, 1838	
,, - -	Edward Pratters - -	26th April, 1839	
,, - -	William Blundell - -	10th Jan., 1840	
,, - -	John Spence - - -	11th Jan., 1840	*Adjutant*
,, - -	Robert Wm. Romer -	15th Jan., 1840	
,, - -	Arthur H. Elton - -	27th March, 1840	
Ensign - -	Thomas E. Holmes -	28th Dec., 1838	
,, - -	John Peter Hall - -	26th April, 1839	
,, - -	Manly G. D. Hall -	23rd August, 1839	
,, - -	Thomas Hamilton -	30th August, 1839	
,, - -	D. T. Armstrong -	22nd Nov., 1839	
,, - -	Robert Halpin - -	10th Jan., 1840	
,, - -	Loftus Hare - - -	31st Jan., 1840	
,, - -	Thomas Dowse - -	27th March, 1840	
Paymaster -	Peter V. Wood - -	2nd August, 1831	*Capt.*, 13th Feb., 1828
Adjutant - -	John Spence - - -	28th Oct., 1836	*Lieut.*, 11th Jan., 1840
Qr.-Master -	Samuel Goddard - -	20th March, 1823	
Surgeon - -	Richard Dowse - -	8th Jan., 1836	
Asst.-Surgeon	H. Drummond, M.D.	13th March, 1835	28th November, 1834
,, ,, -	John Thomas Telfer -	4th Sept., 1835	17th July, 1835

1839-43.

The head-quarters were moved to St. Lucia in January, 1839, leaving detachments in Barbadoes. The St. Kitts detachment proceeded to Dominica. The head-quarters were again in Barbadoes in April, 1840; removed to Trinidad in the following June; and in January, 1841, returned to Barbadoes. In April, 1841, the service companies embarked for Canada; disembarked at Quebec, for Kingston, in June; and in July arrived at London, Canada West, where they stayed two years. The esteem in which the regiment was held by the inhabitants is best shewn by the letter received from the civil magistrates on its departure.* The depôt, which had moved from Brecon to Chester, went back to Ireland in June, 1840.

1843-49.

The service companies left London in September, 1843, for Kingston, remaining there two years, and then embarking in two divisions, in May and June, 1845, proceeded to Quebec.

The depôt, which during the previous four years had been quartered in the north of Ireland, was brought over to England in December, 1844, and stationed at Dover.

While the service companies were at Quebec, the city experienced two of those fearful conflagrations which have recurred so frequently in the course of its history. The first, which destroyed one thousand six hundred and fifty houses and left twenty thousand persons homeless, happened on 28th May, 1845, the night of the arrival of the first division of the regiment. The second, which destroyed upwards of one thousand two hundred dwellings and left fifteen thousand more persons without homes, occurred just a month afterwards. The following copy of the official report of the second fire is preserved among the regimental records:

Report of Major-General Sir James A. Hope, K.C.B., to His

---

* See letter in *Appendix Volume.*

Excellency Lieutenant-General the Earl of Cathcart, K.C.B., Commander of the Forces in North America :

"Quebec,
"1st July, 1845.

"My Lord,

"It is with great grief that I have to report to your Excellency "another dreadful conflagration, that took place in the suburbs of Quebec "on the night of the 20th June, by which 1,200 houses were consumed. "The fire is said to have originated about 20 minutes before 12 o'clock at "night in a hangar behind the house of Mons. Sissy, a notary, at the "beginning of Aiguillon Street, close to the glacis outside of St. John's "Gate, and spread with frightful rapidity along the whole length of that "street, of St. John's Street, and extending to the ruins of the St. Roche "suburbs burned by the former fire. There was no time lost in turning "out the whole of the troops, but the rapidity of the flames for a considerable time defied all human exertion, and continued their frightful progress "along the St. Foy Road (a distance of 1,200 yards) as long (with very "few exceptions) as there were houses to burn, and spreading against the "wind from one narrow street to another in the direction of the St. Louis "Road.

"Unfortunately, a gale of wind was blowing from the east, and the "whole breadth of St. John's suburbs appeared to be on fire at the same "moment.

"During the night Mons. Carron, the Mayor of Quebec, came to me to "point out a spot near the glacis, betwixt the St. John's and St. Louis "Gates, that should be cut off, and a house was immediately blown up, "by which a considerable part of the suburb of St. Louis was saved. It "was therefore to the Mayor's judicious selection of this spot that this "important service was performed, and I have strongly to express to your "Excellency that Mons. Carron exerted himself during the whole of the "night to the utmost, and, owing to the great extent of the conflagration, "was obliged to invest me with full authority to take every step that the "circumstances required.

"Some houses and a chapel were blown up in the intervening space "from where the fire commenced to the spot where it was finally checked, "without producing any good effect, except at one spot, pointed out to me "by Captain Warburton, Royal Artillery, who reported that by blowing "up two houses at the corner of one of the streets leading towards the St. "Louis Road, and taking advantage of some open spaces, a stop might "be put to the flames in that direction. I immediately directed him to "do so, and I am happy to do justice to the intelligence of this young "officer in reporting to Your Excellency that the operation was attended "with success, and I have to add Lieutenant Shakespeare's (of the Royal "Artillery) name as equally deserving of praise, both of these officers "being entrusted with the peculiar duty of placing the barrels of gun-

" powder in the houses, often within a few yards of the conflagration.

" By the blowing up of several houses in rapid succession at the spot " where the St. John's Road extended beyond the St. Louis suburb, taking " advantage of some open spaces, and by clearing away a space betwixt " the burning buildings and the upper part of the suburbs of St. Louis, " the flames, that had raged to this spot with unabated violence, were " checked about 8 o'clock on the morning of the 29th June, and this " direful visitation suddenly and in an instant subdued into the stillness of " exhausted desolation.

" A great part of the St. Louis suburbs, the whole of the St. John's " suburbs, with very few exceptions, five-sixths of the suburbs of St. " Roche, the streets along the foot of the Rock to the point where the fire " was stopped on the former occasion, present one scene of continuous " ruins, having deprived at least 22,000 persons of their houses and homes, " entailing on the poorer part of the population of this devoted city inex- " pressible misery, and sent thousands to wander for shelter and bread, " and to depend for relief and consolation on the benevolence and gene- " rosity of a feeling public. It may be permitted to me under these " appalling circumstances to go a little out of my way, and to beg of your " Excellency to use all your influence that some Legislative measures " should be adopted in the reconstruction of this unfortunate town, that " the houses should be built as much as possible with stone or brick, and " roofed with some material to resist the flames, and that a line should be " drawn round the works, and no houses allowed to be built within this " line but of stone or brick, and securely roofed. Besides almost the whole " of these suburbs being constructed of wood, all the intervening spaces " between the streets, the gardens, yards, etc., were filled with stacks of " deals, connected with wooden palings and every sort of combustible " matter packed together. It therefore ceases to be matter of surprise " that these two dreadful calamities have occurred, and rather to be " wondered at that such a dreadful visitation has not occurred before.

" The spot where the houses were latterly blown up was at the point " where the St. John's suburbs extend beyond the St. Louis ; and owing " to these houses being rapidly blown up, and the utmost exertion of the " troops in clearing a space, a great part of the St. Louis suburbs, the " whole of Clapton Terrace, consisting of villas, and the gentlemen's seats " on each side of the St. Louis Road, were saved.

" I wish it had fallen to my lot to report to your Excellency the services " of the officers and men employed during the progress of both the late " lamentable events, on occasions more honourable and more congenial to " their profession ; but it would indeed be unjust if I did not mention the " names of Captain Boxer, of the Royal Navy, whose exertions during the " whole night, and particularly at the end of the fire in directing the spaces " to be cleared, were of the greatest service ; of Lieutenant-Colonel " Walker, commanding the Royal Artillery ; of Major Watson, command- " ing the 14th regiment ; of Lieutenant-Colonel Thorp, commanding the

" 89th regiment. And I have also to state that I received every assistance
" from the suggestions and activity of Lieutenant-Colonel Ward and the
" officers of the Royal Engineers ; of Lieutenant-Colonel Pritchard, assist-
" ant adjutant general ; of Captain Ingall, deputy assistant quartermaster-
" general ; of Town Major Knight ; and of my aide-de-camp, Captain Hope.
" The conduct of the officers, non-commissioned officers, and men, as I
" expressed on the occasion of the former conflagration, exceeds anything
" I can say in their praise. Though this dreadful misfortune occurred in
" the night, when the men are frequently scattered in various directions,
" they were quickly assembled at the sound of the bugle, and never relaxed
" an instant in their exertions ; and I am happy to add that but one person
" was killed, one man's leg broken, and three or four contused by the
" several explosions that took place ; and that no pains were spared to get
" the people out of the way. There have been no casualties among the
" troops.

" I directed that the whole of the tents should be pitched near the old
" French works, and the Splinter-Proof Barracks given up for the accom-
" modation and relief of the sufferers, and to be placed at the disposal of
" the Mayor.

" I have the honour, etc.

" (Signed) J. A. HOPE,
" Major-General."

A Horse Guards letter, dated 12th August, 1845, was subsequently received, expressing the gratification of the Commander-in-Chief, Field Marshal the Duke of Wellington, at "this fresh display of all " the good qualities of British officers and soldiers—cool deliberation " in the direction of the measures to be adopted ; energy in their " execution ; and the zeal, activity, and persevering labour of all to " attain the object in view—that of arresting the progress of the mis- " fortune and saving the lives and property of Her Majesty's subjects." The letter made special mention of Major James Watson, 14th regiment.

On 25th July, 1846, the grenadiers and Nos. 3 and 9 companies, under Brevet Major Newman, embarked at Quebec, in H.M.S. *Belleisle*, for Halifax, Nova Scotia, where they disembarked on 31st July, and occupied the Citadel. The head-quarters, with Nos. 6 and 7 and the light company, under Major Watson, followed, arriving at Halifax on 31st August, where they occupied the North Barracks. Detachments were furnished at Annapolis, Prince Edward's Island, and Cape Breton.

A letter was soon after received from Quebec, signed by twenty-two justices of the peace on behalf of the inhabitants, expressive of their appreciation of the good conduct of the regiment during its stay in that city, and their best wishes for its success.*

Early in June, 1847, the detachments were called in, and the service companies embarked at Halifax, in H.M.S. *Assistance*, for England; landed on 27th June at Plymouth, where the regiment was quartered in the Citadel Barracks, with detachments on Maker Heights, Drake's Island, and at Falmouth. On 27th July a company, under Brevet Major Newman, proceeded from head-quarters to Exeter. It rejoined on 28th December, on relief by the 47th depôt.

In the course of the year the regiment was armed with the new percussion musket of 1842 pattern, the old flint-lock muskets previously in use in the regiment being returned into store. The new musket, which had a front-action lock, had a calibre of 0·753 in., and weighed 10 lbs. 4 oz., with a 39-inch barrel, the calibre, weight, and length being about the same as the latest patterns of the flint-lock musket. The charge was 4·5 drachms, and the bullets 14·5 to the pound.

An Act of Parliament, passed on 27th June, 1847, altered the conditions of enlistment in the army, marines, and the European forces of the East India Company. Instead of engaging for unlimited service, men were to be entered for a first term of ten and not exceeding twelve years, with the option of re-engaging at its expiration for a second term to complete twenty-one years for life pension.

After commanding the regiment for sixteen years, Colonel Matthias Everard, C.B., retired by exchange on half-pay (3rd Buffs), and was succeeded by Major Maurice Barlow,† who purchased his lieutenant-colonelcy in the 44th, and exchanged back to the 14th the same day. The regiment remained at Plymouth through 1848, and in March, 1849, proceeded by march route to South Wales, the first division, under Major James Watson, marching on the 12th for Carmarthen,

---

* See letters in *Appendix*.

† Afterwards General Maurice Barlow, colonel of the regiment. See Succession of Colonels, *Appendix A*.

followed next day by the second division, under Brevet Major Smith, to the same place; the head-quarters, under Lieutenant-Colonel Barlow, marching on the 14th to Newport. Detachments were furnished to Pontypool, Cardigan, and Pates.

1850-53.

In April, 1850, the regiment proceeded by steamer and march route from South Wales to Bristol, and thence to Preston, where it occupied Fulwood Barracks. In May, the same year, it proceeded from Preston to Liverpool, in three divisions, and embarked for Ireland. The head-quarters were stationed at Athlone, where they arrived on 28th June. Two companies were detached to Roscommon and Castlerea. In October two companies proceeded by march route to Castlebar and Westport, a reinforcement of another company to the first-named place being subsequently added.

An alteration was made in the pattern of the accoutrements in this year. The shoulder-belt and breastplate forming part of the old-fashioned cross-belts were discontinued, and replaced by a waist-belt with bayonet frog, and an Union locket-clasp bearing the number of the regiment, surrounded by the county title—" Buckinghamshire."

On 21st and 23rd April, 1851, the regiment marched in two divisions from Athlone to Dublin, and on arrival was quartered in the Linen Hall Barracks. A couple of months later it moved into the Royal Barracks.

On 16th and 17th March, 1852, the regiment moved from Dublin to Limerick, giving detachments at Cahir, Newcastle, Rathkeale, Kilrush, and Killala. In July the head-quarters moved from the New Barracks to the Castle Barracks, and a detachment was sent to Clare Castle. The various detachments were relieved during the year, and the head-quarters returned to the New Barracks, Limerick.

The regiment was in Limerick when intelligence arrived of the death of Field Marshal the Duke of Wellington on 14th September, 1852. It remained in Limerick through that and the following year. On 2nd March, 1853, a special duty party, consisting of Captain Budd, Lieutenant Helyar, and eighty-nine men, proceeded to Lenagh. The party rejoined head-quarters on 28th March. The various detachments were relieved in the course of the year, and those at Cahir and Newcastle were called in. Asiatic cholera was very prevalent at Limerick during the year.

Ominous clouds were at this time gathering in the East, threatening the European peace which England had enjoyed for nigh forty years. The demands of the Czar Nicholas in respect of the custody of the "Holy Places" of Palestine and the sovereignty of Turkish subjects professing the Greek faith had been met by Turkey with an appeal to the Great Powers in May, followed by a declaration of war by Turkey against Russia on 5th October, 1853. The British Fleet, under Admiral Dundas, which had been already sent to the Dardanelles, entered the Bosphorus on 30th October, and preparations began for the despatch of British troops to the same quarter.

In October, 1853, orders were received for the regiment to hold itself in readiness to proceed to Gibraltar, and in December it was divided into service and depôt companies preparatory to embarking at the New Year.

1854.

Commencement of the Russian War.

On 4th January, 1854, the allied fleets of England and France entered the Black Sea, and on 23rd February the despatch of British troops to the East commenced with the embarkation of the Guards. On 28th March, 1854, England and France declared war against Russia.

The 14th, being the first regiment on the roster for foreign service, expected to be amongst those first sent to the seat of war; but, for some reason or other, its destination was changed to Malta. The service companies—Nos. 1 (grenadiers), 2, 4, 6, 7, and 10 (light company)—under command of Colonel Maurice Barlow, proceeded from Limerick to Cork, and on 26th April, 1854, embarked at Cove (now Queenstown) in the sailing transport *Bombay*, an old Indiaman owned by Messrs. Green. They were towed out of harbour next day, passed Gibraltar on 3rd May, and arrived at Malta on 28th May.* Before disembarking, the 14th was ordered to give one

* Sergeant-Major Ryan (afterwards Quarter-Master of the North Tipperary Light

hundred volunteers to the 41st, 47th, and 49th regiments, who at once joined drafts of those regiments lying in the harbour. On landing, the 14th was broken up in detachments—head-quarters, with Captains Dwyer's and Douglas's companies, at Fort Marioch; Captain Budd's company at Fort Tigna; Major John Watson with two companies at Fort Ricasoli; and a company at Fort St. Angelo. The other corps in garrison were the Buffs, 9th, 62nd, 68th, and Malta Fencibles. The detachments were frequently changed during the summer, and on 1st November the head-quarters moved into Floriana Barracks.

In the meantime, after some months of weary waiting at Scutari and Varna, the British troops had landed in the Crimea on 14th September, 1854; the memorable battles at the Alma, Inkerman, and Balaklava had been fought, and the siege of Sevastopol had begun. The first bombardment was over.

The 14th at this time was the only line regiment left behind at Malta, and to it were attached all the sick and wounded arriving from Scutari who were unfit to proceed to England, as well as all the women and children of the regiments that had gone on to the war.* Large quantities of stores and ammunition were in progress of shipment in Malta Dockyard, so that the regiment, which was held in readiness to proceed to the Crimea, was incessantly engaged on fatigues and other duties.

The depôt companies (Nos. 3, 5, 8, and 9), under Brevet Lieutenant-Colonel James Watson, had left Limerick for Mullingar on the departure of the service companies, and during the remainder of the year were scattered over the south of Ireland in detachments, which also served as recruiting stations. The regiment had kept up no local connection in England, and for long previous to the war, as well as for some time after, was largely recruited in Ireland. Two captains were added to the regiment in June-July, 1854, and in

Infantry Militia) was left behind at the last moment, to complete his service in July, 1854, with the depôt companies, having lost his wife, who left a young family. Sergeant John Glancy, afterwards a major in the regiment, was appointed acting sergeant-major, and subsequently confirmed in the appointment.

* Captain C. M. Wilson writes: "I personally embarked seven hundred women, with nine hundred children, in the *Himalaya* at Malta, in 1854. We had had them with us five months."

December the same year four more were added, bringing up the total strength of the regiment to *sixteen* captains. (It should be observed that, when the regiment was strongest during the war, the regimental pay lists and monthly returns never showed more than *fourteen* companies completed—*eight* in the Crimea, *two* at Malta, and *four* in Ireland.) The minimum standard for infantry recruits was lowered at this time to five feet four-and-a-half inches for men not over thirty years of age. Recruiting parties were stationed at Lincoln, Leicester, Shrewsbury, Windsor, and Glasgow.

1855.

The Crimea.

The 72nd Highlanders having arrived from England, every available man of the 14th service companies embarked for the Crimea in the steamer *Emu* on 10th January, 1855. Having been reinforced by drafts from home, per *Alipore* and *Blake*, on 1st and 5th December, the service companies had been reorganized as *eight* companies, the designations of which, in the December, 1854, pay list, stood as follows: No. 1 (grènadiers,) Captain Hammersley; No. 2, Captain Trevor; No. 4, Captain Budd; No. 6, Captain Douglas; No. 7, Captain Segrave (joined from depôt); No. 10, —; No 11, Captain Sir J. E. Alexander (absent on the staff); light company, Captain Dwyer. The non-commissioned officers and men of these eight companies numbered—including sick and absent—six hundred and ninety-two, of whom five hundred and forty-one were between the ages of 18 and 30. Four hundred and sixty had under five years' service, one hundred and seventy-two between five and ten years', and the rest over ten years' service.

The embarked strength was two field officers, seven captains, ten subalterns, four staff, fifty-two sergeants, fifteen drummers, and six hundred and four rank and file. The following officers were embarked:

Lieutenant-Colonel M. Barlow, commanding; *Major* John Watson; *Captains* John Dwyer, William Douglas, W. C. Trevor, F. Hammersley, Thomas Segrave, and W. H. Hawley; *Lieutenants* W. Heywood, C. M. Wilson, C. Matthews, C. Fairtlough (adjutant), R. H. Vivian, R. H. Inglefield, T. P. Cosby, W. S. Blunt, R. H. Graham, and G. Brydges; *Paymaster* J. P. Hall; *Quarter-Master* J. O'Connor; *Assistant-Surgeons* T. M. Blickley and J. M. Hyde.

ROLL OF OFFICERS 1st JANUARY, 1855.

(From the Monthly Army List.)

Colonel	Sir James Watson, K.C.B., *General*	24th May, 1837
Lieut.-Colonel	Maurice Barlow, *Colonel*	25th Dec., 1847
Major	James Watson, *Lieutenant-Colonel*, dep.	10th Jan., 1840
,,	John Watson	28th March, 1854
Captain	Sir J. E. Alexander, *Lieutenant-Colonel*, staff	18th June, 1830
,,	Ralph Budd, *Major*	1st June, 1841
,,	John Dwyer	22nd Jan., 1846
,,	William Douglas	18th Sept., 1846
,,	W. Cosmo Trevor	29th Nov., 1850
,,	F. Hammersley	25th April, 1851
,,	C. C. Newman, dep.	12th Oct., 1852
,,	Henry Townshend, dep.	22nd Oct., 1852
,,	Jul. Chetham Strode, dep.	16th Dec., 1853
,,	J. McCulloch O'Toole, dep.	6th June, 1854
,,	Plomer J. Young, dep.	25th March, 1853
,,	Thomas Segrave	21st July, 1854
Lieutenant	W. H. Hawley	15th Feb., 1850
,,	Charles E Grogan	28th April, 1848
,,	John Barlow	23rd May, 1851
,,	Fred. Smythe, *Paymaster*, dep.	19th Dec., 1851
,,	William Heywood	16th Jan., 1852
,,	William H. Bower	13th Feb., 1852
,,	J. Echlin Matthews	27th August, 1852
,,	E. D'H. Fairtlough, *Adjutant*	19th Oct., 1850
,,	C. Monck Wilson	16th Dec., 1853
,,	Dawson S. Warren, dep.	16th Dec., 1853
,,	Gage H. Dwyer, *Adjutant*, dep.	28th March, 1854
,,	H. Hope A'C. Inglefield	6th June, 1854
,,	R. Hussey Vivian	6th June, 1854
,,	Thomas Prittie Cosby	11th August, 1854
,,	Walter Frederick Blunt	18th August, 1854
,,	Reginald H. Graham	15th Dec., 1854
,,	George Brydges	15th Dec., 1854
,,	Angus W. Hall, dep.	15th Dec., 1854
Ensign	Ed. W. Saunders, dep.	28th March, 1854
,,	William Dods, dep.	27th June, 1854
,,	Alexander Gordon, dep.	11th August, 1854
,,	F. Gerard Armstrong, dep.	18th August, 1854
,,	J. D. Bradley, dep.	25th August, 1854
,,	Ramsay Harman, dep.	1st Sept., 1854
Paymaster	John Peter Hall	11th April, 1851 *Capt.* 15th Feb., 1850
Adjutant	Ed. D'H. Fairtlough, *Lieutenant*	16th August, 1854
Qr.-Master	John O'Connor	11th March, 1853
Surgeon	William Denny	7th August, 1846
Asst.-Surgeon	William Renwick	4th March, 1853
,, ,,	Thomas M. Blickley	6th Jan., 1854
,,	John M. Hyde	1st Dec., 1854

The *Emu*, her decks covered with materials for huts from Malta Dockyard,* anchored off Seraglio Point, Constantinople, on 14th January, and lay there in heavy snow until the 17th, when she resumed her voyage, and arrived off Balaklava at 3.30 a.m. on 18th January, 1855. She entered the harbour next day, and was berthed just inside, below the lower Genoese fort, with her head to the sea, ready to run out in case of a Russian attack, which was daily expected. In the harbour all this time were over two thousand troops, which had not been landed on account of the scarcity of shore rations.

On 27th January a case of cholera and one of small-pox occurred on board the *Emu*, and on 30th January two hundred and fifty men were landed from her, the light company being first on shore, and encamped just outside Balaklava. Two died of cholera on board, and two fresh cases occurred.

On 2nd February, 1855, the rest of the service companies disembarked from the *Emu*, and the whole encamped on a spur of the hills to the left of the plain, just beyond the head of the harbour. The men, wholly untrained in encamping, at first made many blunders and endured much discomfort. Insufficiently clad as they were, they suffered bitterly from the cold the first night; but on the morrow there was an issue of winter clothing—sheepskin coats with the wool inside.† Balaklava itself was a wilderness of

---

* "Our quarter-deck was completely covered with materials for huts prepared in "Malta Dockyard, as the admiral, Sir Houston Stewart, told us, 'expressly for "the 14th,' but, as it turned out, we never had the use of them, as in the crippled "state of the land transport they were too heavy for conveyance to the front. So we "had, in the first place, the inconvenience of all this lumber, and no room to move, "and in the end no benefit from it."—Colonel and Paymaster Hall's MS. Notes.

† "On 3rd February warm clothing was issued. Croatian sheepskin coats and "jackets with the wool on, and usually a flower or other device painted on the "skin side, worn outward. Being of various cut, patterns, and sizes, they produced "an effect more picturesque than uniform. However, they kept the wearer warm. "Buffalo robes for sleeping on were also issued, and the officers were supplied "gratis with grey tweed jackets lined with rabbit-skin, coming just over the hips. "It was afterwards said that these were intended for the non-commissioned officers "and men, sealskin coats having been sent out for the officers, but which, in the "absence of instructions as to whether they were to be supplied gratuitously or on "payment, had been kept in store until an answer was received at the end of "March, when the weather had improved, and few officers cared to buy them at "£8 each, the price fixed."—Paymaster Hall's MS. Notes.

mud and slush. On every side were bales and packages of all descriptions, some of them sorely needed at the front, but which could not be sent up for want of transport. Sick and frost-bitten men were constantly arriving from the front, many of them brought down by the French ambulance, which was in better order than our own. Steamers were beginning to arrive with navvies and plant for the projected railway from Balaklava to the camp before Sevastopol. On 7th February, Lieutenant Matthews was sent home sick; and on 8th February the battalion had its first experience of the war, when Sir Colin Campbell had his brigade, including the 14th Regiment, in position on the heights to the eastward of the port before daylight, in expectation of a Russian attack. Nothing was seen of the enemy, and the brigade returned to camp soon after daylight. An attack was expected the following day.

On 9th February a draft, consisting of Brevet Lieutenant-Colonel James Watson, in command; Captain C. Newman; Lieutenants D. S. Warren, G. H. Dwyer, A. Hall, E. W. Saunders, and W. Dodds; Ensigns F. G. Armstrong and J. D. Bradley, with six sergeants and one hundred and forty-three rank and file, which had arrived from Cork, in the *Princess Royal*, and had been transhipped to H.M.S. *Terrible*, landed, and marched into camp. Heavy rain and snow prevailed at this time.

On 2nd March, 1885, Assistant-Surgeon William Renwick * died of typhus on board the *Walmer Castle* in Balaklava harbour, whither he had been removed a few days before. He was a very zealous officer, who had done good service in Limerick during the cholera outbreak in 1853. He had preceded the regiment to the Crimea some time, but rejoined it on its arrival.

The regiment was much employed on fatigue duties in and about Balaklava.† After performing much laborious and dirty work, it was

* Assistant-Surgeon William Renwick, M.D., appointed 4th March, 1853.

† "Three hundred pairs of ammunition boots were served out to the 14th, but "the thick, heavy clay sucked the soles off, and for a week back the men have "been going about without any soles, *ergo* with their feet on the ground, which "with the thermometer at 30°, is not agreeable locomotion."—MS. Notes.

moved up to the front, by two companies daily, between 10th and 15th March, and was brigaded with the Royals, 4th King's Own, 39th, and 50th regiments, under Major-General Henry Barnard, forming the first brigade of the third division, commanded by Lieutenant-General Sir Richard England, K.C.B., K.H. The third and fourth divisions formed part of the left attack. The 14th was placed in a double line of huts already constructed, of which the 39th occupied the right, and the 14th the left wing, and at once commenced taking its turn of duty in the trenches, which it continued to do until the fall of Sevastopol.* Its losses throughout the siege were comparatively small, which acquired for it the nickname, "the bomb-proofs." †

On Easter Sunday, 8th April, 1855, the allied camps were busy with preparations for the morrow, on which was to commence what was known as the "second bombardment" of Sevastopol. Showers were frequent during the day, and heavy rain fell all night, flooding the trenches, so that in some of the batteries the men were standing in eight inches of water. A dark night, with a heavy gale, followed. Thick fog and drizzling rain heralded in the 9th April, and rendered the Russian works invisible; but about 5.30 a.m. it cleared partially, and fire opened from the British right and left attacks and French works. Eight days of incessant firing followed, in which thirty-two thousand five hundred and sixty-eight shot and fifteen thousand two hundred and eighty-six shells, many of them of the largest calibre ever used, were thrown into the place; but the Russian fire was not silenced, and their works were no sooner injured than they were repaired again at night. Rumours of an assault were rife in camp; but a Council of War decided that the damage done to the Russian defences was insufficient to justify one. On 22nd

---

* Captain C. M. Wilson writes: "It took 1,100 men to guard our portion of "the trenches. It was the custom in our division to have a roster of corps in the "brigade office, and when men were wanted, the brigade-major detailed the whole "of the duty men of the regiment first on the roster, making up the balance from "the next on the roll; the uncalled-for balance of the second regiment being first "for duty on the next occasion, and so on."

† See *Appendix Volume.*

April daylight parades were stopped, and six hundred men added to the daily guard of the trenches of the left attack. On 23rd April, supernumerary non-commissioned officers and men of the regiment were sent back to Malta, per *Melbourne*. On 26th April the first casualties occurred in the 14th—a man was killed and another wounded by the fall of a pile of muskets belonging to the Royals while parading for trench duty. Another 14th man died of wounds received in the trenches. As the weather became milder, steps were taken to restore order and routine, which had been somewhat neglected during the long and bitter winter. Winter clothing was ordered to be laid aside, except for trench duty at night, and officers were directed to appear on all occasions in their proper uniforms, with black handkerchiefs round the neck. On 1st May, Lieutenant Alexander Gordon, who had been appointed to an ensigncy in the regiment from sergeant, Scots Fusilier Guards, on 11th August, 1854, but had been doing duty with the commissariat up to the end of March, 1855, was appointed adjutant, *vice* Fairtlough. The regiment had two men killed on 4th–6th May, but was not in the trenches on 11th May, when the Russians made a determined sortie by the Woronzoff road, which was repulsed with loss by the guards of the trenches of the left attack.

On 14th May, Brevet Lieutenant-Colonel Sir James Alexander, who had thrown up his staff appointment in Canada to join the regiment, embarked at Cork, with a regimental draft of five officers, six sergeants, two drummers, and one hundred and sixty-two rank and file for Malta. This draft, together with the supernumeraries sent back from the Crimea, formed the two companies which were stationed as a "reserve depôt" at Fort Ricasoli during the remainder of the Crimean campaign. On arrival at Malta, Sir James Alexander was at once ordered on to the Crimea. Major Ralph Budd, who succeeded to regimental majority on the retirement on full-pay of Brevet Lieutenant-Colonel James Watson* on 14th May, 1855, took

---

* Lieutenant-Colonel James Watson, eldest son of General Sir James Watson, was the latest Bhurtpore officer to leave the regiment. He died at the family seat, Wendover, Bucks, 23rd October, 1874.

command of the two reserve companies at Malta, which he held until after the fall of Sevastopol.†

The constant arrival of reinforcements had by this time brought up the allied strength to one hundred and eighteen thousand men, whereof twenty-eight thousand were British. The attack on Sevastopol was prosecuted with renewed vigour. On 23rd May the Cemetery Ridge had been captured by the French; and on 6th June the "third bombardment" commenced. On 7th June the French took the Russian work known as the Mamelon, and the British right siezed and held the Quarries and other counter-approaches in spite of repeated attempts of the Russians to retake them. On this day the 14th Regiment had one man killed in the trenches. On 11th June, Colonel Barlow took his turn in the trenches as acting brigadier. On 14th June the bombardment again ceased. On Sunday, 17th June, the "fourth bombardment" commenced. The morrow—the thirtieth anniversary of Waterloo—had been selected for a combined assault on the Russian works.

It was arranged that the French should attack the Malakoff, and that the British should assail the Redan as soon afterwards as might appear advisable. As day broke, the French commenced operations, their columns as they came within range suffering most severely from the musketry and guns of the works which had been silenced the night before. Perceiving this, Lord Raglan directed the columns for the Redan—portions of the light, second, and fourth divisions, under Lieutenant-General Sir George Brown—to move out of the trenches to the attack at once. A murderous fire of grape and musketry met them, and despite the heroic efforts of the assailants, the attempt failed, with heavy loss. Whilst the attack on the Redan was in progress, Sir Richard England was directed to send Major-General Barnard's brigade of the third division down the Woronzoff Ravine, with a view to supporting the attacking columns on his right, whilst the other brigade of the division, under Major-General Eyre,

---

† Major Budd, according to the regimental pay-list, was struck off the strength of the companies in the Crimea from 6th June, 1855. He is said to have left the Crimea before that date.

attacked the Russian advanced works at the head of Navy Creek. This service Eyre's brigade performed with great gallantry under a most galling fire. They maintained the position they had taken through the day, and in the evening withdrew unmolested, leaving a post in the cemetery at the head of the creek. Barnard's brigade, in which was the 14th, was not brought into action. It was, however, under fire for several hours, and suffered some loss. The 14th, which furnished two captains, four subalterns, and two hundred and ten non-commissioned officers and men to the storming-party of four hundred men, under Lieutenant-Colonel Waddy, 50th regiment, formed at the head of the brigade, and had three men killed (Privates Michael Lynch, William Thompson, and John Mulock) and four men wounded (one dangerously: Corporal Thomas Brown, thigh broken). The Russians attempted a sortie the same night, and the division was kept under arms, but did not march off. So ended the memorable 18th June, 1855.

Sir James Alexander thus describes the affair:—*

"We rose at midnight on the 18th (17th-18th). There was a hum of "voices all over the camp. We fell in while it was dark, and a strong "body of stormers being told off, we moved away towards the right, "crossing in front of Cathcart's Hill. The dust was suffocating, and the "night sultry. We marched in sections down the ravine, and whilst "objects were indistinctly visible, found ourselves with portions of the 4th, "39th, and 89th, in the Woronzoff Road, here commanded by the enemy's "guns. We were directed to ascend to the right and occupy the rocks "above, and we did so, like birds clustering there. Colonel Munro, "of the 39th, commanded the reserve. He was well used to warfare "in India, and moved about with zeal and intelligence among his "charge. We were joined on the hillside by Colonel Norcott, with "some of his Rifle Brigade. Colonel Waddy, of the 50th, headed the "stormers. He had acquired a high character for daring, and, in fact, "exposed himself more than most men on all occasions in the trenches. "He now turned out in his shell-jacket, without flask or haversack for "refreshment, but in one hand carrying a pistol, and in the other a "naked sword (the scabbard left at home), which he had captured in "combat in the East. There was no mistake about the intentions of 'the gallant Waddy, 'Do or Die' seemed his motto.

"Soon, balls, great and small, began to pitch amongst us; and as the

* *Passages in the life of a Soldier*, vol. ii., p. 64, *et. seg.*

"day broke, the roar of the combat became louder and louder. We were "near the Great Redan, which we saw lower down the ravine of the Woronzoff Road; the Malakoff was on our right, and shot and shell from it "flew over us, and ploughed up the ground on the left of the road. All "the while, our batteries of the Left Attack briskly and incessantly fired "at the Redan. The first casualty we observed was a sapper, who was "sitting down waiting the order to move nearer the enemy's works. He "was struck by a round shot that came lobbing down the hillside from "the Malakoff, and taking him at the back of the neck, broke his spine "and knocked him some yards down the hill. . . . .

"Grape-shot now lashed amongst us, and occasioned casualties among "our stormers who were advanced with the scaling-ladders. Now and then "there was a cry of 'Round-shot, look out!' which was avoided by being "on the alert to throw ourselves behind a rock—and then, up again.

"Below us, we saw sailors carrying slowly on stretchers, wounded messmates up from the Redan, whilst the shot knocked up the dust and "gravel around them. A sapper corporal came along the hillside from "the direction of the Redan. I asked him what was the news of the "assault from the trenches of the Right Attack. 'Bad news,' he said. "'The French had failed at the Malakoff; our people had tried the Redan, "and had been forced to retire with great loss. . . . . .' A party of "sappers, mixed up with our men, suffered greatly on this occasion. As "the stormers of the 14th were turning the corner of a rock, the Sergeant-"Major of the sappers was shot dead by a round shot, and, I think, eight "others were killed and wounded.

"Our general, Sir Henry Barnard, knowing the folly of attempting the "800 yards of open ground between us and the Barrack Battery, crossed "as it was by lines of annihilating fire from the Redan and other heavily "armed works, held his men in hand, and thus saved the utter and "inevitable destruction of his brigade, without the possibility of doing any "good, that is, without the previous fall of the Malakoff and Redan. The "Redan now clearly saw us in red masses against the rocks, and the "Russians were observed cutting away part of their parapet to get another "gun to bear upon us. We were accordingly desired to descend the hill-"side, cross the Woronzoff Road and ascend to the second parallel of the "Left Attack. We did so leisurely. We had some difficulty in getting "along the parallel from a crowd of ammunition-mules we found on the "top of the ascent, the enemy crossing us all the while with round shot. "The men were directed to keep as much in the bottom of the trench as "possible, but some, more careless than the rest, and sometimes it "happened out of mere bravado, remained high and unprotected by the "parapet. A round shot took a party of three exposed in this way, though "Major Dwyer and myself were directing them to keep low down, knocking them off their legs, and striking two of their pouches into the air like "crows. One poor fellow, whose bowels were carried clean out of him, "remained on his knees for a minute, wiping his face, in ignorance of "the mortal wound he had received, then fell over, and was covered with "a great-coat by his comrades. I was much pleased to observe on this "occasion the zeal and fearless conduct of Assistant-Surgeon Hyde,* who "moved among the wounded, doing all he could to assist them. . . . .

"It was understood that there was to have been three hours of severe "bombardment on our part and that of the French before they attacked "the Malakoff, and it falling to them the English were to go on the Redan. "The powerful work of the Malakoff not falling from the attacks on the "Redan failed. There was great slaughter there besides what occurred "elsewhere; thus, out of 120 sailors, carrying the scaling-ladders, 80 fell, "and our loss, altogether, amounted to 90 officers and 1,400 men.

"It seemed that, after the dashing way we took the Quarries, and the "French the Mamelon, on the 7th June, we were a little too proud, and "wanted humbling. We seldom have all our own way in this world, and "are doomed to frequent disappointments. And so we retired along our "trenches, 'bent' somewhat, but 'not broken;' and, carrying our "wounded on stretchers, passed the fine fellows of the Royal Artillery, "stripped to shirt and trousers, lying beside their guns and mortars, after "their stupendous exertions, grim-looking, reposing for a while after the "late severe tussle. With Colonel John Watson and the last men of the "regiment, I descended into the Valley of Death."

The weather at this time was exceedingly hot, and cases of cholera were frequent in the camps at Sevastopol and Balaklava. For some time the 14th escaped, although cases afterwards occurred in the regiment. On 24th June, General Estcourt, the chief of the staff, died; and on 28th June, Lord Raglan, who was much broken by the reverse of the 18th June, succumbed to the malady. General Simpson succeeded to the command-in-chief, and Major-General Barnard became chief of the staff in place of General Estcourt, on which Colonel Barlow was given the first brigade of the third division, and Brevet Lieutenant-Colonel John Watson † succeeded to the command of the 14th Regiment. Lieutenants Heaton and Harman, with a draft of fifty men, joined the service companies on 29th June.

The monthly return of the regiment for 1st July, 1855, shows twenty-four officers, fifty-seven sergeants, sixteen drummers, and seven hundred and thirty-seven rank and file present at headquarters before Sevastopol.‡ A company of marksmen, consisting

* Afterwards Deputy Surgeon-General J. M. Hyde, retired.

† Colonel John Watson retired on full pay. Died January, 1888.

‡ The Pay-lists and Monthly Returns of this period show eight companies in

of eight of the best shots from each company, was under the command of Lieutenant C. M. Wilson. Apprehensions were felt as to the water supply in camp, and parties of men were employed during May-August in constructing tanks, some for drinking purposes and some for washing. Lieutenant R. H. Vivian was employed for some time on this duty as an acting engineer. Captain William Douglas was appointed acting major of brigade in the absence of Major Daniell, 38th regiment. Lieutenant Wilson was at first appointed, on the sudden illness of Major Daniell, but it was represented that there would be a difficulty in officers taking orders from a lieutenant, and Captain Douglas was put in his place.

The siege continued. On 2nd August, Sir Richard England resigned command of the third division, and was succeeded by Sir William Eyre. On the 8th, Brevet Lieutenant-Colonel Sir James Alexander assumed command of the regiment, *vice* Lieutenant-Colonel John Watson. The "fifth bombardment" of Sevastopol, which was considered to have been very successful, began on the 17th and lasted till 21st August. Captains O'Toole and Townshend, Lieutenant A. A. Le Mesurier, and Ensign A. Molony arrived with a draft of one hundred and sixty-seven men on 20th August,

---

the Crimea only, the original numbers and the designations of which are given as follows:

No. 1, Grenadiers, Captain Hammersley
2, Captain Trevor
4, Captain Haywood
6, Captain Douglas
No. 7, Captain Segrave
10, Captain Newman
11, Captain Hawley
Light Company, Captain Dwyer

An officer who served with the regiment at Sevastopol states from memory that there were ten duty companies, which about this time were numbered and commanded thus:

No. 1, Grenadiers, Captain Hawley (?)
2, Captain Trevor
3, Captain Segrave
4, Captain Haywood
5, Lieutenant C. M. Wilson, for Captain Hammersley, absent on the staff. Lieut. Wilson also commanded the company of marksmen.
6, Captain Douglas
7, Captain Newman
8, Captain O'Toole
No 9, Captain Young, Old No. 9, which was to have gone as an additional company to the Crimea, was detained on account of the illness of its captain, Strode, who died 16th February, 1855. No. 11, one of the new companies, was sent in its place, and received number and "call," No. 9.
10, Light Company, Captain Dwyer

and on 29th August Lieutenant-Colonel John Watson and Lieutenants Dwyer and Harman left sick. On 27th August there was an investiture of the Order of the Bath at head-quarters, at which five hundred picked men from the third and fourth divisions, under command of Brigadier-General Barlow, formed the square. Captain Barlow arrived out from the depôt as aide-de-camp to his uncle, Brigadier-General Barlow.

The monthly return of the service companies before Sevastopol on 1st September show twenty-seven officers, fifty-seven sergeants, nineteen drummers, fifty-four corporals, and eight hundred and fourteen privates effective before Sevastopol, and three officers, eleven sergeants, six drummers, sixteen corporals, and two hundred and fifteen privates sick. The return shows the equipment of the regiment with the Enfield rifle for the first time as complete, and every man in possession of sixty rounds of Enfield rifle ammunition and seventy caps. The first return from the companies at Malta bears this date, and shows Captain Young's and Captain Grogan's companies (old Nos. 3 and 8) there, with Major Budd in command. The depôt returns show four companies (Nos. 12, 13, 14, and 15). The companies in the Crimea are designated in the returns as before—Nos. 1 (grenadiers), 2, 4, 6, 7, 10, 11, and light company.

On 1st September, 1855, the British sap was within one hundred and fifty yards of the salient point of the Redan, and on the morning of the 5th the "final bombardment" commenced. The assault was fixed for the 8th September, the French attacking the Malakoff, whilst the British second and light divisions again assaulted the Redan. About noon the French got into the Malakoff, and held it against the repeated attempts of the Russians to retake it; but the British were less successful. The Redan was entered; but after a most determined and sanguinary contest, lasting for nearly an hour, it was found impossible to hold it, and the troops had to retire into the trenches. The third division got under arms at 11 a.m., and moved up to the attack; but the 14th remained in reserve in rear of the Greenhill Battery, and was not engaged.

The troops were greatly depressed at this fresh reverse and the prospect of more weary work in the trenches. But the end was at hand. When darkness fell, fires were seen rising from all parts of the town, attended by violent explosions, which were frequent during the night. Sunday morning broke with a brilliant sunrise dimming the watch-fires, and revealed the whole south side of Sevastopol shattered and forlorn, and the last of the brave defenders crossing the harbour to the north. Sevastopol had fallen, and the long siege was at an end.*

A regimental draft, under Captain Young, Lieutenant Vernede, and Ensigns Costin and Bright, marched into camp the same day, Ensign Costin having come into camp the day before.

On 20th September, the first anniversary of the Alma, some Crimean medals were issued, but the supply was limited, and three only fell to the lot of the 14th. On the 21st a warrant was published granting sixpence a day to all non-commissioned officers and men in the Crimea. On 30th September an order was notified that a clasp would be given for the siege with the medal, and directed the word "SEVASTOPOL" to be inscribed on the colours.

The remainder of the year was spent by the regiment in camp outside Sevastopol. On 3rd October, Major Budd rejoined from Malta. On the 8th, Lieutenant G. Brydges left; on the 9th, Sergeant-Major Fred. Rance † was appointed to an ensigncy; and on the 10th, Captain Hammersley ‡ was appointed to the Quarter-Master-General's staff.

---

* "In the forenoon, after church, I went down the Sailor's Ravine to the town, "and met numbers of men, soldiers and sailors, English, French, and Sardinians, "returning with loot of a most miscellaneous description—household furniture, "musical instruments, church decorations, vestments, poultry and pigs, military "clothing and appointments. Here, was an English staff-officer with a large "book bound in crimson velvet under his arm; there, a French soldier in a "Russian drummer's uniform, mounted on a cow, which he had laden with "various articles of furniture, etc. Inside the town I explored several curious "batteries among the houses at the back of the Redan, and returned by the "Woronzoff road. Here I met Captain Lowndes, of the 47th, in charge of a "party of that regiment, with orders to make all English Soldiers give up their "loot, while our Allies were allowed to carry theirs away."—Paymaster Hall's MS. Notes.

† Now Major Fred. Rance, late 36th Foot.

‡ Major-General Hammersley, late Director of Gymnasia, Aldershot.

On 12th November, General Sir James Simpson, G.C.B., left, and was succeeded in the command of the army by Lieutenant-General Sir William Codrington, G.C.B. On 29th November, Captain Townshend died of cholera at Bajukdere, on his way from Scutari. On the same date Brigadier-General Barlow was appointed to the first brigade of the second division.

Up to the end of November the weather continued tolerably fine, with occasional frosts, and but little snow. In December it became colder, although mild when the wind was in the south. But as the month advanced, the severity increased—on 19th December the thermometer fell to three degrees Fahrenheit—and many men were frost-bitten. The army was employed in road-making during the winter months, each brigade having a section allotted to it. A brigadier was told off daily to superintend. Towards the end of December the transport horses began to fail, and all hands were employed in bringing up hutting.

1856.

The regiment was still in the camp before Sevastopol at the beginning of the year. Variable weather, with a good deal of snow, characterised the months of January and February. On 6th February the regiment was inspected by Major-General Sir William Eyre, commanding the division. On 18th February the right wing, under command of Major Budd, moved down to Balaklava and occupied ground to the eastward of the town. The head-quarters, under Brevet Lieutenant-Colonel Sir James Alexander, followed next day. The regiment was assigned to an independent brigade, under Brigadier-General C. Warren, C.B., 55th regiment, consisting of the 14th, 39th, 82nd, and 89th regiments. General Sir Colin Campbell was in command of an army corps to the eastward of Balaklava. The regiment was in huts, but bad ones.*

On 29th February there was a meeting of the chiefs of the staff of the allied and Russian armies at the Traktir Bridge to arrange an armistice; but there was a hitch about the use of the harbour.

---

* Paymaster Hall's MS. Notes.

On 3rd March, a General Order announced that fire would cease between the armies, and on 14th March the armistice was duly signed. On 2nd April news arrived that a Treaty of Peace had been signed on 30th March, 1856.

On 13th April there was a grand review of the Russian army on Mackenzie's Farm Heights, to which the allied commanders-in-chief were invited; and on the 17th a similar review of the allied armies, at which the Russian general, Luders, was present with the French in the morning. He then lunched at General Codrington's quarters, and in the afternoon saw the English army, which in its smart appearance presented a marked contrast to the French, at which, it was rumoured, Marshal Pelisser was by no means pleased.

A Horse Guards Order, dated 28th April, 1856, directed that the minimum standard for recruits for line Infantry out of India be raised to five feet six inches for men and lads between the ages of seventeen and thirty. Limited enlistments, for *two* years only, which had been sanctioned the year before, were ordered to be discontinued.

The evacuation of the Crimea shortly afterwards commenced, and the 14th proceeded to Malta in the following divisions:

Per ss. *Etna*, 16th May, 1856.—*Major* R. Budd; *Captains* W. Hawley, W. Hayward, and J. McC. O'Toole; *Lieutenants* D. Warren, R. H. Vivian, T. P. Cosby, W. T. Blunt, E. W. Saunders, H. Vernede, A. A. Le Mesurier, and H. W. Heaton; *Ensigns* J. M. Bright and J. Glancy; *Assistant-Surgeon* Hyde.

Per *Thames*, 29th May, 1856.—*Brevet Major* W. Douglas; *Captain* Young; *Lieutenants* J. D. Bradley, A. Malony, and C. Costin; *Ensign* F. Rance; *Assistant-Surgeon* T. H. Blickley.

Per *Robert Low*, 31st May, 1856.—*Captain* C. Newman; *Lieutenant* Dods.

Per ss. *Ottawa.*—*Brevet Lieutenant-Colonel* Sir James Alexander; *Brevet Major* J. Dwyer; *Captain* W. C. Trevor; *Lieutenants* C. M. Wilson, H. H. A'C. Inglefield, A. Hall, and F. G. Armstrong; *Paymaster* J. P. Hall; *Adjutant* A. Gordon; *Quarter-Master* J. O'Connor; *Surgeon* W. Chalmers; *Assistant-Surgeon* W. H. Price; head-quarters, etc.

Arriving at Malta in May-June, 1856, the regiment disembarked and went into quarters, giving detachments at St. Lucien's Tower, Marsa, Sirocco, and St. Francisco.

In July, 1856, an officer instructor of musketry was first allowed for the regiment, Lieutenant C. Costin performing the duty until the arrival of a certificated officer from Hythe.

In October, 1856, the depôt companies, under Brevet Major W. Douglas, proceeded from Mullingar to Fermoy, and with the depôts of the 28th and 38th regiments formed a depôt battalion there.

Crimean Honours.

The word "SEVASTOPOL" was inscribed on the regimental colours, and the English and Turkish Crimean medals were given to every officer and man of the regiment who served in the Crimea previous to the fall of Sevastopol (and afterwards, if actually engaged with the enemy after that date). In addition the following honours were awarded:

C.B.—Colonel Maurice Barlow.

Knights of Legion of Honour—Brevet Major Dwyer; Sergeant John Macdonald.

3rd Class of the Medjidie—Colonel Maurice Barlow.

5th Class of the Medjidie—Major and Brevet Lieutenant-Colonel Sir J. E. Alexander; Major Budd; Brevet Majors Dwyer and Douglas; Captains Hammersley and Maycock.

French Military Medal.—Two hundred and twenty-six of these medals were given for distribution to distinguished non-commissioned officers and soldiers of the British army. The following awards were made in the 14th Regiment: Sergeant Thomas Cooper: volunteered with twenty men to attack a Russian Rifle pit in the cemetery, left attack, 18th June, 1855. Sergeant John Macdonald: carried off, under fire, from the open ground in front of the trenches a wounded man of the 39th regiment. Sergeant Thomas Brown: served in the Crimea from 10th January, 1855, to 6th January, 1856; constant and zealous duty in the trenches; wounded 18th January, 1855. Private Robert Harrison: zealous attention to his duty in the trenches. Private Thomas Caby: volunteered to go out under a heavy fire to bring in a wounded man 1st Royal regiment. Private Patrick Carty, a clean, well-conducted soldier in the trenches, where he was wounded.

Sardinian Medal for Valour.—Four hundred of these medals were given for distribution in the British army. In the 14th they were awarded to Colonel Maurice Barlow, C.B.; Brevet Lieutenant-Colonel Sir James Alexander; Brevet Major W. C. Trevor; Captains J. G. Maycock and W. Young; Sergeant Hopkins.

General Sir James E. Alexander, C.B., K.C.L.S., F.R.S.
Commanded 1st/14th in the Crimea, raised 2nd/14th
in 1858 & Commanded it in New Zealand 1860-62.

1857.

On 27th January, 1857, Colonel Maurice Barlow, C.B., exchanged to half-pay, unattached, with Colonel Peter Farquharson, who retired the same day, Brevet Lieutenant-Colonel Ralph Budd* succeeding to the command of the regiment.

The head-quarters at Malta moved on 30th April from the Isola Barracks to Fort St. Elmo. During the summer the regiment suffered greatly from ophthalmia, over a hundred men being in hospital at one time with the complaint. A smart shock of earthquake split the walls of the officers' mess, and did other damage.

The establishment of the regiment underwent several changes in the course of this year. On 18th June the number of non-commissioned officers and rank and file were reduced to the following: one sergeant-major, six staff-sergeants, one drum-major, twelve colour-sergeants, forty-eight corporals, twenty-four drummers, and seven hundred and ninety-two privates. In August the distribution and strength were as under:

	Service Companies.	*Depôt.*
Companies	8	4
Staff-sergeants	6	0
Colour-sergeants and Sergeants	40	10
Corporals	40	10
Drummers	21	4
Privates	760	120

In July, 1857, the depôt companies moved from Fermoy to Cork, furnishing a detachment, under Captain Hammersley, at Spike Island, and detachments at Rocky Island and Carlisle Fort. On 26th November a draft, consisting of Lieutenant G. H. Dwyer, Ensigns Watson and Bryan, and thirty men, joined the service companies from the depôt.

* General Ralph Budd. Died 8th March, 1889. He entered the regiment as ensign 10th March, 1825, became lieutenant 16th March, 1826; captain, 1st June, 1841; brevet major, 20th June, 1854; major, 15th May, 1855; and was lieutenant-colonel commanding from January, 1857, to 12th July, 1867. He was in the Crimea from 10th January to 5th June, 1855 (medal and clasp and 5th Class of Medjidie), and again after the fall of Sevastopol.

1858.

Re-formation in Two Battalions.

The outbreak of the Indian Mutiny in May, 1857, led to the addition of second battalions to a number of line regiments during the succeeding autumn and winter, and to a fresh reduction in the standard of recruits, which, by a Horse Guards Order of January, 1858, was lowered to five feet four inches for men and lads between seventeen and thirty for infantry at home, and to five feet three inches for recruits, not under eighteen, for infantry in India. Among the regiments augmented was the 14th, which, in January, 1858, was directed to form a SECOND BATTALION—the second the regiment has possessed. The establishment of the new battalion was fixed as follows: one lieutenant-colonel, two majors, eight captains, ten lieutenants, six ensigns, one paymaster, one adjutant, one quartermaster, one surgeon, one assistant-surgeon, nine staff-sergeants, eight colour-sergeants, thirty-two sergeants, thirty-two corporals, seventeen drummers, and seven hundred and eight privates.

Lieutenant-Colonel E. W. D. Bell,* V.C., from 23rd Royal Welsh Fusiliers, was appointed lieutenant-colonel, and Major John Dwyer and Brevet Major William Douglas were gazetted to the majorities. The head-quarters were fixed at Naas.

1858.—First Battalion.

On 28th August, 1858, the service companies of the 1st battalion, under command of Major W. C. Trevor, embarked at Malta in H.M.S. *Perseverance* for Corfu. On arrival, the head-quarters and five companies were transferred to Cephalonia, occupying Argostoli,

---

* Colonel Edward William Doddington Bell, V.C., commanding 2nd battalion 23rd Fusiliers. Since dead. Second lieutenant, 23rd Fusiliers, 15th April, 1842; lieutenant, 17th November, 1843; captain, 18th December, 1848; brevet major, 12th December, 1854; major, 23rd March, 1855; brevet lieutenant-colonel, 26th December, 1856; lieutenant-colonel, 14th Regiment, 8th January, 1858. Transferred back to 23rd Fusiliers, March, 1858. Captured and secured with his own hands the first Russian gun taken at the Alma; brought his regiment out of action there. Particularly distinguished himself in command of a working party at the siege of Sevastopol on 2nd April, 1855. (Victoria Cross, Knight of Legion of Honour, 5th Class of Medjidie, medal and clasp.)

and giving detachments at Lesuire and Fort George. The remaining five companies encamped at Zido until a passage could be provided for them, when four companies proceeded to Zante, and one company (the grenadiers, under Captain Grogan) to St. Maura, with a detachment of thirty men in Ithaca. The battalion remained in the Ionian Islands during the remainder of the year.

1858.—Second Battalion.

On 1st April the establishment of the 2nd battalion was increased to twelve companies, and the number of rank and file raised to nine hundred and fifty. On 23rd April Lieutenant-Colonel Sir James Alexander,* assumed command of the battalion, *vice* Bell, re-transferred to the 23rd Royal Welsh Fusiliers.

On 28th May the battalion marched from Naas to Newbridge, and proceeded by rail to Waterford, there to be stationed. On 9th June it was inspected by Major-General Eden, C.B., who expressed himself as highly satisfied. The battalion at this time only numbered three hundred and ninety-five non-commissioned officers and men.

In August it was ordered to Kilkenny in aid of the civil power, and furnished detachments to Thomastown and Callan. The conduct of the men on this occasion was most creditable, and a letter of congratulation was received from the commander of the forces in Ireland (Lord Seaton) expressing his satisfaction thereat.

Having been summoned by telegram to Dublin, the battalion proceeded from Kilkenny by special train, and took over quarters in the Richmond Barracks.

It was inspected by General Lord Seaton on 6th December, who expressed his high approval of its appearance, and his opinion that as much progress had been made as could have been expected since its formation.

On 31st December colours were presented to the 2nd battalion at Dublin by the Countess of Eglinton, in the presence of the Lord Lieutenant of Ireland, the Commander of the Forces, and other

---

* General Sir James Alexander, C.B., K.L.S. Joined the 14th regiment, from Captain 42nd Highlanders, in 1840, and finally left it, as lieutenant-colonel, in 1862. For particulars of services, see *Appendix Volume.*

distinguished persons. The battalion paraded under the command of Brevet Colonel Sir James Alexander. His Excellency the Lord Lieutenant inspected the battalion, and the colours were consecrated by the Chaplain to the Forces, the Rev. Robert Halpin, who had once served as a commissioned officer in the regiment. Her Excellency formally presented the colours to Ensigns Butler and Harington, and addressed the battalion as follows:

"Sir James Alexander, Officers, Non-commissioned Officers, and "Privates of the 14th Regiment,—It is with feelings of pleasure and "pride that I present these colours to so distinguished a corps. If the "history of our country did not already tell us, we have but to look "at the glorious names inscribed here to be reminded that this gallant "regiment has upheld the honour of Great Britain in all quarters of the "world. Tournay and Java, Bhurtpore and Corunna, Waterloo and "Sevastopol, tell of deeds which warm the hearts, not only of our country-"men, but of those whose sex preclude their participation in such glories; "and in giving these colours into the keeping of such a regiment, com-"manded by one who has distinguished himself, not only with the sword, "but with the pen, I well know that they will be borne in triumph wher-"ever the call of duty leads you."

In reply, Colonel Sir James Alexander said:

"I beg leave to return thanks for the distinguished honour conferred "upon the corps, in the colours being presented by the Countess of "Eglinton, the wife of the Lord Lieutenant of Ireland. I hope that "everyone belonging to the regiment will remember this day, and be "proud of it and of the colours which are now unfurled for the first time "in public. The colours should be a bond of union to a regiment. They "have been compared to a spire of a village church, round which the "people desire to live and die. These colours, now taken up with honour, "should stimulate to a proper discharge of duty by having the emblems of "Royalty emblazoned on them, and also the record of the former service "of the regiment of which the battalion forms a part, and they ought "never to be abandoned as long as there is a hand left to uphold them.

"Her Excellency has noticed the name of Corunna. On that occasion "the old 2nd battalion fought on that well-contested field, in presence of "our most esteemed and distinguished Commander of the Forces, Lord "Seaton. The 3rd battalion, composed of very young soldiers, like those "here present, had the proud distinction of placing Waterloo on their "banner.

"There is still room for other distinction on these colours, surmounted "as they are by the Lion of England; and although it is wicked and "sinful to wish for war, I trust that if on any future occasion this regi-"ment is called upon by our beloved Sovereign to fight the battles of our

Lieutenant and Adjutant
J. Glancy

Ensign
W. B. Lindsay

Ensign
W. J. Willis

Lieutenant
J. J. Carberry

Lieutenant
Kenrick Hill

Lieutenant-Colonel
Sir J. E. Alexander

Lieutenant
Iver McIver

Major
John Dwyer

Ensign
F. W. Harrington

Ensign
R. Langtry

Captain
E. W. Saunders

Captain
H. Cowell

Ensign
J. S. Johnson

Officers, 2nd Battalion, at Dublin, 1858.

"country, these young soldiers will rush to the conflict with as much zeal "and gallantry as those warriors who have preceded them when they "acquired for the corps the honourable title of the 'Old and Bold,' per-"petuating also by the quick-step of the regiment the remembrance of a "gallant achievement—the surprise of a French post, on which they "advanced playing '*Ça Ira.*' It is a remarkable fact that the benediction "was pronounced to-day by a gentleman who was formerly an officer in "the regiment. He had for some time a commission in it, and it may "not be known that he was actually engaged in the suppression of the "rebellion in Canada, and succeeded in capturing a rebel standard. He "also wears a Crimean medal.

"There is another person on the ground who was with the regiment "when it assaulted one of the breaches at Bhurtpore, and he is now "occupying a respectable position in this city. I hope that I will be "excused adverting to those matters interesting in the history of a regi-"ment, and I trust that this battalion will maintain a character for order "and discipline in quarters as well as for gallantry in the field. I beg "now, in the name of the regiment, to return thanks to the distinguished "individuals who have honoured us with their presence on this occasion."

General Lord Seaton, commanding the forces in Ireland, desired that the regiment should be informed that he was well pleased with the manner in which the ceremony was performed and with the appearance and steadiness of the men.

1859.—Second Battalion.

In April the 2nd battalion was completed to its full establishment, recruits being obtained principally from the Liverpool district and from the midland counties of Ireland. In May it proceeded by rail to the Curragh Camp, and after a summer spent there, left in the middle of October for Mullingar, where it was quartered in the barracks.

1859.—First Battalion.

In October of this year the 1st battalion was moved to Corfu. The head-quarters and six companies, under Major W. C. Trevor, occupied the citadel. The Zante detachment—four companies—under Major Holworthy, rejoined head-quarters, and a detachment of four companies, under Captain Blunt, proceeded to Vigo. On 17th December a detachment of thirty-four men, under Ensign Hutchison, was furnished at Paxo, and the Ithaca detachment, under Lieutenant Barlow, rejoined head-quarters.

ROLL OF OFFICERS 1ST JANUARY, 1859.

(From the Monthly Army List.)

Rank	Batt.	Name	Date
Colonel - -		Sir James Watson, K.C.B., *General* - -	24th May, 1837
Lieut.-Colonel -	1	Ralph Budd - - - - - - -	27th Jan., 1857
" " -	2	Sir J. E. Alexander, *Colonel* - - -	30th March, 1858
Major - -	1	Wm. C. Trevor - - - - - - -	27th Jan., 1857
" - -	1	Ed. J. Holworthy - - - - - -	17th Feb., 1857
" - -	2	John Dwyer - - - - - - -	8th Jan., 1858
" - -	2	Wm. Douglas - - - - - - -	8th Jan., 1858
Captain - -	1	F. Hammersley, *Major*, Staff - - -	25th April, 1851
" - -	1	Plomer J. Young - - - - - -	25th March, 1853
" - -	1	Wm. H. Hawley - - - - - -	29th Dec., 1854
" - -	1	Chas. Ed. Grogan - - - - - -	29th Dec., 1854
" - -	1	John G. Maycock, Staff - - - -	29th Dec., 1854
" - -	1	Martin Petrie, Military College - - -	5th May, 1854
" - -	1	Dawson S. Warren, dep. - - - -	10th April, 1857
" - -	1	Thos. P. Cosby - - - - - -	24th July, 1857
" - -	2	Wm. Heywood, Military College -	15th May, 1855
" - -	1	Walter F. Blunt - - - - - -	2nd Oct., 1857
" - -	1	Angus W. Hall - - - - - -	23rd Oct., 1857
" - -	2	R. H. Vivian - - - - - - -	9th Jan., 1857
" - -	1	Ed. D'H. Fairtlough - - - - - -	8th Jan., 1858
" - -	2	Hugh M. Lloyd - - - - - -	5th March, 1858
" - -	2	Drury R. Barnes - - - - - -	16th March, 1858
" - -	2	Ed. W. Saunders - - - - - -	26th March, 1858
" - -	2	Wm. Dods - - - - - - -	4th June, 1858
" - -	2	J. M. M'Kenzie - - . - - -	30th July, 1856
" - -	2	H. T. Vernede - - - - - .	30th July, 1858
" - -	2	Ed. D. Fenton - - - - - -	16th Oct., 1857
" - -	2	Alex. Strange - - - - . - -	1st Oct., 1858
" - -	2	Henry Cowell - - . - . -	1st Oct., 1858
" - -	2	Gage Hall Dwyer - - - - - -	2nd Oct., 1858
" - -	2	And. A. Le Mesurier - - - - -	5th Oct., 1858
Lieutenant -	1	Alexander Gordon - - - - - -	29th Dec., 1854
" -	2	Fred. D. Armstrong - - - - - -	17th Feb., 1855
" -	1	John D. Bradley - - - - - -	9th March, 1855
" -	1	Ramsey Harman - - - - -	9th March, 1855
" -	1	Hy. Wm. Heaton, *Instructor of Musketry* -	9th March, 1855
" -	1	Arthur Melony, dep. - - - - -	30th Nov., 1855
" -	1	Chas. Costin, *Adjutant* - - - - -	1st Feb., 1856
" -	2	R. A. L. Furneaux - - - - - -	8th Feb., 1856
" -	2	G. J. N. Beamish - - - - - -	17th Feb., 1857
" -	2	Iver McIver - - - - - - -	24th Feb., 1857

Lieutenant	1	J. T. Casson	23rd Oct., 1857
"	1	Peter Barlow	30th Oct., 1857
"	2	John Glancy, *Adjutant*	8th Jan., 1858
"	2	J. Oct. Machell	8th Jan., 1858
"	2	Kenrick Hill	8th April, 1857
"	2	G. H. Cope	7th May, 1857
"	2	James Anderson, *Instructor of Musketry*	2nd Oct., 1857
"	1	G. H. Morgan	27th Nov., 1857
"	2	J. S. Phelps	9th Jan., 1858
"	1	Hy. A. Burton	26th March, 1858
"	1	Stephen Watson	15th June, 1858
"	1	Geo. L. Bryce	15th June, 1858
"	1	John Wilson	15th June, 1858
"	2	John L. Carbery	22nd June, 1858
"	2	A. R. Keogh	1st Oct., 1858
"	1	Ed. J. Briscoe	2nd Oct., 1858
"	1	Fred. A. Atkinson, dep.	5th Oct., 1858
"	1	Hy. J. Harington	10th Dec., 1858
"	1	Jas. T. Edwards	10th Dec., 1858
"	1	John B. Frizell	10th Dec., 1858
Ensign	1	Joseph Laing	27th March, 1858
"	2	Jas. Stephen Johnston	13th April, 1858
"	2	Robert Langtry	14th April, 1858
"	2	Franc Le Breton Butler	16th April, 1858
"	2	Wm. Bayford Lindsay	17th April, 1858
"	2	John Lawrence	18th April, 1858
"	2	Wm. J. Willis	23rd April, 1858
"	2	F. W. Harrington	28th May, 1858
"	1	Dennis Creagh, dep.	6th August, 1858
"	1	Hy. A. Williams, dep.	6th Oct., 1858
"	1	Hy. Metcalfe, dep.	7th Oct., 1858
"	1	Hy. M'L. Hutchison, dep.	12th Nov., 1858
"	2	Stainsby H. Pigott	26th Nov., 1858
Paymaster	1	William Macdonnell	19th March, 1857 22nd April, 1853 *Lieut.* 25th June, 1852
"	2	J. C. V. Minnett	13th March, 1858
Inst. of Musk.	1	H. W. Heaton, *Lieutenant*	20th May, 1858
" "	2	J. Anderson, *Lieutenant*	19th July, 1858
Adjutant	1	Chas. Costin, *Lieutenant*	11th Sept., 1857
"	2	J. Glancy, *Lieutenant*	19th Feb., 1858
Quarter-Master	1	W. A. Armstrong	4th Dec., 1857
" "	2	Jas. Spry	31st Dec., 1857
Surgeon	1	G. S. King, M.D.	29th June, 1855
"	2	John E. Carte, M.B.	18th Feb., 1853
Asst.-Surgeon	1	T. M. Blickley, M.B.	6th Jan., 1854
" "	1	J. M. Hyde	3rd Nov., 1854
" "	2	Thos. Bennett	22nd Jan., 1858
" "	2	A. T. Carbery	16th Nov., 1858

1860.—First Battalion.

Early in February the 1st battalion was ordered to proceed to the West Indies. The Paxo detachment was recalled on 2nd February, and on the 15th the head-quarters and six companies, under Lieutenant-Colonel Ralph Budd, embarked in H.M.S. *Perseverance*, and sailed for Jamaica, where they arrived and disembarked at midnight on 30th March. The head-quarters and four companies marched to Newcastle to be stationed, and Nos. 2 and 5 companies, under Major W. C. Trevor, to Stronghill, where they occupied the barracks.

The remaining four companies, which had been left behind at Corfu, embarked, under the command of Major E. Holworthy, on the *Indiana*, for the West Indies on 19th March, and disembarked on 19th April at Trinidad, where they were stationed in the barracks at St. James'. The light company, under Captain Young, proceeded to St. Lucia, disembarked there on the 30th, and encamped at Fort Morne Fortuné.

The Stronghill detachment was recalled to head-quarters on 3rd July.

A sergeant-instructor of musketry was added to the establishment of both battalions during this year.

1860.—Second Battalion.

Under instructions from the Horse Guards, dated 1st April, 1860, the 2nd battalion formed a depôt of two companies, which was placed under the command of Captain Lloyd. After a six months' stay in Mullingar, the battalion proceeded, in two divisions, on 9th-10th April, to the Curragh Camp, there to be stationed. The depôt rejoined from Mullingar on 24th April.

A very satisfactory confidential report was communicated to the battalion by order of H.R.H. the Field Marshal Commanding-in-Chief.

The 2nd battalion having been placed under orders for foreign service from 8th April, 1860, was augmented to the following establishment :

Mounted Officers :—Major W. L. Trevor *(in front)* ; Captain C. E. Grogan *(on the right)*.

Paymaster W. Macdonnell · Quarter-Master J. Moore · Lieutenant J. Laing · Ensign H. A. Williams · Ensign W. Close · Ensign H. McL. Hutchison · Ensign Riley · Ensign B. W. C. Firman · Captain A. A. Le Mesurier · Captain T. P. Cosby · Ensign W. T. Blois · Captain A. Gordon

The 1st Battalion at Newcastle, Jamacia, 1861.

Twelve companies—viz., ten service companies and two depôt companies—

1 Lieutenant-Colonel	7 Staff-Sergeants
2 Majors	70 Colour-Sergeants and Sergeants
12 Captains	70 Corporals
14 Lieutenants	24 Drummers
10 Ensigns	1,330 Privates
6 Staff	

On 8th September, 1860, the head-quarters and five companies of the 2nd battalion, under command of Colonel Sir J. E. Alexander, C.B., proceeded by rail from the Curragh to Cork, and embarked in the *Robert Low* for New Zealand,* where what has since been known as the "Second Maori War" had commenced.

A second division, under Major John Dwyer, embarked, in like manner, in the *Boanerges* on 10th September. A third division, consisting of a single company, under Captain Vivian, followed on 15th September, and embarked in the *Seville.*

War had broken out with the Maories in the Taranaki district of New Zealand in consequence of disputes respecting the sale of a certain block of land, claimed as tribal land. Various native outrages had been committed, and several encounters between the native and the colonial forces had taken place by the time the divisions of the 14th arrived at Auckland, on 29th November, 22nd December, 1860, and 26th February, 1861. The head-quarters at first occupied barracks at Auckland, with detachments of eight officers and two hundred and ninety-six men, under Major Dwyer, at Wellington, and five officers and two hundred and twelve men, under Captain Barnes, at Napier.

1861.—Second Battalion.

New Zealand War of 1860-61.

On 4th January, 1861, two companies of the battalion, consisting of seven officers and two hundred men, under Major Douglas,

* The order was received when the battalion was returning from a grand field day at the Curragh. A German air the band was playing at the time was afterwards known in the regiment as the "New Zealand March."

proceeded to Taranaki, to join the field-force employed against the Maories. The latter, well-armed with double-barrelled guns and rifles, and expert in bush-warfare and the erection of improvised defences, proved themselves no mean antagonists; but a succession of defeats convinced them of the futility of further resistance, and on 19th March, 1861, the campaign of 1860-61 came to an end.

On 18th May the head-quarters of the battalion left Auckland for the camp at Otahuhu, where the companies which had been detached on field-service rejoined. Colonel Sir J. E. Alexander promulgated to the battalion the good reports of the detachment received from Major-General Pratt, C.B., commanding the field-force, who stated that, although quite young soldiers, they had fought as bravely, worked as hard, and behaved as well as the men of older battalions. The officers employed on this service were Major Douglas, Captains McIver, Saunders, and Strange, Lieutenants Hill, Phelps, and Frizell, and Ensigns Lawrence and Curtis.

All the troops, the 57th excepted, were now withdrawn from the Taranaki district, which remained under martial law. In December the 2nd battalion 14th shifted its camp from Otahuhu to Pokeno, to assist in making the great road south-west from Drury to Havelock. The battalion, with a detachment of the 12th Foot, occupied the advanced post. The other corps employed—1st battalion 12th, 40th, 65th, and 70th regiments—were posted in rear along the road to Drury, where were the head-quarters of Major-General Cameron, C.B., commanding the forces.

1861.—First Battalion.

In March, 1861, the head-quarters and two companies proceeded from Newcastle to Spanish Town, to do duty during the stay of H.R.H. Prince Alfred, now Duke of Edinburgh. A company proceeded to Kingston at the same time. At the termination of the royal visit, the head-quarters returned to Newcastle, where, at the end of April, the following detachments were furnished: a company, under Lieutenant Laing, was sent to Port Royal, and a company, under Lieutenant Gordon, was detached to Kingston, and twenty men, under Ensign Hutchison, to Spanish Town. The Port Royal and Kingston detachments rejoined head-quarters in May, and the Spanish Town detachment was removed to Port Royal, whence it returned in July.

Quarter-Master-Sergeant James Conroy received the medal for meritorious conduct, with an annuity of £10, on discharge to pension on 6th August, 1861.

Grenadier and light companies were abolished this year, and the companies ordered to be distinguished by a letter instead of a number, and to stand on parade according to the seniority of their captains. A new pattern shako was introduced, similar in shape to the one in use, but without a peak in rear, and with cloth substituted for the felt body and leather crown of the old pattern.

On 12th August, 1862, the veteran colonel of the regiment, General Sir James Watson, K.C.B., whose connection with the 14th dated from 1787, died, in his ninetieth year, and the seventy-ninth of his military service. He was succeeded by Lieutenant-General, afterwards Sir James Wood.*

A Horse Guards letter, dated 26th August, 1862, conveyed to the general officer commanding in Jamaica the approval of the Commander-in-Chief of the Confidential Report of the Inspection of the 14th Regiment, which was highly satisfactory in every respect. The "excellent rifle practice of the battalion was specially noticed by His Royal Highness."

In December, three companies, under Captain Grogan, proceeded from Newcastle, Jamaica, to Trinidad, where on the 20th they relieved a similar detachment, under Captain W. H. Hawley, in St. James' Barracks, which rejoined head-quarters at Newcastle, Jamaica, on the 30th.

1862.—Second Battalion.

In January, 1862, the establishment of this battalion was reduced from fourteen hundred rank and file to nine hundred and fifty.

---

* For services, see *Appendix A*.

After six months' incessant labour in the bush, the 2nd battalion was ordered into winter quarters at Otahuhu. The head-quarters proceeded on 18th June, leaving a detachment of one hundred and forty men, under Captain McIver, at Pokeno, and after a most difficult and fatiguing march, in drenching rain, the road for the most part of the second day's march being knee-deep in mud, arrived at the camp, where the battalion was hutted.

The nature of the duties performed by the battalion during this campaign may be gleaned from the following extracts from a General Order, dated head-quarters, Auckland, 18th June, 1862:—

"The completion of the military road from Drury to Pokeno having "enabled the general commanding to move the troops into winter "quarters, he desires to express to the several corps and departments "employed, his sense of the zeal and goodwill with which they have "performed this laborious and harassing duty. The Lieutenant-General "feels much indebted to Colonel Mould for the professional ability with "which he has directed the work, and to his zealous superintendence, and "that of the officers acting under his orders, is to be attributed in a great "measure the success of the operation. The extent of the bush which has "been filled and cleared, the quantity of stone broken and quarried for "mud metal, the numerous elevations and cuttings on the slopes and "summits of the hills, sufficiently attest the labour undergone by the "troops during the latter portion of the period.

"The operation was carried on in a most inclement season of the year. "Too much praise cannot be awarded to the energy and perseverance of "those by whom this important and useful work has been successfully "extended.

"The Lieutenant-General desires also to express his satisfaction at "the uniform good conduct of the troops whilst employed on this "duty."

A letter from His Excellency the Governor of New Zealand to the Hon. Lieutenant-General Cameron, C.B., commanding troops in New Zealand, which was published in General Orders, contains the following:—

"I have the honour to acknowledge the receipt of your letter of the 25th "ult., reporting the completion of the Military roads to the Waikato as far "as Pokeno.

"I gladly avail myself of this opportunity of expressing my own

Corporal	Colour-Sergeant	Private	Private	Private	Captain	Private	Sergeant	Corporal
J. Halse	G. Imber	L. Hapgood	J. Rabbitt	C. Gordon	J. D. Bradley	D. Griffiths	W. Armstrong	F. Foxwell

Captain Bradley and group of No. 6 Company, at Trinidad, West Indies, 1862.

"thanks, and that of the Government, to yourself and the Officers and "N. Commissioned Officers and Privates of the Force employed by you "on the great work, for the ability with which the road was planned, and "for the energy and zeal with which it has been so rapidly executed, as "well as for the operations having been continued on it to so advanced a "period of the unusually severe rainy season. I shall not fail to report to "Her Majesty's Government my sense of the great service the Military "Force has rendered to the Colony in this respect."

Colonel Sir James E. Alexander, on 5th August, 1861, resigned the command of the 2nd battalion, in which he was succeeded by Lieutenant-Colonel Charles Wilson Austen,* and published the following Battalion Order:—

"Colonel Sir J. E. Alexander, in now leaving the head-quarters of the "2nd battalion 14th Regiment, for family reasons, has to express his great "regret at parting with a battalion in which he has served from its first "formation at Naas, where he joined it on St. George's Day; and, up to "this time he can truly say that generally he has had much satisfaction in "the command with which he was honoured by H.R.H. the General "Commanding-in-Chief, that of one of the new battalions, and one of two "selected for foreign service, when active operations were going on in the "field. That portion of the battalion which was employed in the late "Maori War did good service before the enemy, and other portions "underwent all the trial and hard work on the military road to the "Waikato.

"The detachments well performed the duties that were required of "them. He trusts that the best feeling will prevail in the battalion of "good comradeship among all ranks, of cordiality towards other corps, of "an earnest desire to maintain the good character of the battalion, and at "the same time, and under all circumstances, to render a cheerful and willing "service, the best test of a good soldier. Sir James Alexander will always "take a deep interest in the fortunes of the 2nd battalion 14th Regiment,

* Lieutenant-Colonel Charles Wilson Austen, ensign, 83rd Foot, 14th December, 1838; lieutenant, 15th December, 1840; captain, 1st December, 1848; major, 16th May, 1856; lieutenant-colonel, 26th October, 1858. Served with the 83rd during the Indian Mutiny, including the affair at Sanganu, 8th May, 1858; defeat of the Gwalior rebels at Kolaria; and commanded the head-quarters of the 83rd in the attack on the rebels at Sickur, on 21st January, 1859 (medal.) Appointed to the 14th, 10th June, 1862. Died of wounds received when in command of the 2nd battalion 14th Regiment, at Rangariri, on 20th November, 1863.

" and if at any time hereafter he can be of the least use to any member of
" the battalion, he will consider it his duty, and it will always give him
" pleasure to render any service which may be in his power."

On 26th November the head-quarters of the 2nd battalion marched from the camp at Otahuhu to Shepherd's Bush, where the battalion was employed, until 12th February following, in repairing the great south road between Drury and Pokeno, afterwards returning to the camp.

1863.—Second Battalion.

The New Zealand War of 1863-66.

Unfortunately, the peace with the natives proved but a brief truce. On 4th May, 1863, hostilities were resumed by a native attack on a military escort, which was fired on from the bush, and eight men killed. In consequence of this, an encounter took place at Kuit Kara, where twenty-four Maories were killed, and the Province of Auckland again became the scene of war—the protracted war of 1863-66.

The Maories having chosen one of their chiefs as "king" previous to the outbreak, now declared their intention of driving the settlers out of the North Island. General Duncan Cameron, C.B., issued a proclamation, declaring all natives rebels if they did not submit in a week. The Maories returned with their belongings to a place called Koheroa, and took up a strongly fortified position on a mountainous ridge about two miles therefrom, and close to the Waikato river, whence they bade defiance to General Cameron.

On 8th May the 2nd battalion 14th was ordered into the field, and on the following morning thirteen officers and two hundred and forty-five men marched for Drury, and thence, on 10th July, to Queen's Redoubt, where No. 9 company, consisting of three officers and one hundred and forty-six men, joined head-quarters.

On 17th July the battalion attacked the enemy, strongly posted at Koheroa, and drove them from their rifle-pits into the adjacent swamps. The battalion had one man killed and two men mortally wounded, and the commanding officer, Lieutenant-Colonel Austen, and ten men wounded.

On the night of 31st July the battalion marched to the rebel villages of Paperoa and Paperota, which it attacked next morning and carried without loss. On 14th August the battalion advanced on the enemy's position at Mere Mere, and at a distance of two thousand four hundred yards from it proceeded to erect a redoubt. During the progress of the work the sentries and working-parties were much annoyed by the fire of the enemy from a flax and fir swamp which they had occupied. The redoubt was completed on 26th September, and, leaving a detachment to occupy it, the battalion marched on to Koheroa.

On 1st November the battalion embarked in the gunboat *Pioneer*, and proceeded to the rebel position at Mere Mere, which was discovered to have been evacuated that morning. The position was occupied by the battalion.

On 20th November the battalion moved to Rangariri (literally, "Angry Heavens"), and took part in the attack on the enemy's works at that place—a fine specimen of the field entrenchments in which the Maories displayed so much skill. The works consisted of a long line of high parapet with double ditch, extending from the Waikato to Lake Waikare, the centre of this line being strengthened by a square redoubt of very formidable construction, the ditch being twelve feet wide and—reckoning from the sole of the ditch to the crest of the parapet—eighteen feet deep. The strength of the work was not known until it was attacked, as the profile could not be seen from the river or the ground in front. Behind the left centre of this main line, and at right angles thereto, was a strong line of rifle-pits facing the river, and obstructing the advance of the troops in that direction. About five hundred yards behind the front position was a high ridge, the summit of which was fortified by rifle-pits.

Annexed is a copy of a despatch from Captain Strange, on whom the command of the battalion devolved, giving particulars of the action, in which the 14th had five men killed, two officers wounded (died of their wounds), and nine men wounded.

" In compliance with instructions received, I have the Honour to report " for the information of the Lieutenant-General Commanding, that the " 14th Regiment started from Mere Mere about 10 a.m. for Rangariri. " On arrival in front of the enemy's position, two companies were thrown " into skirmishing order opposite the right of their works, for the purpose " of preventing their escape by a swamp on their flank, the remainder " being in support. At the cessation of the cannonade, on the order to " advance, the skirmishers and supports advanced under a heavy fire, " availing themselves of such cover as the ground afforded, until within " about fifteen yards of the enemy's works, and lay down, continuing the " firing to keep down that of the enemy, who in considerable number " occupied the ditch and parapet of a strong redoubt. Several men " afterwards occupied the ditch to be in readiness to prevent any of the " enemy from escaping in the event of their being carried by escalade, " upon whom a heavy fire was kept up by the enemy.

" This party remained in the ditch until after dark, and was then with- " drawn, a few men having been detained to form a mine under the " direction of the C.R.E. I beg to add that Lieutenant-Colonel Austen " was severely wounded early in the action, and Captain Phelps danger- " ously while directing the fire of the men close to the enemy's works. " There were also five men killed and one corporal and eight men " wounded.

" In conclusion, I have much pleasure in bringing to the notice of the " Lieutenant-General the coolness and gallantry of Lieutenant and Adju- " tant Glancy, whose conduct was most conspicuous during the whole of " the attack, and who was the first officer to enter the ditch; and of " Ensign Green, who ran along the parapet, desiring the men to follow " him, until he descended into the ditch. He there joined the party " under Lieutenant Glancy, both these officers, with men of several regi- " ments, remaining in a very dangerous spot, where I found them, until " withdrawn.

" I am happy to say that the officers under my command, viz., Lieu- " tenant Langtry, Ensigns Swanson, Calwell, and Staff Assistant-Surgeon " Kellet (who rendered a great deal of service under a heavy fire), as well " as Sergeant-Major Mills and Quarter-Master-Sergeant Bellew, were all " conspicuous for their coolness and bravery. Quarter-Master-Sergeant " Bellew attended to a wounded officer (Lieutenant Gresson, 65th Regi- " ment) and bandaged his wounds under a heavy fire. I consider that " the whole of the non-commissioned officers and men behaved well."

Lieutenant-Colonel Austen and Captain Phelps died shortly afterwards of their wounds, deeply regretted by the whole regiment. Colonel Austen's wounds, though severe, were not thought dangerous,

The Storming of Waikato Pa by the 2nd Battalion.

but they refused to heal.* Captain Phelps,† a fine young man, who had been brought up as a surgeon, was perfectly aware of the fatal nature of his own wound, which was in the groin, and begged the medical officers to leave him and attend to men who had a better chance of recovery. Force had to be resorted to to induce him to have his wound dressed.

Captain Strange and Lieutenant and Adjutant Glancy, as soon as promoted to a company, received brevet majorities for their services at Rangariri.

1863.—First Battalion.

On 1st April, 1863, the establishment of the 1st battalion was fixed as below:

TEN SERVICE COMPANIES.

Field Officers, 3; Captains, 10; Lieutenants, 11; Ensigns, 9; Staff, 5; Sergeants, 48; Drummers, 21; Corporals and Privates, 680.

TWO DEPÔT COMPANIES.

Captains, 2; Lieutenants, 3; Ensigns, 1; Sergeants, 10; Drummers, 4; Corporals and Privates, 120.

On the departure of the 3rd West India Regiment from Jamaica, the 1st battalion furnished the following detachments from Newcastle:

On 2nd July, a company under Captain Bradley to Port Royal; on 11th July, two companies under Captain H. A. Burton and Ensign Blois to Up Park Camp; also a company under Ensign Van Heythusen to Up Park Camp on 26th September. In November the company under Lieutenant Barlow, stationed at St. Lucia, rejoined; the detachment at Up Park Camp was relieved; two companies under Major W. C. Trevor rejoined head-quarters at Newcastle, and a company under Captain Burton proceeded to Port Royal to be stationed.

* Poor Colonel Austen, whose services are given in a footnote on p. 189, was a very heavy smoker. Sir James Alexander makes the curious statement that the nicotine in his system interfered with the healing of his wounds.

† Captain Phelps became an ensign, 57th Foot, in 1855, and joined the 14th as a lieutenant on the formation of the 2nd battalion. He had served at first in the Crimea as an assistant-surgeon, 57th, to which he was appointed on 11th August, 1854.

1864.—First Battalion.

In January, this year, Major E. Holworthy,* commanding the detachment at Trinidad, was accidentally drowned while bathing. Major W. C. Trevor† succeeded to the command of the 2nd battalion, *vice* Austen, died of wounds, and Captains Hawley and Grogan obtained the vacant majorities in succession to Major Trevor, promoted, and Major Holworthy, deceased.

The 1st battalion being ordered home, the detachments were called in. The Trinidad detachment, under Major Hawley, embarked in H.M.S. *Orontes* on 11th April, and arrived at Jamaica on the 18th. On the 19th the head-quarters, under Colonel Budd, marched from Newcastle to Kingston, and embarked in the same vessel, and on the 24th the Port Royal detachment of two companies, under Captain Burton, came on board. The battalion, consisting of three field officers, seven captains, twelve subalterns, three staff officers, and six hundred and fourteen non-commissioned officers and privates, sailed the same day for Portsmouth, having, during its service of four and a quarter years in the West Indies, lost only nine men from disease. It disembarked at Portsmouth on 19th May, and proceeded by rail to Aldershot, where it was quartered in the Permanent Barracks.

---

* Major Holworthy exchanged into the 14th Regiment from a depôt battalion in 1857. He had previously served over twenty years in the Ceylon Rifles.

† Lieutenant-General William Cosmo Trevor, C.B., retired; reserve of officers. Ensign 14th Regiment, 11th March, 1842; lieutenant, 22nd January, 1846; captain, 29th June, 1856; major, 6th January, 1857; lieutenant-colonel, 8th December, 1863; colonel, 8th December, 1863; major-general, 16th September, 1878; lieutenant-general, 25th July, 1884. Served in the Crimea from 10th January, 1855, including the siege and fall of Sevastopol and the assault of 18th June, 1855 (brevet of major, medal and clasp, and Sardinian and Turkish medals). Commanded the 2nd battalion 14th Regiment in the New Zealand War of 1864-66; was in command of a mixed force of imperial, colonial, and native troops in front of the Wereroa Pa, on which occasion fifty prisoners were taken and possession obtained of the pak; accompanied General Chute, with part of the 14th, through the Wanganui and Taranaki campaign, including the assaults on Putako, Otapana, and Waikato, and the operations round Mount Egmont (mentioned in despatches, C.B., and medal).

1864.—Second Battalion.

New Zealand War.

On 23rd May the 2nd battalion marched for Otahuhu, whence a detachment of two hundred men, under Captain Furneaux, proceeded on the 28th to Auckland, and embarked in the steamship *Alexandria* for Napier, to be stationed. The detachment reached its destination on 4th June. The head-quarters of the battalion moved to Auckland on 13th June, and also proceeded in the steamship *Rangotera* to Napier, where they disembarked, and went into quarters on the 19th of the same month.

The head-quarters of the 2nd battalion moved again in November from Napier, in H.M.S. *Falcon*, to Auckland, where the following detachments rejoined: No. 7 Company, under Brevet Major Strange, from Mankoia Heads; No. 8, under Captain Armstrong, from Waikato Heads; and No. 2, under Lieutenant Johnson, from Wellington.

1865.—Second Battalion.

New Zealand War.

In January the 2nd battalion proceeded, in two divisions, from Auckland to Otahuhu; Captain Vivian and three companies on the 9th; and the head-quarters, under Lieutenant-Colonel Trevor, on the 12th. A draft of ten men, under Captain Bryce, joined the 2nd battalion at Otahuhu on 21st January from the depôt, and on 4th February the two companies (six officers and one hundred and forty men), under Lieutenant-Colonel Dwyer, which had been stationed at Wellington, rejoined head-quarters.

A detachment of two officers and one hundred and twenty-four men, under Captain Buck, proceeded to Mount Trafford on 14th February; and on the 15th three companies (five officers and two hundred and twenty-four men), under Brevet Major Strange, proceeded to Auckland, and relieved a detachment of the 68th Regiment there.

On 1st March, 1865, the battalion, under command of Lieutenant-

Colonel Trevor, marched from Otahuhu to Auckland, where No. 2 Company joined head-quarters. Thence the battalion proceeded in H.M.S. *Brisk* to join the field force under Lieutenant-General Sir Duncan Cameron, K.C.B. The field strength of the battalion was nine officers and three hundred and sixteen men, which was augmented on 7th March by fifty men of No. 6 company from Mount Trafford.

Captain Vivian, with three officers and one hundred and five men, occupied Stewart's Redoubt; and Lieutenant Green and twenty-five men were detached on 15th March to Maxwell's Farm, where they remained until 19th May, when they rejoined head-quarters.

On 6th June the head-quarters, under Lieutenant-Colonel Trevor, marched from Wanganni to Makumaru, where they were stationed until 21st July, when they marched to Wereroa, and occupied a "pa" which had been evacuated by the enemy.

Confinement to the pa was exceedingly irksome; but much risk was incurred in entering the surrounding bush. On one occasion Captain Bryce, Lieutenant and Adjutant Butler, and Ensign Symonds were out tracking footprints in the bush, and one of the party had turned aside after game, when they were suddenly fired on from the bush. Lieutenant Butler, who was badly hit, and scarcely able to walk, could not have escaped had not Captain Bryce, himself wounded, supported him and kept the Maories at bay by firing at them from time to time. All three eventually reached the camp, but their lives would have been sacrificed had it not been for the coolness and gallantry of Captain Bryce.*

1865.—First Battalion.

On 28th January, 1865, the depôt companies under Captain T. P. Cosby joined the head-quarters of the battalion at Aldershot. The

---

* Sir James Alexander relates the following incident as having occurred during the campaign: "A sergeant of the 14th was on outpost duty, and returning from "patrol found that his blanket had disappeared. As the night was not warm, he "went in search of a covering, and seeing some men in blankets lying on the "ground, he lay down beside them, and gradually drew off a couple of their "blankets to make himself comfortable. After dozing for some time, he found "he was lying beside some dead Maories, and speedily left his silent bedfellows."

establishment of the battalion was reduced by two sergeants and fifty men from 1st April.

Vacating its quarters in the Permanent Barracks, Aldershot, the Battalion encamped on Cove Common on 19th May, and in July proceeded with the flying column under Major-General Hodges, C.B., to Midrow Hill, near Sandhurst, where it encamped for field exercise and instruction.

In the course of the manœuvres, Lieutenant-General Sir J. L. Pennefather, commanding the troops at Aldershot, expressed his satisfaction at the rapidity with which the battalion got under arms on the "Assembly" being suddenly sounded, and moved to the outposts to repel an attack. At the conclusion of the manœuvres the battalion again pitched its camp on Cove Common. From Aldershot it proceeded to Portsmouth by march route, in three divisions—the first division (four companies), under Major Grogan, starting on 25th July; the head-quarters and four companies, under Colonel Budd, on 31st July; and the last division (four companies), under Captain Maycock, on 10th August. The battalion was quartered in the Clarence Barracks, and furnished a detachment of one company, under Captain Dixon, to Fort Tipnor.

1866.—Second Battalion.

End of the New Zealand War.

The head-quarters of the 2nd battalion, under Lieutenant-Colonel W. C. Trevor, with Captains Vivian and Furneaux, Lieutenants Keogh, Swanson, and Caldwell, and Ensigns Wood and Churchward, marched on 3rd January from Wereroa as the infantry contingent of the field force, then formed under the command of Major-General Chute, for further operations against the Maories. The battalion was employed in the capture of pas at Okutuku, Putahi and Otapawa, in which engagements it lost one sergeant and two privates, killed; and had two officers (Lieutenants Keogh and Swanson) and six privates wounded. In the attack on Putahi the Hauhau flag was hauled down by Private Michael Coffy, of the 2nd battalion 14th. The battalion proceeded inland, by the Mount Egmont route, in pursuit of the enemy, and returned by Taraki and the sea coast to Wereroa on 8th February, having, during the time it was in the bush, endured the severest privations.

The services of the 2nd battalion 14th were specially noted by the Major-General Commanding, in Camp Orders, as follows:—

"Camp, near Ohinomutu,
"4th January, 1866.

"The Major-General Commanding witnessed with great satisfaction the "soldierlike and gallant conduct of the troops (2nd battalion 14th Foot, "3 officers, 105 men), engaged in front of that most formidable enemy's "position, Okutuku Pa, taken this day, and begs to thank Captain Vivian "and the detachment 2nd battalion 14th Regiment . . . . . . for "their services. The noble conduct of Captain Vivian, Lieutenant "Keogh, Ensign Caldwell, and the detachment of the 2nd battalion 14th "Regiment in charging and occupying the Pa, under a raking fire from "the enemy, not only elicited the Major General's warmest approbation, "but surprised the spectators, who were all under fire."

"Camp, in front of Putahi,
"7th January, 1866.

"The Major-General is at a loss to find words sufficiently to express "his thanks to the field force engaged against the formidable Pa of Putahi "this morning, after a harassing march of four hours, through dense bush "and forest, and up and down almost perpendicular ravines. The "Major-General was proud to see the force first advance against the Pa, "nearly inaccessible to troops, with as much coolness and precision as if "on their private parades, and afterwards charge it in the most gallant "and spirited manner when within about eighty yards of it. Such "conduct proves that no troops can cope with those of Great Britain. "The Major-General begs most earnestly to thank Lieutenant Carr, "commanding detachment R.A.; Lieutenant-Colonel Trevor, commanding "2nd battalion 14th Regiment; Major Rock, commanding detachment "2nd battalion 18th Regiment; Captain Johnson, commanding detach-"ment 50th Regiment; Major Von Temskey, commanding Forest "Rangers; Major McDounell, commanding Native Contingent; and the "Officers, N.C.O.'s and Privates under their respective commands, for "their brave and soldierlike conduct; also Dr. Gibb, Deputy-Inspector-"General of Hospitals, and his Staff, for their zeal, energy, and attention "on this occasion."

"Head-quarters, Camp Katamari,
"13th January, 1866.

"Through the gallantry displayed by the troops in the assault on Otapawa "this morning, a Pa, strongly stockaded and celebrated for the natural "strength of its position, fell in a few minutes before the gallant charge "of the troops, and another important success has in the course of a few "days, been obtained over the enemy. The Major-General Commanding "thanks the officers, non-commissioned officers and men engaged, for the "spirit and determination displayed on this occasion, which could not fail "to ensure success, and it will afford him the greatest satisfaction to bring "under the notice of H.R.H. the conduct of those to whom he is so much "indebted for the success of the operation."

With the capture of these strongholds, the campaign was virtually brought to a close. Operations were thenceforward carried on by the Colonial troops, unaided.

On 6th February, 1866, Major-General Chute published the following General Order on the breaking-up of the field force:—

"The Major-General begs to thank the officers, non-commissioned "officers and men of the 14th Regiment, the Detachment R.A., and "Native Contingent comprising the field force, now about to be broken "up, for the cheerfulness with which they have undergone the hardships "of a short but arduous campaign in the Wanganui Taranaki district, "and through the forests east of Mount Egmont; also for the gallantry "displayed on all occasions when they have had the good fortune to "encounter the enemy. The discipline and good conduct invariably "evinced by the regulars reflects great credit on officers commanding "corps."

For services rendered in this campaign, "NEW ZEALAND" was added to the honours borne on the regimental colours. Lieutenant-Colonel Trevor was created a C.B., Captains Vivian and Strange were promoted to brevet majorities, and the New Zealand medal was granted to the officers and men engaged.

On 15th October the head-quarters of the 2nd battalion, under Lieutenant-Colonel Trevor, C.B., marched to Auckland, and embarked in the *Monarch* for Melbourne.

Three companies, under Brevet Lieutenant-Colonel Dwyer, embarked for Hobart Town, Tasmania, and two companies, under Brevet Major Vivian, for Adelaide.

On arrival at their several destinations, the head-quarters and detachments were quartered in barracks.

1866.—First Battalion.

In January, this year, the battalion was ordered from Portsmouth into the Northern District. On 28th January, head-quarters and five companies, under Major Grogan, left for Sheffield, and two companies, under Captain Ranier, for Tynemouth. On the 30th, five companies, under Major Hawley, proceeded by rail to Weedon.

On 1st April, 1866, the establishment of the battalion was reduced from twelve companies to ten.

The stay of the battalion in the north was brief. On account of the unsettled state of Ireland in the autumn of the year, it was hurried off, at short notice, by rail from Sheffield to Liverpool, and thence in two divisions by steamers to Dublin. On the day of its arrival it was sent to the Curragh Camp, there to be stationed.

1867.—First Battalion.

The Fenian conspiracy—a treasonable organization, chiefly of the labouring classes in Ireland, under Irish-American leaders—was at this time becoming very prominent. Secret drilling, and outrages directly traceable to the organization, were of frequent, indeed daily, occurrence. On 12th February, in Kerry, parties of armed peasants attacked several police-stations and fired on the police. The day after this outrage, at 4 a.m., the battalion received orders to send off three hundred and fifty men by the 5.30 a.m. train from the Curragh Stand-house to Killarney in aid of the civil power.

The party proceeded under the command of Major Hawley, and on arrival at their destination a detachment of three officers and one hundred and thirty men immediately marched for Killorglin, twelve miles distant. The remainder were located in the most suitable buildings in the town. The same day (13th February) the head-quarters and every available man (except the quarter-master and twelve men, who were left behind to pack the regimental baggage) proceeded, at one hour's notice, by rail from the Curragh Stand-house to Cork, to reinforce that garrison. The 1st battalion was employed in aid of the civil power in the counties of Cork and Kerry up to the middle of April, and furnished detachments under the orders of the local magistrates as occasion required.

On 16th April the head-quarters and two companies, under Colonel Budd, proceeded to Tralee to be stationed, and detachments

The 1st Battalion at Malta, 1867

of the regiment were located at Mallow (three companies under Major Hawley), Kenmare (two companies under Major Grogan), Killarney (two companies under Captain Le Mesurier), and Mill Street (one company under Brevet Major Foster).

The undress blue frock coat worn by officers was discontinued in April, and replaced by a blue patrol jacket with braided front, etc. The muzzle-loading Enfield rifles were returned to store, and breech-loading Snider rifles were issued to the regiment instead. On 6th June the 1st battalion furnished a detachment of three companies, under Major Hawley, to Ballincollig. Two companies proceeded from Killarney, and one from Mallow.

In June, 1867, the 1st battalion was placed under orders for the Mediterranean, and was concentrated at Cork. The head-quarters arrived from Tralee, under Colonel Budd, on the 19th, and the detachments from Ballincollig, Kenmare, and Mallow on the 14th and 15th. By an Act passed on 20th June, the first period of service of soldiers was made twelve instead of ten years. The pay of the non-commissioned officers and privates was by Royal Warrant increased by twopence per diem, and an additional penny per diem was granted to all soldiers re-engaged to complete twenty-one years' service.

Towards the end of June the embarkation of the 1st battalion was postponed until the autumn, and two companies, under Captain Lloyd, proceeded to Youghal to be stationed. Detachments were also furnished early in July, one company, under Lieutenant Hall, to Bantry, and one company, under Captain Cosby, to Dungarvon. Colonel Ralph Budd, who had commanded the 1st battalion from 27th January, 1857, was on 12th July transferred to the recruiting staff, and was succeeded in the lieutenant-colonelcy by Major and Brevet Lieutenant-Colonel John Dwyer,* from the 2nd battalion; Captain Petrie was promoted to the majority.

---

* Colonel John Dwyer, retired; since dead. Ensign, 35th regiment, 19th July, 1831; lieutenant, 21st November, 1834; exchanged to 14th, 1835; captain, 22nd July, 1846; major, 25th September, 1857; brevet lieutenant-colonel, 17th May, 1864; lieuttenant-colonel, 17th July, 1867; retired on full pay, 8th May, 1872. Served in the Crimea from 10th January, 1855, including the siege and fall of

Early in August a depôt was formed, consisting of two companies, which left for Youghal, under the command of Captain Le Mesurier, shortly before the embarkation of the battalion for the Mediterranean, and all the detachments were recalled to Cork. The depôt subsequently proceeded to Chatham and joined the depôt battalion there.

On the morning of 19th August the head-quarters and ten companies, under the command of Major W. H. Hawley, embarked on board H.M.S. *Himalaya* at Queenstown, and left that day for Malta. On arrival there on the 29th, five companies, under Major Grogan, proceeded to Pembroke Camp, and the next day the head-quarters and remaining companies disembarked and encamped on the glacis of Fort Manoel, in which they were quartered on the departure a few days later of the 1st battalion 60th Rifles.

A severe form of cholera made its appearance in the detachment at Pembroke Camp a few days after its arrival there; the disease also affected the head-quarter companies, in which a few cases occurred. Active measures were, however, taken to stamp it out; the Pembroke Camp detachment was moved to Gozo, and within a fortnight of the first appearance of the disease the last fatal case occurred.

During the epidemic the battalion lost nineteen non-commissioned officers and privates and one soldier's wife. The five companies which had been moved on sanitary grounds to Gozo returned to Pembroke Camp on 29th October.

1867.—Second Battalion.

On 15th May a detachment of the 2nd battalion, four officers and one hundred and sixty-seven men, under Captain Saunders, left Hobart Town, Tasmania, in H.M.S. *Virago*, for Perth, Western Australia, and early in August a detachment of two officers and one hundred and seventy-two men, under Brevet Major Vivian, pro-

---

Sevastopol and assault of 18th June, 1855 (medal and clasp, Knight of the Legion of Honour, 5th class of Medjidie, and Turkish medal). Served in New Zealand war of 1864-65, and commanded the 2nd battalion 14th Regiment in the campaign of February-September, 1864 (medal).

ceeded from Adelaide, South Australia, in the ship *Haversham*, to Hobart Town, Tasmania, to be stationed.

1868.—First Battalion.

The head-quarters and four companies of the 1st battalion proceeded from Fort Manoel to Pembroke Camp on 14th April to be stationed. Two drafts joined the battalion during this year from the depôt, one (under Lieutenant Pigott) of forty-six men on 9th May, and the other (of forty-six men, under Lieutenant Laing) on 9th October.

In September the 1st battalion received orders of readiness for India, and on 19th October the battalion, under Lieutenant-Colonel Dwyer, consisting of thirty-four officers and seven hundred and nineteen non-commissioned officers and men, embarked at Malta on board H.M.S. *Crocodile* for Alexandria, where a draft of thirty-seven men, under Captain Cope, which had arrived from the depôt, joined the service companies. At Alexandria the battalion disembarked, and proceeded by rail to Suez, where it again embarked in H.M.S. *Euphrates* for Bombay. The battalion disembarked at Bombay on 14th November, and at once left by rail for Deolali, where it arrived and encamped the next morning.

The following officers landed in India with the battalion :

Lieutenant-Colonel Dwyer.
Majors Hawley and Grogan.
Captains Warren, Cosby, Le Mesurier, Dixon, Cope, Briscoe, Thompson, Lawrence, and Metcalfe.
Lieutenants Lemon, Laing, Lindsay, Blois, Patton, Carlyon, Firman, Pigott, Hosack, Tombs, and Trench.
Ensigns Richardson, Purkis, Morris, Robinson (H. M.), Robinson (T. M.), and Reed.
Paymaster Carden.
Lieutenant and Adjutant Hutchison.
Quarter-Master Moore.
Surgeon Price.
Assistant-Surgeons Cherry and Randall.

From Deolali the battalion proceeded in four divisions on the 19th, 20th, 24th, and 28th November by rail to Nagpore, thence by bullock train one hundred and eighty miles to Jubbulpore, and onward by rail to Cawnpore, where the battalion arrived early in December, and was stationed in relief of the 101st Bengal Fusiliers.

1868.—Second Battalion.

The head-quarters, under Lieutenant-Colonel Trevor, C.B., remained during the year at Melbourne. The detachment at Hobart Town was under Major Maycock. Detachments under Captains Saunders and Morgan were furnished to Perth and Freemantle, Western Australia.*

1869.—First Battalion.

The sick and weakly men of the 1st battalion proceeded, under Captain Le Mesurier, from Cawnpore to Kasauli, a hill sanitorium, towards the end of March, this year, to remain during the hot season.

Towards the end of July some fatal cases of cholera occurred amongst the men of the 1st battalion in the Echelon Barracks at Cawnpore. The companies affected—"B," "C," and "K"—were promptly moved into the cholera camp at Riapore, about twenty miles distant. Early in August the disease attacked "E" and "G" companies, which were also moved into camp, and later on, "A" company and the band.

The heat in camp was intense, but the health of the men was good, and no fatal cases of the disease occurred after the men quitted the barracks.

About the end of September the band and the companies in

---

* The following particulars of the battalion are taken from the Monthly Returns for this year:

Nationalities: English, 368 non-commissioned officers and men.
Scotch, 42 ,, ,, ,, ,,
Irish, 444 ,, ,, ,, ,,

Heights: 6 ft. and over, 37; 5 ft. 11 in., 52; 5 ft. 10 in., 121; 5 ft. 9 in., 135; 5 ft. 8 in., 187; 5 ft. 7 in., 178; 5 ft. 6 in., 131; 5 ft. 5 in., 8; under, none.

camp returned to their barracks. On this occasion the battalion lost nine men from cholera.

In December, Lieutenant Ridgeway and one hundred and eight men of the 1st battalion joined the services companies at Cawnpore from the depôt, and the detachment under Captain Le Mesurier rejoined from the hill sanitorium at Kasauli. On the 29th of the month the battalion furnished a detachment of two companies, under Captains Briscoe and Metcalfe, to Allahabad, to garrison the fort, consequent upon the departure of the 58th Regiment. Captain Metcalfe died at Allahabad (in 1870), a short time after his arrival there, from the effects of a gunshot wound in the leg, received accidentally whilst shooting in the neighbourhood.

1869.—Second Battalion.

The detachment of the 2nd battalion, under Captain Saunders, stationed at Perth, Western Australia, left that place early in March in the ship *Rover*, for Hobart Town, where it arrived on the 7th of that month, in relief of the detachments there stationed under Brevet Major Vivian, which embarked in the same ship for Adelaide, South Australia, where they arrived on the 29th of the month. In the same month the 2nd battalion furnished a detachment to Sydney, New South Wales, whereof two officers and fifty-five men, under Captain Bradley, proceeded from Melbourne in the *Alexandria;* Lieutenant Noyes and forty-five men, from the same place, in the *Victoria;* and two officers and seventy-five men, under Lieutenant Daly, in the *City of Hobart*, from Tasmania.

The establishment of the 2nd battalion was reduced by two ensigns and augmented by ninety privates, from 1st April The shell jacket worn by the non-commissioned officers and privates on fatigue and undress parades was discontinued from 1st April, and was replaced by a Norfolk frock of scarlet kersey, with regimental facings and buttons.

1870.—First Battalion.

From 1st April, 1870, the establishment of both battalions was reduced from ten to eight companies. The medical officers were removed from the establishment of regiments, but remained attached for duty.

On 9th August, this year, Lieutenant-General Maurice Barlow, C.B.,* was appointed to the regiment in the room of General Sir William Wood, K.C.B., K.H., deceased.

About this time attention was drawn to our system of Army organisation, and to the need of providing a reserve of men to be available for active service in case of emergency. An Act was passed, and came into operation on 9th August in this year, under which Army enlistments were to be made for twelve years as before, but on the condition that the first six only were to be with the colours, or active Army, and the remaining six years in the Army Reserve. Soldiers on transfer to the reserve were to return to their own homes, and to receive pay at sixpence a day, but to be liable to be recalled to active service with the colours in case of war or other special national emergency.

Early in September the detachment of the 1st battalion at Allahabad was increased to a half battalion by two companies, under Lieutenants Hosack and Richardson, from Cawnpore.

1870.—Second Battalion.

On 3rd March, 1870, the detachment at Hobart Town, Tasmania, left that place in the steamship *Hero* for Melbourne, joining head-quarters there on the 8th of the same month.

On 19th March, the battalion, numbering twenty-one officers and two hundred and eighty-three non-commissioned officers and privates, under command of Lieutenant-Colonel Trevor, C.B., embarked in the ship *Walmer Castle* for England.†

The battalion had received much attention and hospitality during

* For services, see *Appendix A.*

† Major Alexander Strange died during the passage home. He entered the army in the 25th Foot in 1848, and was promoted from that regiment to a company in the 14th Regiment on the formation of the 2nd battalion. He was mentioned in despatches for "conspicuous forwardness" with the skirmishers in the attack on Kobrema, 17th July, 1863 (brevet of major and medal). Captain Heywood also died during the voyage home.

its four years station in the Australian colonies. Social gaities, race meetings, and cricket and football matches had enlivened its stay, and it bade farewell to the Australian shores with a grateful sense of the kindness which it had experienced at the hands of the colonists.

The battalion disembarked at Gravesend on 22nd June, and proceeded by rail to Colchester Camp, there to be stationed. Soon after its arrival, the depôts of both battalions, which had been attached to the 2nd depôt battalion at Chatham, arrived at Colchester, and were attached to the battalion.

A Horse Guards Order, No. 101, of this year, awarded the silver medal to No. 969, Sergeant T. B. Ryle, 2nd battalion 14th regiment, as the best shot in the British Army for the year 1869-70.

On 27th September the 2nd battalion proceeded from Colchester to Sheffield by rail, and was quartered in the barracks.

In this year the 2nd battalion was supplied with the new valise equipment. The weight carried by the soldier in marching order was slightly reduced, and the forty-round ammunition pouch and shoulder belt, as well as the knapsack, were rendered obsolete. The quantity of ammunition carried by the soldier on service was increased. Provision was made for forty rounds in two small pouches slung to the waistbelt, and for the remainder in the valise. The new equipment admitted of being readily put on or taken off by the soldier unaided, and the weight carried was more equally distributed, and less severely felt by the soldier.

1871.—Second Battalion.

On 1st February, 1871, the establishment of the 2nd battalion was fixed at forty corporals and five hundred and sixty privates.

On 30th May the head-quarters and three companies of the 2nd battalion left Sheffield by rail for Chester, where they were quartered in the castle. The same day three companies, under Major Maycock, proceeded by rail from Sheffield to the North Fort, Liverpool, for duty; and on 9th August the remaining companies proceeded from Sheffield, under Major Long, by rail to Bradford, there to be stationed.

Early in September a company of the 2nd battalion, which had in March been detached at Weedon, under Brevet Major Vivian, proceeded to Bradford on relief by a company under Captain Keogh. This company also left Weedon for Bradford on 28th September. On 3rd November, Major Long,* of the 2nd battalion, died at Bradford. Captain Warren, of the 1st battalion, succeeded to the majority.

1871.--First Battalion.

In January of this year a draft of one hundred and four men, under Lieutenant Whitbourne, joined the 1st battalion at Cawnpore, from the depôt; and the half battalion stationed at Allahabad, under Captain Furneaux, returned to head-quarters at Cawnpore on being relieved by the 104th Fusiliers. Quarter-Master-Sergeant William Hopkins was discharged to pension, receiving a medal for meritorious service.

In March a party of convalescents and young soldiers of the last arrived draft proceeded, under Captain Williams, from Cawnpore to the hill sanitorium at Kasauli for the hot season. The party returned to the plains and rejoined the 1st battalion at Cawnpore towards the end of October.

On 28th October the head-quarters and five companies of the 1st battalion, under Major Hawley, left Cawnpore, by rail, for Calcutta. The battalion travelled by night and halted for the day at the rest camps (Allahabad, Dinapore, and Muddapore), and, arriving at Howrah on 1st November, crossed the Hoogly and marched to Fort William, there to be stationed, in relief of the 2nd battalion 19th Regiment.

The agitation for army reform still continued, and the demand for improvement in the professional training of officers, and for their being drawn from a wider field, led to the introduction of a Bill for the abolition of purchase in the Army. The Bill passed the

---

* Major Charles Poore Long entered the army in 1851 in the 13th Light Infantry, with which he served in the Crimea, including the battle of Tchernaya and the siege of Sevastopol (medal and clasp, 5th class of the Medjidie and Turkish medal. He exchanged from the 13th, as major, with Major R. Douglas.

Captain E. W. Saunders	Lieutenant C. A. Morris	Lieutenant H. K. Ridgeway	Assistant-Surgeon J. Scanlan	Paymaster W. J. Carden

Captain R. S. Lennon	Captain Furneaux	Lieutenant-Colonel W. H. Hawley	Lieutenant H. McL. Hutchison	Lieutenant R. G. F. Pigott	Lieutenant J. Hosack	Captain J. Laing	Assistant-Surgeon J. Randall	Lieutenant J. Reid	Sub-Lieutenant W. S. Hewett

Sub-Lieutenant C. W. Ravenshaw	Surgeon-Major J. E. Moffatt	Lieutenant W. S. Purkiss	Major C. E. Grogan	Lieutenant R. W. Richardson	Captain R. F. de Lacy Wooldridge

Officers, 1st Battalion, 1871.

Commons, but was thrown out in the House of Lords. A Royal Warrant was, however, immediately published, abolishing the purchase and sale of commissions in the army from 1st November, 1871. At this time the regulation value of an ensigncy was £450; that of a lieutenantcy, £700; that of a company, £1,800; that of a majority, £3,200; and that of a lieutenant-colonelcy, £4,500. In addition to the foregoing sums, it was customary, in order to induce retirement, and thereby accelerate promotion for the officers succeeding to a step, to pay the retiring officer a bonus, varying in amount according to circumstances, which was termed the "over regulation."

By an Act of Parliament, passed shortly after the abolition of purchase, the State guaranteed the value of the commissions held by all officers on 1st November, 1871, also the average amount of "over regulation" money which it was customary to pay in the regiment for promotion.

The rank of ensign was abolished on 1st November, and that of sub-lieutenant substituted. Appointments to this latter rank were to be obtained by open competition, and to be temporary, for two years, at the expiration of which period the officers, provided they passed certain professional examinations, were to be gazetted lieutenants.

The period during which a lieutenant-colonel might retain the command of a battalion was limited to five years; and service in the rank of major was restricted to the same period, but subsequently extended to seven years.

1872.—First Battalion.

In December, 1871, two companies, under Captain Lemon, proceeded frnm Cawnpore to Fort William, Calcutta, and rejoined the head-quarters, and in February, 1872, the remaining companies, under Major Grogan, left Cawnpore for the same place, on being relieved by the 1st battalion the 8th King's Regiment.

A draft, consisting of two sergeants and one hundred rank and file, under Captain Saunders and Lieutenant Firman, arrived at Fort William, Calcutta, on 16th March, from the depôt, and joined the battalion.

On 8th May, Major W. H. Hawley* was promoted to the lieutenant-colonelcy of the 1st battalion, in succession to Colonel J. Dwyer, retired on full pay.

1872.—Second Battalion.

In June, two companies, under Captain Burton, proceeded from Chester, by rail, to Liverpool, and were quartered in the North Fort, and the companies there stationed, under Captain Morgan, rejoined the head-quarters at Chester. In the following month Captain Harington's company returned to Chester from the North Fort, and were relieved by "E" company, under Lieutenant Lapenotiere, from Chester.

The establishment of the 2nd battalion was altered, with effect from 1st April, to the following numbers:—

1 lieutenant-colonel	3 staff officers
2 majors	48 sergeants
10 captains	48 corporals
16 lieutenants and sub-lieutenants	18 drummers
	480 privates

On 20th July a company, under Captain Morgan, proceeded from Chester to Bradford, to be stationed, in relief of Brevet Major Glancy's company, which rejoined head-quarters at Chester. In August, two companies proceeded from Chester to Bradford, under the command of Captains Bradley and Harington, in relief of "B" and "A" companies, which, under Captain Bryce and Lieutenant Gordon, rejoined head-quarters. Brevet Major Glancy's company returned to Bradford in September, and Captain Wilson's company proceeded from thence to Chester.

1873.—First Battalion.

Localization of the Forces.

On 1st April, 1873, the military districts in Great Britain and Ireland were divided into sixty-six infantry and twelve artillery sub-

* Major-General William Hanbury Hawley; reserve of officers. Ensign, 10th July, 1846; lieutenant, 20th June, 1848; captain, 1st April, 1855; major, 8th June, 1870; lieutenant-colonel, 8th May, 1872; colonel, 8th May, 1877; retired, as major-general, 2nd October, 1886. Served in the Crimea from 10th January, 1855, including the siege and fall of Sevastopol and assault of 18th June, 1855. (Medal and clasp and Turkish medal.)

Captain A. F. de B. Dixon — Lieutenant J. Hosack — Lieutenant J. Reid — Paymaster W. J. Carden — Captain A. E. Blair — Captain J. Laing — Sub-Lieutenant W. S. Hewett — Quarter-Master J. Moore — Lieutenant H. McL. Hutchison — Sub-Lieutenant H. B. Urmston — Lieutenant B. W. L. Firman

Captain A. F. de L. Wooldridge — Captain E. W. Saunders — Surgeon-Major J. E. Moffatt — Major C. E. Grogan — Lieutenant-Colonel W. H. Hawley — Major D. S. Warren — Lieutenant R. F. G. Pigott — Captain Furneaux — Lieutenant C. A. Morris

Lieutenant R. W. Richardson — Lieutenant C. D. Ferrier — Assistant-Surgeon J. Randall — Sub-Lieutenant C. W. Ravenshaw

Officers of the 1st Battalion at Fort William, 1873.

districts. To each infantry sub-district were assigned two battalions of the line, one of which was to be ordinarily at home and the other abroad. These battalions were ordered to be linked together, for the purposes of enlistment and service.

In each infantry sub-district a brigade depôt was ordered to be located, under the command of a lieutenant-colonel, the depôt to be composed of two companies from each of the line battalions assigned to the sub-district. The line battalions, militia battalions, the brigade depôt, the rifle volunteer corps, and the infantry of the army reserve constituted the infantry sub-district brigade, all of which were ordered to be under the command of the officer commanding the brigade depôt. The infantry brigades were numbered by lot; the 14th Regiment was designated the 10th brigade, and allotted to West Yorkshire. The brigade depôt was formed on 23rd April at Bradford, to which place the depôt of the 1st battalion proceeded from Chester under Captain Lemon, and was there joined by two companies of the 2nd battalion, under Brevet Major Glancy. Colonel Broadley Harrison was appointed to the command of the brigade depôt, and the 4th West Yorkshire Militia and the 3rd, 7th, and 9th West Yorkshire Rifle Volunteers were allotted to the sub-district.

During the whole of 1873 the 1st battalion remained at Calcutta, with a detachment at Cawnpore. The duties were severe. A number of men selected by the surgeon were sent to Darjeeling during the hot weather.

The establishments of the brigade depôt and home battalion were fixed as hereunder from 14th June, 1873:

	Brigade Depôt.	*Home Battalion.*
Lieutenant-Colonels	1	1
Majors	1	1
Captains	4	8
Lieutenants and Sub-Lieutenants	4	13
Staff	0	3
Sergeants	10	46
Corporals	8	40
Drummers	4	16
Privates	100	488

ROLL OF OFFICERS, 1ST JUNE, 1873.

(From the Monthly Army List.)

Rank	Bn.	Name	Date
Colonel		Maurice Barlow, C.B., *Lieutenant-General*	9th August, 1870
Lieut.-Colonel	1	Wm. H. Hawley,	8th May, 1872
,, ,,	2	Jas. Sinclair Thomson, *Colonel*	30th April, 1873
Major	1	Chas. Ed. Grogan	14th Jan., 1864
,,	2	John G. Maycock, dep.	30th Oct., 1866
,,	1	Dawson S. Warren	3rd Nov., 1871
,,	2	Thos. P. Cosby	8th May, 1872
Captain	2	R. H. Vivian, *Major*	9th Jan., 1858
,,	1	Ed. W. Saunders, *Major*	26th March, 1858
,,	1	And. A. Le Mesurier, *Major*	5th Oct., 1858
,,	1	Ramsay Harman, *Major*, dep.	1st July 1859
,,	2	Hy. A. Burton	6th Feb., 1863
,,	2	Monteford S. Morgan	23rd Sept., 1862
,,	2	John D. Bradley	4th August, 1863
,,	1	Rich. A. L. Furneaux	28th August, 1863
,,	2	John Glancy, *Major*, dep.	14th Jan., 1864
,,	2	Geo. Leslie Bryce, dep.	14th June, 1864
,,	1	Aug. F. De B. Dixon, *Major*	22nd April, 1859
,,	2	John Wilson	12th Sept., 1865
,,	2	Geo. Harwood Cope	21st March, 1866
,,	1	Honble. J. D. Drummond, Staff	16th Jan., 1863
,,	1	Rich. S. Lemon, dep.	12th June, 1870
,,	1	De Lay R. F. Wooldridge	1st April, 1870
,,	1	Joseph Laing	3rd Nov., 1871
,,	2	Fred. W. Harington	8th May, 1872
,,	1	Æneas G. Blair	1st July, 1862
Lieutenant	1	Hy. A. Williams, dep.	6th Feb., 1863
,,	1	Hy. M'L. Hutchison, *Adjutant*	4th August, 1863
,,	2	Roger Hall	28th August, 1863
,,	1	W. Thornhill Blois, dep.	14th Jan., 1864
,,	2	Aubrey Lisle Patton	14th June, 1864
,,	2	Tredenham F. Carlyon	18th April, 1865
,,	1	Bertram W. C. Firman	12th Sept., 1865
,,	2	Gerard Van Heythuysen, *Inst. of Musketry*	6th Feb., 1866
,,	2	Wm. Mills	30th Oct., 1866
,,		*Rich. H. Atkinson*	30th March, 1867
,,	1	E. G. F. Pigott, *Inst. of Musketry*	29th May, 1867
,,	1	John Hosack	6th July, 1867
,,	2	Arthur Walter Noyes	14th August, 1867
,,	2	Powell W. Symonds	14th Sept., 1867
,,	2	Herbert Lovell Woodland, *Adjutant*	20th June, 1868
,,	1	Wm. Adams Ridgway	12th June, 1870
,,	1	Robt. Wm. Richardson	3rd Sept., 1870

Lieutenant -	2	Wm. M. G. Lapenotiere	14th Jan., 1871
,, -	1	Wm. Syer Purkis	3rd May, 1871
,, -	1	Chas. Alex. Morris	28th Oct., 1871
,, -	2	Patrick Crosbie, dep.	28th Oct., 1871
,, -	2	Somerset J. Butler	28th Oct., 1871
,, -	2	Alfred Ruttledge, dep.	28th Oct., 1871
,, -	1	Thos. Middleton Robinson	28th Oct., 1871
,, -	1	Jas. Reed	28th Oct., 1871
,, -	2	Chas. Stewart Gordon	28th Oct., 1871
,, -	1	Chas. Daniel Ferrier	8th June, 1870
,, -	1	Harry Howlett Young	28th Oct., 1871
,, -	1	W. S. Hewett	30th Dec., 1871
,, -	2	Thos. L. Penno	30th Dec., 1871
Sub-Lieutenant	1	Edw. H. Molesworth, dep.	24th Feb., 1872
,, ,,	1	Chas. W. Ravenshaw	8th May, 1872
,, ,,	2	Thos. Rich. Mills	8th June, 1872
,, ,,	2	Gerald Grant-Dalton	8th June, 1872
,, ,,	2	S. C. de Trafford	10th June, 1872
,, ,,	1	Hy. Brabazon Urmston	13th Nov., 1872
,, ,,	1	Hy. Rich. Marrett, dep.	13th Nov., 1872
Paymaster -	1	Wm. Jos. Cardew	4th Sept., 1867
,, -	2	Geo. Ed. Earle	4th Sept., 1867 16th Sept., 1868 *Lieut.* 2nd March, 1866 *Hon. Capt.* 2nd Sept., 1862
Inst. of Musk.	2	G. Van Heythuysen, *Lieutenant*	1st August, 1870
,, ,,	1	R. G. F. Pigott, *Lieutenant*	19th Sept., 1870
Adjutant -	1	H. M'L. Hutchison, *Lieutenant*	4th Nov., 1864
,, -	2	H. L. Woodland, *Lieutenant*	17th March, 1869
Qr.-Master -	1	John Moore	25th March, 1862
,, -	2	John Mills	26th April, 1863
Surgeon-Major	2	R. Webb	7th July, 1869
,, ,,	1	J. E. Moffatt	29th Dec., 1869

No. 10 Sub-District—County, York, West Riding.

Brigade Depôt, Bradford.

Corps:—1st & 2nd Battalions 14th (Buckingham Regiment.)
4th West York Militia (133rd); head-quarters, Leeds; facings, white.
5th Administrative Battalion West Riding, Yorkshire, Rifle Volunteers.
29th West Riding, Yorkshire, Rifle Volunteers.
34th West Riding, Yorkshire, Rifle Volunteers.

In 1873 the badge of the WHITE HORSE and the motto "NEC ASPERA TERRENT" were authorised by Horse Guards Order* to be added to those borne on the regimental colour.

About this time the glengarry cap, with regimental metal badges, replaced the round forage cap, with tuft and chin-strap, and regimental number in brass, previously worn by the sergeants and rank and file.

1873.—Second Battalion.

Three companies were withdrawn from Bradford on 12th February, and proceeded, two—under Captain Bradley—to Chester, and one—under Captain Morgan—to the North Fort, Liverpool.

On 30th April, Colonel W. C. Trevor, C.B., exchanged to the 54th Regiment, and Colonel James Sinclair Thomson† was gazetted to the lieutenant-colonelcy of the 2nd battalion.

On 15th May a company under Lieutenant Patton rejoined head-quarters, Chester, from the North Fort, Liverpool, and in June a company under Lieutenant Butler proceeded thither, in relief of Brevet Major Vivian's company, which rejoined at Chester.

During the same month, companies under Captain Cope and Lieutenant Patton proceeded from Chester to Liverpool to relieve Captains Morgan's and Burton's companies, which rejoined head-quarters.

A division having been directed to assemble at Cannock Chase in August for field manœuvres, the 2nd battalion 14th Regiment was directed to join it, and an advanced party, consisting of a company under Brevet Major Vivian proceeded there, by rail, from

---

* See *Appendix Volume.*

† Lieutenant General James Sinclair Thomson, retired. Ensign, 54th Regiment, 25th November, 1842; lieutenant, 13th June, 1846; captain, 9th July, 1852; major, 21st September, 1860; lieutenant-colonel, 27th July, 1866, exchanged to 14th Regiment; colonel, 27th July, 1871; major-general, 1st April, 1882; lieutenant-general, 2nd December, 1882. Served on board the *Sarah Sands* with the head-quarters 54th Regiment when that vessel was burnt in the Bay of Bengal, and the conduct of the regiment was the subject of a special Horse Guards Order. Served in Oude in 1858-59 under Sir Colin Campbell (medal.)

Chester. Next day the head-quarters of the battalion and four companies, under Colonel Thomson, left Chester by rail, and having been joined at Crewe by three companies under Captain Bradley from Liverpool, the whole proceeded to Cannock Chase and encamped.

At the conclusion of the manœuvres in September, the 2nd battalion was ordered to Aldershot. A detachment under Lieutenant Hall proceeded from Liverpool to Aldershot as an advanced party on the 9th of the month; and on the 12th the head-quarters and eight companies, under Colonel Thomson, left Cannock Chase, by rail, for Aldershot, and on arrival, the same day, joined the third brigade in the North Camp.

From 1st October, the daily stoppage of 4½d. at home, and 3½d. abroad, from the pay of the non-commissioned officers and privates for the ration of bread and meat was abolished, and the issue of the money allowance in lieu of beer (1d. per diem) was discontinued.

1874.—First Battalion.

On 9th January four companies of the 1st battalion, under Captain Furneaux, left Calcutta, by rail, for Benares, to be there stationed, and on the 12th of the same month the head-quarters and remaining companies, under Lieutenant-Colonel Hawley, left Calcutta, by rail, for Cawnpore, whence they proceeded by route march, *viâ* Lucknow, nine marches to Sitapur. The battalion was ordered to halt, *en route*, at Lucknow for divisional manœuvres, at the conclusion of which General Sir Henry Tombs, V.C., commanding, was pleased to express himself as being highly pleased with the battalion, and especially with the steadiness and precision with which it performed the various movements.

In March a draft (seventy rank and file, under Sub-Lieutenant Molesworth) joined the 1st battalion at Sitapur from the 2nd battalion.

On 7th June, Quarter-Master-Sergeant Thomas Moriarty was discharged to pension, and subsequently received the medal for meritorious service, with an annuity of £10.

On 19th June the badge of the WHITE HORSE was authorized to be worn on the forage caps of the officers by a Horse Guards Letter of that date.

1874.—Second Battalion.

In connection with the summer drills at Aldershot, the 2nd battalion, under Colonel Thomson, encamped at Woolmer from 22nd to 24th June, at Chobham from 6th to 9th July, and at Frensham from 22nd to 30th July.

In December the 2nd battalion was supplied with the Martini-Henry breech-loading rifle. Authority was granted for the badge of the ROYAL TIGER to be worn on the collar of the clothing of the men of the regiment.

About this time a button of universal pattern was supplied with the clothing, in place of the regimental button bearing the number of the regiment, the badge of the Tiger, and "Waterloo," and a Union locket of universal pattern was issued for the soldiers' waist-belts, instead of the regimental pattern bearing the number "14" and the county title "Buckinghamshire Regiment."

On 16th December a detachment of the 1st battalion, under Sub-Lieutenant Urmston, left Sitapur, by march route, for Benares, to join the half battalion at that station.

1875.—First Battalion.

In March, Lieutenant Penno and eighteen rank and file joined the 1st battalion at Sitapur from the 2nd battalion. In July, Major D. S. Warren was appointed an assistant adjutant-general in India, and was removed to the seconded list. Brevet Major Saunders was promoted to the vacant majority in the 1st battalion.

General James Webber Smith, C.B., a veteran 95th officer, was appointed colonel at the death of General Maurice Barlow, C.B.

In General Order No. 63 of this year, No. 1,282, Private J. Gardiner, of the 1st battalion, was awarded the prize of £20 and silver medal as the best shot in the army for the year 1874-75.

1875.—Second Battalion.

On 27th July, after nearly two years at Aldershot, the 2nd battalion, under Colonel Thomson, proceeded by rail to Portsmouth, embarked in H.M.S. *Simoom*, and sailed to Devonport, where the battalion was quartered in the Raglan Barracks.

In December the battalion proceeded to Plymouth; head-quarters and two companies occupied the Citadel Barracks, and the remaining six companies the Millbay Barracks.

1876.—First Battalion.

On 1st January the head-quarters of the 1st battalion and right half battalion, under Lieutenant-Colonel Hawley, left Sitapur, by march route, for Lucknow, where the battalion arrived and encamped on the 5th of the month, for duty during the visit of H.R.H. the Prince of Wales.

A draft of one sergeant and eighty-six rank and file, under Captain Lemon and Lieutenant de Trafford, which had been furnished by the 2nd battalion, joined the 1st battalion on its arrival at Lucknow.

His Royal Highness the Prince of Wales, on arrival at Lucknow, consented to present new colours to the 1st battalion, and directed that the left half battalion, then at Benares, should be present on the occasion. The half battalion received orders by telegram on 9th January, and within an hour left, by rail, for Lucknow, and the following morning, at 9 o'clock, one hour before the presentation parade, arrived there, under the command of Brevet-Major Dixon, and joined the head-quarters.

On 10th January the 1st battalion, under Lieutenant-Colonel Hawley, paraded with all the troops in garrison at Lucknow, and was reviewed by H.R.H. the Prince of Wales, who was pleased to express his high approbation of the soldierlike appearance and steadiness of the battalion. The new colours were then consecrated by the garrison chaplain, and handed by Major Saunders and Brevet Major Dixon to His Royal Highness, who presented them to the two senior lieutenants, W. S. Purkis and C. A. Morris, and then addressed the battalion in the following words:

"Colonel Hawley, Officers, Non-commissioned Officers, and Men of "the 1st battalion 14th Buckinghamshire Regiment,—To-day I have the "high privilege of presenting new colours to this distinguished corps, "which in nine years will complete its second century of service. It "bears on its colours the honoured names of 'Tournay,' 'Corunna,' "'Java,' 'Waterloo,' 'Bhurtpore,' 'India,' 'Sevastopol,' and 'New Zea- "land,' showing that from its existence it has rendered distinguished "service in almost all parts of the world. I confide these colours to your "keeping, in the fullest assurance that wherever they are carried they will "preserve the high reputation the regiment has sustained. Every regi- "ment in Her Majesty's service is expected to do its duty, but I feel sure "that none will do its duty better than the Fourteenth."

Colonel Hawley thanked His Royal Highness for the honour he had done the battalion, and the new colours were escorted to their place in the ranks. His Royal Highness was pleased to accept the old colours, which were forwarded to him and deposited in Sandringham House; and to announce to Colonel Hawley his intention of taking steps to obtain the necessary sanction for the regiment to be styled the "PRINCE OF WALES'S OWN."

The left half battalion, under Brevet Major Dixon, left Lucknow, by rail, for Benares on 11th January, and the next day the headquarters and right half battalion, under Lieutenant-Colonel Hawley, left, by march route, for Sitapur, and arrived there on the 16th of the same month.

From 1st April the pay of all non-commissioned officers and soldiers was increased by twopence per diem during the first twelve years of army service, but the issue of the increase was ordered to be deferred until the soldier was in due course discharged or transferred to the Reserve.

On 6th June it was announced in the *London Gazette* that Her Majesty had been pleased to command that the regiment should be styled the "FOURTEENTH (BUCKINGHAMSHIRE) PRINCE OF WALES'S OWN REGIMENT," and be permitted to bear the PRINCE OF WALES'S PLUME on its second colour.

1876.—Second Battalion.

In July the 5th Army Corps, to which the 2nd battalion 14th Regiment was attached, was ordered to assemble at Salisbury; the

Presentation of Colours to the 1st Battalion by H.R.H. the Prince of Wales, at Lucknow

battalion, under Lieutenant-Colonel Thomson, left Plymouth, by rail, on 5th July, and encamped at Honnington Picking, the place of rendezvous, where it remained until the termination of the manœuvres on the 28th of the month, and then returned, by rail, to Plymouth.

In November the 2nd battalion 14th Regiment was ordered to Ireland. Under Lieutenant-Colonel Thomson, it embarked at Liverpool on 14th November, in H.M.S. *Orontes*, and, after a fearfully rough passage and the excitement of a fire on board, disembarked at Belfast on the 17th, and was quartered in barracks there.

The battalion furnished a detachment of twenty-nine non-commissioned officers and privates, under Lieutenant P. Crosbie, to Carrickfergus Castle.

The rank of sub-lieutenant was, in regard of fresh appointments, abolished from 1st December, and that of second lieutenant introduced in its stead. Commissions as second lieutenants were permanent; but candidates were required, after passing the open competitive examination, to join the Military College at Sandhurst for instruction in military subjects for one year, and to pass a further examination in those subjects in order to qualify for a commission. The establishment of the 2nd battalion 14th Regiment was ordered to be increased, with effect from 1st November, to fifty-eight sergeants and drummers and eight hundred and twenty rank and file.

1877.—First Battalion.

On 1st January, this year, the 1st battalion 14th Regiment paraded at Sitapur, with the troops in garrison, to celebrate the proclamation of Her Majesty the Queen as Empress of India. A silver medal, commemorative of the occasion, was presented on the parade to the regiment, and conferred upon Sergeant-Major Daniel Griffin. The 1st battalion, head-quarters and four companies, at the end of January moved into camp at Sitapur; and on 3rd February, on being relieved by the 92nd Highlanders, marched, under Lieutenant-Colonel Hawley, for Ranikhet. From Moradabad, owing to the limited space in the camping-grounds in the hills, the march was continued in three detachments; the whole arriving at Ranikhet on

3rd, 4th, and 5th March, and occupying the barracks vacated by the 1st battalion 19th Regiment.

On 2nd March the left half battalion, under Brevet Lieutenant-Colonel Grogan, left Benares, by rail, for Moradabad, and from thence marched in two detachments to Ranikhet, where they arrived on the 18th and 20th of the same month.

1877.—Second Battalion.

In January an improved pattern of Martini-Henry rifle was issued to this battalion.

On 12th May new colours were presented to the 2nd battalion at Belfast by Lady Templetown. The battalion paraded under Colonel Thomson, and the new colours, having been consecrated by the Lord Bishop of Down, were handed by Major Cosby and Brevet Major Morgan to Lady Templetown, who presented them to Lieutenants Noyes and Crosbie, and addressed Colonel Thomson as follows:

"Colonel Thomson,—As wife, daughter, and sister of soldiers, no one "can feel more deeply impressed than I do with the importance of the "ceremony you have done me the honour of asking me to undertake, and "I can assure the 2nd battalion 14th Regiment that it is with very great "pride and pleasure I present them to-day with their new colours. Your "regiment stands high in its numbers, high in its reputation, and high "also in the honours recently attained in its new designation. I wish to "add I feel very happy in the presence, on this occasion, of the distin-"guished General now in command of this district. In presenting you "these beautiful colours, just blessed by the Lord Bishop of the Diocese, "I commit to your charge the precious emblems of the honour and "renown of your gallant regiment, knowing that whenever and wherever "you may be called upon you will guard and carry them forward through "the dangers and difficulties of every conflict, and hold them firm on to "the end. In bidding you farewell, I beg you to accept my sincere wishes "for your honour, welfare, and success."

Colonel Thomson, addressing Lady Templetown, then replied:

"On behalf of the regiment, I wish to return you my sincere thanks for "your kindness in presenting us with our new colours, and for the kind "remarks you have been pleased to make with reference to the 2nd "battalion 14th Regiment.

"I feel this more particularly, knowing, as I do, that your ladyship met "with a severe accident recently, from the effects of which you are still "suffering, and I consider your kindness in coming here to-day very great "indeed. If the time comes, as come it may before long, when the colours "are brought face to face with the enemy, I am confident that the men "now around us, the Prince of Wales's Own, will be as steady under fire, "guard them as bravely, and carry them as gallantly and gloriously forward "to victory as ever the men of the 14th did in days of yore, when they "earned for themselves the 'Old and Bold,' and helped to win those "many battles and sieges now inscribed on their colours. But there is "room for more, and if war does come, I feel persuaded that they will add "more to their number, and help to win one more crown to the glory and "triumph of Old England. Again I thank your ladyship, and you, my "lord bishop, for your great kindness in taking part in the ceremony."

In May the 2nd battalion 14th Regiment left Belfast for the Curragh Camp, in two divisions—five companies, under Major Cosby, on the 14th, and the head-quarters and remaining companies, under Colonel Thomson, on the 18th. The Carrickfergus detachment rejoined the head-quarters at Belfast on the 17th.

On 4th June the 1st battalion 14th Regiment was armed with the Martini-Henry Rifle.

In August, Lieutenant-Colonel Hawley was removed from the command of the battalion to the half-pay, under the operation of the five years' rule. Brevet Lieutenant-Colonel Grogan* was promoted to the lieutenant-colonelcy, and Brevet Major Le Mesurier obtained the majority in succession to Lieutenant-Colonel Grogan.

1878.—Second Battalion.

In January, this year, the 2nd battalion was supplied with an improved pattern bayonet for the Martini-Henry Rifle (weight, one pound; length of blade, one foot nine and a-half inches.)

---

* Major-General Charles Edward Grogan, retired. Appointed cornet, 7th Hussars, 16th April, 1847; transferred to 12th Lancers, 3rd September, 1847; lieutenant, 28th April, 1848; exchanged to 14th Foot, 17th January, 1851, in which regiment he became captain, 29th December, 1854; major, 14th January, 1864; and lieutenant-colonel, 15th August, 1877. Exchanged to the 8th King's, 13th September, 1879, and commanded the 2nd battalion of that regiment in the Afghan campaign of 1879-80, and afterwards the Lower Kurrum brigade (medal). Brevet Colonel, 6th January, 1879. Retired, 2nd July, 1885.

About this time, the attitude of Russia had excited popular feeling in England to the verge of frenzy. War was regarded as inevitable, and the streets of the large towns echoed with so-called "jingo" songs. In March it was considered necessary to mobilize the army and militia reserves, and the excitement was at its height.

The 2nd battalion, being in the First Army Corps, was ordered to be made up with reserve men to the war establishment, and to be provided with regimental transport, consisting of twelve waggons and twenty-seven horses, for the care of which, and for regimental transport duties generally, one officer, one sergeant, and twenty-six men were detailed.

Colonel J. S. Thomson retired in view of an appointment to the command of a brigade depôt, and Brevet Lieutenant-Colonel D. S. Warren,* from the staff in India, was promoted to the lieutenant-colonelcy.

From 1st April, 1878, paymasters ceased to be regimental officers. The Army Pay Department was formed of volunteers from the regimental and control paymasters and army officers holding the rank of captain.

The establishment of the 2nd battalion was augmented by four second-lieutenants, one transport sergeant, and one hundred and eighty rank and file from 1st April, 1878, the numbers being twenty-officers, fifty-seven sergeants, sixteen drummers, and one thousand rank and file.

On 22nd and 25th April, one sergeant and two hundred and fifty-seven rank and file of the 1st class Army Reserve joined the 2nd battalion at the Curragh Camp from the districts of Cork, Clonmel,

---

* Major-General Dawson Stockley Warren, C.B., retired. Ensign, 14th Regiment, 19th December, 1851; lieutenant, 6th December, 1853; captain, 10th April, 1857; major, 3rd November, 1871; brevet lieutenant-colonel, 1st October, 1877; colonel, 1st October, 1881; retired major-general, 16th November, 1885. Served with the regiment in the Crimea from 10th January, 1855, including the assault of 18th June and siege and fall of Sevastopol (medal and clasp and Turkish medal.) Served on the staff in India. Served with the 2nd battalion in Afghanistan, including the affair at Mazeena and the Kama expedition. (Mentioned in despatches, C.B., and medal.) Assistant adjutant-general in the Soudan in 1885 (medal and clasp and Khedive's star.)

The 2nd Battalion at the Curragh, 1878

and Birr. At this time the 2nd battalion numbered one thousand three hundred of all ranks, and the 1st battalion upwards of one thousand, as fine a regiment as ever stepped in the British army.

On 31st July the Army Reserve men were demobilized and sent to their homes; and in September the 2nd battalion was placed under orders for India, and the establishment reduced by one sergeant and one hundred and eighty privates. The regimental transport horses, waggons, and equipment were handed over to the Control department.

On 14th September, Major Young was gazetted to a majority in the regiment, from the 49th regiment, in exchange with Major and Brevet Lieutenant-Colonel Vivian.

The 2nd battalion, under Lieutenant-Colonel Warren, consisting of twenty-four officers, thirty six sergeants, and six hundred and ninety-eight drummers and rank and file, left the Curragh Camp, by rail, in two divisions, on 3rd and 4th October, for Cork, and on the 4th and 5th embarked on board H.M.S. *Crocodile* for Bombay, where the battalion disembarked on 8th November, and proceeded by rail to Deolalee.

From Deolalee the battalion proceeded on the 12th and 13th, by rail, to Lucknow, and occupied the left infantry barracks there on 16th and 17th November. The officers named below embarked with the battalion, Lieutenant-Colonel Warren commanding:—

Major	Cosby	*Lieut.*	Ruttledge, *I.M.*	*2nd Lieut.*	Vowell
„	Young	„	Gordon, *Adjt.*	„	Carter
		„	Rae	„	Lester
Captain	Lemon	„	Parker	„	Christie
„	Harington	„	Walker	„	Wemyss
„	Hall	„	Geaves	„	Mitchell
„	Van Heythuysen	„	Schuyler	„	Fry
„	Noyes	„	Kitchener	„	Barchard

Paymaster Franklin, A.P.D., attached.

1878.—First Battalion.

The Two Battalions.

Captain Laing and Lieutenants Pigott and Richardson, one sergeant, and fifty rank and file, joined the 1st battalion at Ranikhet on 17th January from the 2nd battalion.

In the autumn the battalion was placed under orders for Aden, and directed to reduce its effectives by giving volunteers to other corps and transfers to the 2nd battalion. Five hundred and fifteen sergeants, drummers, and rank and file, mostly men under the Army Enlistment Act of 1870, were thus transferred, and seventy-seven men volunteered to other corps.

On 22nd and 24th November, the 1st battalion, under Lieutenant-Colonel Grogan, left Ranikhet, by march route, for Moradabad, where it encamped on 4th and 5th December, and proceeded to hand over its camp equipment, native establishment, and regimental bazaar to Lieutenant Rae for the battalion. The transfer effected, the battalion left Moradabad, by rail, on 7th December, for Deolalee, and early the following morning arrived at Lucknow, where the 2nd battalion was stationed, thus enabling the two battalions to meet for the first time since the formation of the existing 2nd battalion.

As the 1st battalion train entered Lucknow, the officers and men received a most enthusiastic greeting from those of the 2nd battalion, the greater part of whom had assembled at the station. Their band played the regimental quick-step, "*Ça Ira*," as the train came in, and all ranks of the younger battalion most hospitably greeted and entertained their more seasoned comrades. After a brief stay, prolonged to the utmost, the train with the 1st battalion moved off on its journey, amid the ringing cheers of the 2nd battalion, repeated again and again until they died away in the distance.

The 1st battalion arrived at Deolalee on 14th-15th December, and on 18th-19th left by rail for Bombay, and there embarked on board H.M.S. *Malabar* for Aden. The battalion landed at Aden on 28th December, and took over the barracks which had been vacated by the 1st battalion 8th King's; head-quarters and four companies at the Crater Position Camp; three companies, under Major Le Mesurier, at the Isthmus; and one company, under Captain Richardson, at Steamer Point. During the stay of the battalion at Aden the detachments were relieved every two months by the companies from head-quarters.

In this year the 1st battalion won the prize—a cup worth one hundred pounds—given by His Excellency Sir Frederick Paul Haines,

Private Holmes	Lance-Corporal Brosman	Lance-Corporal Sweeney	Private Davis	Corporal Cook	Corporal Powell	Private Woods	Private Clarke	Private *Hamilton*	*Lance*-Corporal Busby
Colour-Sergeant Lowing	Private O'Neil	Private Sherring	Lance-Corporal Clarke	Sergeant Saunders	Sergeant Brown	Lance-Corporal McClelland	Private *McCern*	Private Cafferty	Colour-Sergeant Kenyon

Winners of the Prize Cup,

Presented by His Excellency General Sir F. P. Haines, G.C.B., Commander-in-Chief in India, to the Best Regimental Team in the Musketry Competition of the year 1877-78, in Bengal.

commander-in-chief in India, to that battalion serving in Bengal, in which a team of twenty men, selected under certain conditions, would, in firing seven rounds at six hundred and eight hundred yards, make the highest score with the Martini-Henry rifle. Thirty battalions competed for the prize, which the 1st battalion won with a score of seven hundred and forty-five points, forty-seven in excess of that obtained by any other corps. The cup, in accordance with the conditions on which it was presented, has been deposited in the officers' mess.*

Brevet Lieutenant-Colonel Cosby† completed seven years' service as a regimental major, and a total service of twenty-seven years, on 23rd November, and was placed on half-pay from that date under the retirement rules, and promoted to lieutenant-colonel. Brevet Major Burton succeeded to the majority.

1879.—First Battalion.

General Sir Alfred Horsford, G.C.B., was appointed colonel on 5th January, this year.

The 1st battalion was in the autumn ordered to return from Aden to England, and on 17th November the head-quarters and four companies, under Colonel F. Barry Drew, C.B.,‡ who had exchanged from the 8th King's with Colonel Grogan, marched to Steamer

---

* See *Appendix Volume.*

† Major-General Thomas Prittie Cosby, retired. Ensign, 14th Regiment, 23rd November, 1852; lieutenant, 11th August, 1854; captain, 24th July, 1857; major, 8th May, 1872; brevet lieutenant-colonel, 2nd September, 1878; lieutenant-colonel, 2nd September, 1882; retired, 27th August, 1887. Served in the Crimea from 10th January, 1855, including the assault of 18th June and siege and fall of Sevastopol (medal and clasp, and Turkish medal.)

‡ Major-General Francis Barry Drew, C.B., retired. Ensign, 28th Foot, 28th May, 1845, exchanged to 40th Regiment in 1847; lieutenant, 17th August, 1848; adjutant, 11th February, 1848, until promoted to captain, 20th November, 1851; exchanged to 64th Regiment in 1852; to 11th Regiment in 1852; to 94th Regiment in 1855; to a depôt battalion, 1859; brevet major, 1863; retired on half-pay, unattached, 1865; exchanged to 8th King's, 1868; lieutenant-colonel, 7th March, 1877; commanded the 8th King's in the Afghan campaign of 1878, and succeeded to the command of the brigade at the forcing of the Peiwar Kotal (three times mentioned in despatches, C.B., and medal); exchanged to 14th Regiment, 13th September, 1879; retired, 26th April, 1882.

Point and encamped, preparatory to embarkation. The detachments joined the head-quarters the following day. General Loch, commanding the Aden brigade, inspected the battalion on the afternoon before it embarked, and was pleased to say that during his thirty-six years of service he had never seen a better drilled, a better behaved or a better disciplined regiment.

On 25th November the 1st battalion embarked on board H.M.S. *Serapis*, and sailed the same day for Portsmouth, where the battalion was transhipped on 22nd December, and conveyed to Cowes, Isle of Wight, whence it marched to Parkhurst, and was stationed in the barracks.

Thirty-five officers and seven hundred and fifty-six men of the 1st battalion landed in India in November, 1868, of which seven officers and one hundred and forty-two men returned to England with the battalion; of the remainder, one officer and seventy-four men died in India; four officers and two hundred and thirty-three men were invalided to England, and did not return to the country; two officers and two hundred and thirty-eight men left the Service; twenty-one officers and fifty-two men went to other corps and to the 2nd battalion; and seventeen men deserted.

The undermentioned are the officers who proceeded to India with the battalion, and returned to England with it:—Captains Laing, Firman, Pigott, Richardson, and Purkis; Lieutenant and Adjutant Reid; and Captain and Paymaster Carden, Army Pay Department, attached.

The establishment of the 1st battalion was, from the date of return to England, reduced to the following numbers:—

1 Colonel	4 2nd Lieutenants	8 Color Sergeants
1 Lieutenant-Colonel	1 Adjutant	24 Sergeants
1 Major	1 Quarter-Master	40 Corporals
8 Captains	8 Staff Sergeants	17 Drummers
8 Lieutenants		

440 Privates.

The battalion, on arrival at Parkhurst, was issued with the valise equipment and new-pattern helmet with spike.

1879.—Second Battalion.

In the autumn the British Resident, Major Cavagnari, who had been appointed to Cabul under the treaty of Gundamuck, together with his escort of native troops, were murdered in that city by the Afghan soldiery. The army then retiring from Afghanistan was ordered to retrace its steps and march on Cabul, and additional troops were detailed for active service in that country. The 2nd battalion 14th Regiment was ordered to the front, and on 29th December left Lucknow, by rail, for Jhelum, with a strength of twenty-one officers, forty-three sergeants, and six hundred and seventy-five drummers and rank and file. The following were the officers:

Lieutenant-Colonel Warren, commanding.
Major Young.
Captains Harington, Haywood, Van Heythuysen, Noyes, and Butler.
Lieutenant and Adjutant Gordon.
Lieutenants Rae, Graves, Schuyler, and Lester.
2nd Lieutenants Wemyss, Mitchell, Fry, Barchard, Vialls, Vanrenen, and Murray.
Major and Paymaster Franklin (Army Pay Department, attached).
Quarter-Master Bayley.

1880.—Second Battalion.

Afghanistan.

The 2nd battalion 14th Regiment arrived at Jhelum on 2nd January, 1880, and was detained there for carriage until the 5th, when it left, by march route, for Peshawur, where it arrived on the 19th. On the 27th of the same month the battalion crossed the Indian frontier, and encamped at Jamrood at the entrance to the Khyber Pass.

The battalion, with the 41st Bengal Native Infantry and 1st Ghoorkhas, formed the 2nd brigade of the reserve division, under General Ross, C.B.

The battalion was reinforced on 16th March by Captain Ogden, Lieutenant St. George, and sixty-five men from Lucknow, and on the 24th by a draft from England of two sergeants and sixty-one men, under Lieutenant Baird.

On 10th April the 2nd battalion 14th Regiment marched to Ali Masjid, and the following day to Lundi Kotal, an almost impregnable position at the head of the Khyber Pass, where it relieved the 1st battalion 5th Fusiliers. On the arrival of the 1st battalion 18th Regiment, the 2nd battalion 14th Regiment moved on to Pesh Bolak, where it arrived on 5th May, and with a battery of artillery and the 8th Hussars formed the British garrison.

Pesh Bolak is situate on a stony plain, about six miles from the main road, and was a strategical position to hold in check the turbulent Shinwari tribesmen.

On the night of 18th May four companies of the 2nd battalion 14th Regiment, under Lieutenant-Colonel Warren, with a detachment of the 8th Hussars, 5th Bengal Cavalry, and 32nd Punjab Pioneers, and four guns of L 5, R.H.A., the whole under Brigadier-General Gib, marched—without camp equipage or baggage—for Trelai, eight miles distant, where they arrived at daybreak, and thence to Shershai, about nine miles further towards the west, arriving there about 10 a.m. The inhabitants seemed undecided whether to fight or not, but ultimately tendered their submission.

The troops bivouaced on the open plain, and during the night information was received that Mulah Fakir had collected his forces further to the westward. At 5 a.m. on the 20th the force resumed its march, along the right bank of the river which washes the valley, for the villages of Hissarak and Mazeena, near which latter place the enemy, estimated at from four thousand to six thousand, were reported to have taken up a position. After a march of about six miles, beating of tom-toms, planted standards, and desultory firing gave indication of the enemy's position.

They appeared to be occupying the right bank of the river which fertilizes the lower ground on the left bank, where lay many forts amidst the fields. The right bank, on the contrary, was uncultivated, dry, and stony, the ground slightly undulating, and intersected by nullahs. Close to the right bank of the river was a small, elevated plateau, occupied by the enemy, which appeared to command the ground in its neighbourhood. This was evidently the point first to be seized.

The artillery and cavalry, together with Captain Harington's company of the 2nd battalion 14th Regiment, as an escort to the guns, moved to the left and opened fire upon the enemy, who occupied

strong ground on the right bank of the river, which was everywhere fordable, their left thrown forward, with supporting bodies at the centre. The other companies of the 2nd battalion, under Lieutenant-Colonel Warren, moved in attack formation, two lines of companies at double-company distance and two hundred yards interval between the two lines, and advanced upon the mound before alluded to.

The 32nd Punjab Pioneers were in reserve, to the right rear. On the attack getting to within five hundred yards of its objective, shots came from the left bank of the river from a wood slightly in advance of the mound, to clear which Captain Morris was directed to wheel his men by half-companies to the right, and, advancing, to extend his leading half-company, which he performed. Here the first casualties befel the enemy. Captain Gordon's company having moved up from the second line into Captain Morris' place, the attack proceeded. The enemy did not accept the assault, and Morris' company was drawn in, and the regiment re-formed under cover. From the mound large masses of the enemy were seen moving across from the forts on the left bank to reinforce their right flank, and volleys by marksmen, and ultimately volleys by companies, were delivered with great effect at distances ranging from four hundred and twenty yards on the left upon the enemy in position, to six hundred or seven hundred yards on the right upon his advancing supports.

In the low ground directly to our front the enemy held a very strong position, sheltered from the effects of artillery fire among the stone walls and trees.

In the rear of this a ford was seen, behind the right rear of which a high ridge of ground with a ziarát appeared as a special rallying point.

Many standards being here planted, Captain Noyes' company, supported by Captain Morris', was directed to advance over the low ground alluded to. The enemy made a decided stand, and a regular advance by rushes of half-companies was executed by "D" company. The position was ultimately cleared with the bayonet, the men led on by Captain Noyes, who captured the enemy's standard, cutting down the standard-bearer with his own hand. The enemy lost heavily, and "D" company had two men killed and three wounded in the attack.

Captain Noyes also received a severe wound from an Afghan knife, a most formidable weapon some two or three feet in length. The artillery, which had assisted by preparing the various points of attack, were informed that the next objective of the battalion was the ziarát. Their effective Shrapnel fire so shook the enemy, that on the advance being resumed they again evacuated their position. The disposition of the battalion now was—two companies, under Captains Noyes and Morris, in the bed of the river; two companies, under Captains Haywood and Gordon, crowning the ziarát ridge. From the high banks of the river, just beyond this, the retreating enemy again came within effective range of half-company volleys at five hundred to seven hundred yards; whilst the 32nd Punjab Pioneers were seen advancing and driving the enemy before them.

The enemy appeared completely beaten. The battalion advanced, and ultimately crossed to the left bank of the river, where, at about 12 noon, in conjunction with the rest of the force, it rested through the hottest period of the day under the trees, one company being pushed forward on the left flank as a precautionary measure.

Captain Harington's company still remained with the guns on the right bank. At about 2.30 p.m. the enemy, having collected again, began to fire upon our left flank.

A freṣh advance was ordered, and, with but slight resistance on the part of the enemy, the entire valley was cleared, some heights in the vicinity of Mahrey being crowned during the advance; the 32nd Punjab Pioneers had worked on the right flank of the 2nd battalion 14th Regiment.

At about 6.30 p.m. the battalion retired through the valley, burning the crops as they went, and took up their quarters for the night in some forts. During the night the enemy tendered their submission. Next morning the force returned to Shershai, the duties of rear guard being confided to the 2nd battalion 14th Regiment, and, contrary to the usual custom of the country, no molestation was attempted.

The loss inflicted on the enemy was about one hundred and twenty killed, two hundred wounded, and some forty prisoners. Our losses were—2nd battalion 14th Regiment, two men killed and one officer and three men wounded; Royal Artillery, one officer severely and one bombardier slightly wounded; 32nd Punjab Pioneers, one sepoy wounded. One doolie-bearer of the 2nd battalion was killed, and one missing.

The despatches published in the *Gazette of India* gave great praise to Brigadier-General Gib for the services he had rendered in the successes in the Mazeena Valley from 18th to 23rd May, 1880.

As the despatches are too long for insertion at full length in these records, one or two extracts may be given which more especially concern the regiment.

"About 7.30 a.m. we found the enemy in a very strong position in a "cultivated valley, which was studded with forts and covered with orchards, "terraced fields, sangas, and watercourses.

"His left rested on the village of Hisarak, and his line extended to "Mazina, about a mile in left, and faced what would have been my right "flank, had I advanced further. As we came up they waved standards, "fired guns, and beat drums. . . . .

"I took the guns along the plain to about the middle of the enemy's "line, and opened fire at about 1,200 yards. I then ordered two com-"panies of the 14th, in skirmishing order, with two companies in support, "and the 32nd in column on the right rear, to attack the enemy's left, and "gradually to push on as he became shaken by the guns.

"The enemy being in great force I determined to use the guns "freely. . . . .

"At one place, near what is marked in the map 'Old Kaffir Fort,' "there was a sanga which the guns could not reach, owing to a high "bank; this position was obstinately held, but carried finally by a charge "led by Captain A. W. Noyes, 14th Regiment, who was first in, killed the "first man who opposed him, and grappled with the second, who was "bayonetted by the men, but not before he had wounded Captain Noyes "in the hand. I beg to bring Captain Noyes' gallant conduct particularly "to the notice of the Major-General.

. . . .

"About 1 o'clock the enemy were in full retreat.

. . . .

"The 2-14th Regiment, though composed mostly of very young soldiers, "behaved with great steadiness, coolness, and gallantry, and were well "kept in hand by the commanding officer, Lieutenant-Colonel D. S. "Warren, assisted by his company officers. The action was one well "calculated to produce wild firing, but there was none, and the percentage "of rounds fired, viz., 8·25 per man, was marvellously low in a long day's "fight."

The brilliant march of the force under General Sir Frederick

Roberts from Cabul to Candahar, and the subsequent defeat of the Afghan army at that place practically concluded the war.

Hostilities ceased; and after a while the troops evacuated Afghanistan. The 2nd battalion 14th Regiment on its way back, at the end of August, was quartered at Nowshera. From Nowshera, "B" company, sixty-seven strong, under Captain Butler, was despatched on 29th October to garrison the Fort of Attock. At Nowshera, too, in November, the depôt, under Captain Lemon, which had been left behind at Lucknow when the battalion took the field, rejoined.

By command of Her Majesty, "AFGHANISTAN" was added to the regimental honours, in commemoration of the gallant conduct of the 2nd battalion 14th Regiment in the campaign of 1879-80. The officers, non-commissioned officers, and men present in the campaign received the Afghan medal.* Lieutenant-Colonel D. S. Warren was mentioned in despatches and made C.B. Major A. W. Noyes was specially mentioned for his gallant conduct at Mazeena on 25th May. Captain Hutchison, who was employed on staff during the campaign, received a brevet majority, and Lieutenant Kitchener was specially mentioned in despatches for services with the transport.

On 21st November, 1880, the present colonel, General Alfred Thomas Heyland, C.B.,† was appointed to the regiment on the transfer of Sir Alfred Horsford to the Rifle Brigade.

Under the Localization of the Forces Scheme of 1873, new barracks had been erected at York to accommodate the 6th and 10th brigade depôts (depôts 25th King's Own Borderers and 14th Regiment), together with the permanent staff of the militia battalions appointed to these brigades, and the battalions themselves when assembled for training and exercise.

These barracks were completed early in May, this year; and on 6th May the 10th brigade depôt proceeded from Bradford to York, under Captain Reid, and occupied the new barracks.

Colonel Lloyd, commanding the Sixth Sub-District, assumed command of both brigade depôts and sub-districts. A few months later, Colonel Chichester, C.B., was appointed to command the brigade

---

* A silver medal, worn with a dark-red and green striped ribbon.

† For services see *Appendix A*.

depôts and sub-districts, and Colonel Lloyd was transferred to Guildford.

1881.—First Battalion.

On 26th June a detachment of the 1st battalion 14th Regiment (four sergeants and seventy-six men, under Captain Firman and Lieutenant Cox) proceeded from Parkhurst to Marchwood, there to be stationed. This detachment rejoined the head-quarters at Parkhurst on 20th July. On 19th July the 1st battalion furnished a guard of honour at Osborne House, East Cowes, under Captain Van Heythuysen, during the stay there of Her Majesty the Queen. The detachment returned to head-quarters after the departure of Her Majesty on 26th August. In October a draft of one sergeant and one hundred and two rank and file, under Lieutenant Watts, left the 1st battalion at Parkhurst for India, and joined the 2nd battalion at Nowshera the following month.

In this year some changes were made in the officers' uniform: an improved pattern of shoulder-cord was approved for the tunic and mess jackets, shoulder-straps were directed to be worn on the great coats and patrol jackets, the badges of rank were transferred from the collar to the shoulder-strap, and a drooping gold-embroidered peak replaced the horizontal forage-cap peak previously worn.

The 1st battalion was reinforced by several drafts of men during the latter part of 1880 from the depôt at York, and in December of that year a guard of honour of two sergeants and forty-five rank and file, under Captain Firman and 2nd Lieutenant Salmonson, proceeded by route march to East Cowes, during the time Her Majesty should remain at Osborne.

Her Majesty having graciously expressed a desire to see the 1st battalion, many years having elapsed since she had inspected a regiment at Parkhurst, several attempts were made to march to Osborne, but the weather was always unpropitious. At last, by special command, the battalion was paraded, on 12th February, 1881, at Parkhurst for Her Majesty's inspection. It was drawn up in line, with a strength of fifteen officers and four hundred and thirty-four non-commissioned officers and men. The Queen, accompanied by the

Princess Beatrice, was received with a royal salute. After driving down the line and seeing the regiment march past and advance in "review order," Her Majesty was pleased to express to the commanding officer, Colonel F. Barry Drew, C.B., her extreme satisfaction at the way the men manœuvred and her approval of their appearance under arms. "Billy," the regimental deer, headed the band on this occasion, and was specially noticed by the Queen. The battalion left Parkhurst on 16th and 17th February for Portland in two divisions. The first party of two hundred non-commissioned officers and men, under Captain Purkis, with Lieutenant Heigham, 2nd Lieutenant Salmonson, and Quarter-Master King, with the heavy baggage of the regiment, marched to Cowes, and embarked in H.M.S. *Medina* for Southampton, proceeding from thence by special train to Portland. This party took over the Verne Barracks from the 56th Regiment.

The second division, consisting of head-quarters and the remainder of the battalion (with the exception of "A" company, which, under the command of Captain Firman, with 2nd Lieutenant H. J. Crofton, was left behind to hand over barracks to the 66th Regiment), marched to Cowes, and embarked in H.M.S. *Sprightly* for Southampton, proceeding thence by special train. The strength of the party was nine officers, two hundred and fifty-two non-commissioned officers and men, thirty-four women, and twenty-one children, the whole under the command of Colonel F. Barry Drew, C.B. Captain Firman's company, forty-three strong, joined head-quarters on the 21st.

The barracks in which the battalion was quartered were spacious, and excellently found in every particular, with a magnificent mess-house, married quarters, gymnasium, and even a racquet court. The situation on the top of so steep and high a hill was, however, a drawback. The musketry was carried out on the Chesil Beach, which joins Portland with the mainland, the battalion furnishing a picquet over the convicts, which was its chief duty. A detachment of thirty non-commissioned officers and men was sent to the Nothe Fort at Weymouth from within a few days of the arrival of the battalion at Portland until January, 1882, when it was relieved by a detachment from the depôt 39th Regiment.

Inspection of the Ist Battalion by the Queen, at Parkhurst, 1881

The officers on this detachment were: Lieutenant Wemyss, Lieutenant Mitchell, Captain Schuyler, and Lieutenant Swaine, who succeeded each other in the above order.

The Territorial System.—Change of Title.

On 1st July, 1881, the memorable changes in title, organisation, and dress, consequent on the introduction of the Territorial system, were adopted in the infantry regiments of the line.

The practice of distinguishing line regiments by numbers was discontinued, and the 14th (BUCKINGHAMSHIRE) REGIMENT became THE PRINCE OF WALES'S OWN (WEST YORKSHIRE) REGIMENT. The territorial district assigned to it—styled the 14TH REGIMENTAL DISTRICT—was centred at York.

The territorial regiment consisted of four battalions, two line and two militia, as hereunder:

1st battalion, the old		1st battalion 14th Regiment.
2nd "	"	2nd battalion 14th Regiment.
3rd "	"	2nd West York Light Infantry Militia.
4th "	"	4th West York Militia.

Subsequently, the volunteer corps of the Regimental District were added to the territorial regiment, as volunteer battalions:

1st	Volunteer Batt.,	the old	1st West York Rifle Vols. (head-quarters, York).
2nd	"	"	"	2nd West York Rifle Vols. (head-quarters, Bradford.)
3rd	"	"	"	7th West York Rifle Vols.

The time-honoured *buff* facings gave place to *white*, which became the facing of all English line regiments not entitled "royal." White facings and gold rose-pattern lace were, accordingly, adopted by the line battalions of the regiment, and also by the militia battalions, which had previously worn white facings and silver lace. The shape of the cuff was changed from pointed to banded. The Royal Tiger badges on the collars gave place to the Prince of Wales's plume. The brass numeral "14" on the men's shoulder straps was replaced by "W. YORK," in white letters.

The loss of the historic number was bitterly felt by almost every regiment in Her Majesty's Army, and universal regret was expressed at the further loss of individuality in so many regiments through the alterations in the facings and other details.

ROLL OF OFFICERS OF THE PRINCE OF WALES'S OWN (WEST YORKSHIRE) REGIMENT,

(From the Monthly Army List for August, 1881—the first appearance of the Regiment under its Territorial title.)

Honorary Colonels: { Heyland, General A. P., C.B., *1st & 2nd Battalions.*
Van Straubenzee, H., *3rd Battalion.*

1st and 2nd Battalions.

Lieutenant-Colonels (4):

2 Warren, D. S., C.B.
1 Drew, T. B., C.B.
2 Cosby, T. P.
1 Saunders, E. W.

Majors (8):

1 Le Mesurier, A. A., *L.C.*
2 Burton, H. A.
2 Dixon, A. F. de'B., *L.C.*
1 Lemon, R.S.
d. 2 Harington, F. W.
1 Hutchison, H. Mc.L.
2 Haywood, W. W.
1 Firman, B. W. C.

Captains (10):

d. 1 Van Heythuysen, G.
2 Ogden, D. A.
1 Pigott, R. G. F.
1 Hosack, J.
2 Noyes, A.W.
2 Morris, C. A.
2 Purkis, W. S.
1 Crosbie, P.
2 Butler, S. J.
1 Ruttledge, A.
1 Robinson, P. M.
2 Reid, J.
2 Gordon, C. S.
2 Ferrier, C. D.
2 Penno, T. W. L.
1 Grant-Dalton, G.
1 Rae, V. R.

Lieutenants (30):

Thurston, J. W. (prob.)
d. 2 Walker, F. D.
Adye, C. G.
d. 2 Geaves, R. L.
1 Schuyler, E. E. S.
2 *Kitchener, F. W., Adjt.*
1 *Mills, E. C., Adjt.*
2 St. George, A. W.
2 Baird, W. R. C.
1 Cox, C. H., *I. of M.*
2 Vowell, H. A.
1 Burke, J. H.
Carter, R. L. B. (prob.)
2 Lester, C. M.
2 Christie, J. H.
1 Lowry, F. J. S.
1 Wemyss, G.
1 Mitchell, E. W.
1 Fry, W.
2 Barchard, C. P.
Hogge, A. F., (prob.)
1 Heigham, C. J. M.
1 Walker, H.
2 Vialls, H. G.
1 Swaine, G. W.
1 MacAdam, F. R. P.
1 Critchley - Salmonson, H. B. S.

Lieutenants (continued):

2 Yale, J. C.
1 Crofton, H. J.
2 Roberts, R. J.
2 Watts, H. E.
2 Roberts, A. N.
2 Loudon, W.C.
2 Williams, A. B. C.
2 Clements, C. H.

Paymaster:

2 Franklin, W., *ret. M.*
1 " "

Instructor of Musketry:

1 Cox, C. H., *L.C.*
2 " "

Adjutants:

1 Mills, E. C., *Lieut.*
2 Kitchener, F. W.

Quarter-Masters:

1 King, W.
2 Scott, R.

3rd Battalion.

Lieutenant-Colonel:

ps. Hay, G. J.

Majors:

Chaplin, R.
Lees, F. G.

Captains (10):

Rickaby, J.
Hodgson, N. H. L.
Dashwood, C. F., *Hon. M.*
ps. Tennant, J. R.
ps. Jackson, W. B.
I'Anson, J.
Tennant, G. G.
ps. Lockley, E. H.
Goldie-Taubman, A. H.
Dashwood, E. P.

Lieutenants (15):

Lees, G. F.
ps. Trafford, H.
Hine-Haycock, R. N., *I. of M.*
Bartlett, E. A.
Sagar-Musgrave, A. M.
Sutton, G. W.
Sullivan, H. E.
Yates, A. du P.
Smith, S. B.
Taylor, H. N.
Shaw, F. C.
Pritchard, R. N. N.
Ballard, E. B.
Orr, J. S. B.
Smyth, C. E.

Instructor of Musketry:

Hine-Haycock, R. N., *Lt.*

Adjutant:

Dixon, H. G., *Capt. K.O. Borderers.*

Quarter-Master:

Crombie, J. *(temp. Q.-M. in Army.)*

Medical Officer:

Ramsay, J., M.D., *Surgn.*

4th Battalion.

Lieut.-Col. Commandant:

Pollard, W., *Hon. Col.*

Majors:

ps. Walford, T. S., *Hon. Lt.-Col.*
ps. Hartley, J., *Hon. Lt.-Col.*

Captains (10):

Irwin, H., *Hon. Major*
Callaway, A.
ps. Maude, W. W.
Fawkes, G. B., *Hon. M.*
Cubitt, C. C., *Hon. M*
Cumming, W., *Hon. M.*
Rogers, G.
ps. Irwin, R. B.
Turnor, W. W.
ps. Gott, W. H.

Lieutenants (15):

Lee, J. T.
Pilkington, H. W.
Pollard, W., Junr.
ps. Bennett, A. C.
Harrison, R. E.
Bulkely, H. C.
ps. Eden, H. H. F., *I. of M.*
Blackett, H.
Murray, H. S.
Stanhope, J.M.S.
Wingfield-Stratford, H.
Pilkington, W. H.
Mill, W. F. G.
Moir, W. B.

Instructor of Musketry:

Eden, H. H. F., *Lieut.*

Adjutant:

Hilliard, W. E., *Capt., S. York Regt.*

Quarter-Master:

Bayley, T. *(temp. Q.-M. in Army.)*

Medical Officer:

Walker, J., *Surgeon.*

The establishment of officers was somewhat changed. In place of one lieutenant-colonel, two majors, and eight captains per battalion, were established two lieutenant-colonels, four majors, and four captains, the two senior majors on parade being mounted. All majors were to command companies. The rank of 2nd lieutenant was abolished, and officers joined as lieutenants, but had to serve three years on the lower rate of pay.

1881.—Second Battalion.

During this year the 2nd battalion remained at Nowshera. The detachment at Fort Attock was relieved in January by "H" company, under Captain Morris, the strength being three officers, eighty-five sergeants, rank and file. This detachment was relieved in October by "C" company, under Lieutenant Christie, and subsequently by Major Harington.

At Nowshera the 2nd battalion started the regimental paper, *Our Journal*, which replaced the *Bucks Chronicle* (the 1st battalion paper while in the East). Lieutenant Lester was the first editor.

Nowshera was not a particularly lively station, nor yet a sporting one, but field sports and games varied the monotony at Fort Attock. In October, "F" company, under Lieutenant Vialls, was detached to Peshawar, but did not make a long stay there, as they were withdrawn in the December following.

1882.—First Battalion.

On 27th February the 1st battalion was moved from Portland to Aldershot, and, under command of Brevet Lieutenant-Colonel Dixon, proceeded in two divisions. The first division consisted of head-quarters and "B," "C," and "G" companies, with the married families, and travelled by special train from Portland to Aldershot, where it took over quarters in "C" and "D" Lines, South Camp. The second division, consisting of the remainder of the battalion, under the command of Major Van Heythuysen, moving in like manner, joined head-quarters at Aldershot on 2nd March.

It was much remarked at the time that the two subalterns* carrying the colours, who marched into Aldershot with the head-quarters, were both decorated with the Humane Society's medal.

During the course of the spring and early summer, various drafts joined from the depôt.

During the summer the Egyptian question assumed a most serious aspect. In July, Alexandria was bombarded. There was great excitement in Aldershot at the time, and the officers' club house was besieged by enquirers anxious to get the latest telegrams from the front. Owing to the outbreak of this war and the concentration of the troops for that expedition at Aldershot, the battalion, whose turn for service was far off, received but short notice of its removal from the station. Late on the evening of 27th July the order was received for the battalion to leave Aldershot by the 8 p.m. train the next day. The baggage was to be on the parade at an early hour the next morning. At 6.45 p.m. the battalion paraded, and was inspected by Lieutenant-General Sir Daniel Lysons, K.C.B., who highly complimented it on its good behaviour since its arrival at Aldershot, and expressed his regret at losing it. The Queen's (old 2nd Foot) placed their mess at the disposal of the officers, and the regiment marched from camp, headed by the band of the Queen's, and escorted by the greater part of that regiment, who—both officers and men—were very friendly and popular with the 14th. The battalion arrived at Bradford early next morning, and marched into the Bradford Moor Barracks about 9.30 a.m. Letter "D" company, under the command of Captain Schuyler, which had been left at Aldershot, consisting of one officer, three sergeants, one drummer, and thirty-one rank and file, proceeded on detachment to the Isle of Man.

On arrival at Bradford, a party of recruits, with Lieutenants Cox and Swaine, were sent to Strensall, near York, for musketry drill; and a party of casuals, under Captain Adye, followed soon after. The camp was formed for the first time during this year, and the detachment had a very wet, cold, and uncomfortable time there. The

* Lieutenants MacAdam and Ward.

battalion also furnished a detachment of two companies, under Captain Purkis, at York.

A draft of two sergeants and one hundred and ten rank and file, under Major Noyes, with Lieutenant Cheyne, left Bradford on 14th December to join the 2nd battalion. The winter was a severe one, and the barracks were much exposed. On the day the draft left, the snow lay several inches thick on the square, and the cold was intense. To the people of Bradford, where the head-quarters of a regiment had only recently been stationed for the first time, the band was a great attraction, and crowds of civilians escorted the regiment on every march out, the regimental pet deer also coming in for a large share of admiration. Field sports, save cricket and rifle shooting, were not to be obtained by the men in Bradford, but a regimental theatre was constructed in a dismantled cavalry stable, and proved a valuable addition to their means of amusement.

1882.—Second Battalion.

On 17th April, this year, the detachment at Fort Attock was again relieved, "A" company proceeding there, and Major Harington taking over command.

On 13th May a further detachment, consisting of "B," "C," and "F" companies, proceeded, by march route, to Cherat, where they remained till 19th October. The strength of this detachment was: Captain Ferrier, Lieutenants R. J. Roberts, A. N. Roberts, and W. C. Loudon, twelve sergeants, six drummers, and two hundred and seventy-six rank and file.

On 22nd October the battalion paraded for presentation of Afghan war medals. Colonel D. S. Warren distributed medals to all officers, non-commissioned officers, and men who crossed the frontier in 1878-80. During this year the theatre at Nowshera proved itself of special value, and brought forth a good deal of histrionic talent in the regiment.

The battalion was inspected on 13th and 14th March, 1882, by Brigadier-General Gordon, who, after a most searching inspection, expressed himself as thoroughly satisfied with the "excellent state of efficiency of the battalion," but also referred to the prevalence of drink, and the consequent crowded state of the guard-room and cells.

Nowshera was always held to be a slow, but not unhealthy, station, and the battalion during its long stay there found it extremely monotonous, and during the last cold weather unhealthy to boot. Cricket, polo, gymkhanas, theatricals, and shooting, however, tended to counteract the general depression, and lent a certain amount of interest to the dull routine of everyday life.

1883.—First Battalion.

In February a large party of recruits, under Lieutenant Cox, with Lieutenant O'Donnell, proceeded by rail from York and Bradford to Fleetwood for a recruits' course of musketry. The battalion remained in Bradford till June, when, the musketry camp at Strensall being reopened, the whole of the battalion, except the head-quarters, band, and a small party for duty, proceeded from Bradford to Strensall. The companies from York also joined this party, and the strange spectacle was seen of a battalion with eight companies on detachment separated from their head-quarters. Subsequently the companies were withdrawn to York, and occupied the Militia Barracks. The whole had been, during their stay at Strensall, under command of Lieutenant-Colonel Whitting. In April of this year the post of musketry instructor was abolished and a revised code of musketry rules issued.

The detachment at York, together with the depôt and permanent staff of the 3rd and 4th battalions 14th Regiment, was inspected most minutely by Major-General Cameron, C.B., on 1st and 3rd September. A most searching inspection was made, and each individual was called upon to drill and be catechised by the general.

On 21st September the companies at York left that station late at night, reaching Liverpool early next morning, and embarked on board H M.S. *Assistance.* The head-quarters left Bradford on the 22nd, reaching Liverpool at 9 o'clock the same morning, and also embarked. The battalion disembarked at the North Wall, Dublin, on 24th September, and proceeded by special trains to their respective destinations.

The following officers accompanied the battalion:

Lieutenant-Colonels Cosby and Whitting.
Majors Pigott, Morris, and Soote.
Captains Ruttledge, Adye, and Joyce.
Lieutenants Cox, Heigham, Swaine, MacAdam, Critchley-Salmonson, Crofton, Ward, O'Donnell, Ridge, Freer, De Berry.
Lieutenant and Adjutant Mills.
Captain and Paymaster Reid.

The head-quarters took over barracks at Castlebar, county Mayo, relieving the 45th Regiment, for which purpose Quarter-Master King had proceeded in advance with a small party. Detachments were found as under: Two companies, "C" and "F," consisting of Captain Ruttledge, Lieutenant Cox, Lieutenant O'Donnell, and Lieutenant Ridge, with six sergeants and sixty-five rank and file, the whole under the command of Major Soote, proceeded by train to Claremorris, billeted the night at that village, and marched the following day to Ballinrobe, eighteen miles distant, there to be stationed, taking over the Cavalry Barracks, which had been unoccupied for some time; the Infantry Barracks, some two hundred yards distant, were occupied by the married families. On 1st October a company, under Captain Joyce, proceeded to Westport, taking over quarters in the old Militia Barracks.

This detachment was recalled on 8th December, when a draft of two sergeants and one hundred and twenty-two rank and file, under the command of Major Moberley, 2nd South Lancashire Regiment, left Castlebar for Cork, *en route* to India to join the 2nd battalion 14th Regiment.

In addition to these detachments, a small musketry detachment was furnished at Athlone, where all casuals and recruits were exercised.

1883.—Second Battalion.

The 2nd battalion 14th Regiment was inspected by Brigadier-General E. Dandridge on 15th and 16th January, when a most searching inspection was made, and, at the conclusion, a most flattering speech by the inspecting officer.

On 3rd February the battalion was relieved at Nowshera by the 1st West Riding Regiment (33rd), and marched to Sialkote and Amritsar, the head-quarters and right half-battalion proceeding to the former and the left half to the latter station, leaving about sixty men behind unable to march, in this fever-stricken station. A large number of both old and new residents (most conspicuous amongst whom was Dr. Kenny, of the Army Medical Department, whose long connection with the battalion had endeared him to all ranks) accompanied the battalion part of the way on the march, to wish it "God speed."

At Attock, on the march down, Captain Penno's detachment, there stationed, rejoined head-quarters. The head-quarters party consisted of twenty-one sergeants and two hundred and seventy-one rank and file, and the left half-battalion of twelve sergeants and two hundred and sixty-one rank and file. The head-quarters were halted at Jhelum for five days, owing to the bridge of boats being swept away, and further on the march, at Wazirabad, Lieutenant Barchard, with his party of invalids, joined the battalion. The battalion separated on 26th February at Wazirabad, leaving Lieutenant Lester and about one hundred and eighty time-expired men in camp, to proceed by rail the following day to Meean Meer, *en route* for Deoalee.

The following officers marched from Nowshera with the battalion :

Lieutenant-Colonel Saunders.
Majors R. S. Lemon and F. W. Harington,
Captains C. D. Ferrier and T. W. L. Penno.
Lieutenants W. R. C. Baird, J. H. Burke, C. M. Lester, H. G. Vialls, J. C. Yale, R. J. Roberts, H. E. Watts, A. N. Roberts, C. H. Clements, W. M. Carpendale, and O. B. S. Shore.
Captain and Adjutant F. Kitchener.

On 20th March, Colonel Warren, C.B., completed five years as a regimental lieutenant-colonel, and was succeeded by Lieutenant-Colonel Saunders.*

* Major-General Edward William Saunders, retired. Ensign 14th Regiment, 28th March, 1854 ; lieutenant, 29th December, 1854 ; captain, 26th March, 1858 ; major, 5th July, 1872 ; brevet lieutenant-colonel, 10th January, 1880 ; colonel, 16th January, 1884. Retired, with rank of major-general, 23rd November, 1887.

The battalion had been sadly reduced in numbers by sickness during its stay at Nowshera, and by the further loss of one hundred and eighty men at Wazirabad, and on its arrival at Sialkote and Amritsar was but a shadow of its former self. However, during its stay at these stations, in spite of the great heat, and with the assistance of a large draft from home, by the end of the year it had considerably improved.

Amritsar was perhaps not one of the healthiest stations in India; but both Sialkote and Amritsar, with their gardens and the opportunities they offered for outdoor games and sports, were as the seventh heaven compared to Nowshera; and the comradeship of the Carabineers and the Battery L.A., and afterwards D.A. at Sialkote, with whom the battalion was on excellent terms, and who joined in making its stay there a pleasant one, contributed not a little to the general improvement in the health and condition of the battalion. Glad as the battalion was to leave Nowshera, the estimation in which it was held by all its comrades in the district was well expressed by the following complimentary remarks: "The Peshawur Valley, "holding two such regiments as the Queen's and the 14th, cannot "be a very bad place."*

1884.—First Battalion.

Lieutenant-Colonel Whitting was detached from head-quarters for the purpose of superintending the musketry drill at Athlone, where the battalion was exercised during this year. On 25th June the head-quarters at Castlebar was inspected by General Sir Thomas Steele, K.C.B., commander-in-chief in Ireland, who expressed his entire satisfaction with the general state of the battalion. Castlebar was not a favourite quarter with the men; the following extract from *Our Journal* will perhaps explain the reason: "Although there is "gas burnt in the town, there is none in the barracks, though the "town main runs past the barrack gates. An officer's kitchen does

Served with the regiment in the Crimea from 10th January, 1855, including the siege and fall of Sevastopol and the assault of 18th June, 1855 (medal and clasp and Turkish medal). Served in the field during the Maori War (medal).

* *Our Journal* (regimental paper), March, 1884.

" duty for a school-room for adults and boys, and a nine-men's room " for infant-school by day and reading and recreation room by night. " There is no skittle-alley, no piece of ground in barracks to play " cricket on; not even a decent piece of ground for quoit-pitching, " and a very poor ball alley."

The detachment at Ballinrobe was very much better situated, and all ranks were on friendly terms with the inhabitants.

Major-General Lord Clarina, commanding the Dublin District, inspected the battalion on 7th August, and expressed his satisfaction at its general appearance, working, and interior economy. He afterwards visited the Ballinrobe detachment. "D" company had replaced "F" company during the summer on this detachment, when "F" had proceeded to Athlone for musketry drill.

In September orders were received for the head-quarters to move to Galway, with detachments at Gort and Oughterard; consequently, on 25th September, an advance party, under Captain Penno and Lieutenant H. B. Critchley-Salmonson, proceeded by rail to Galway, and took over barracks from the 2nd battalion Suffolk Regiment (12th).

On the same day part of "H" company, under Major C. A. Morris, proceeded by rail to Gort, and took over barracks from a detachment of the 2nd battalion Suffolk Regiment.

"E" company also, under Lieutenants H. Walker and G. W. Swaine, marched from Castlebar to Ballinrobe on the 24th, where they remained for the night. The following day they marched to Cong, thence, by steamer, to Oughterard, in relief of a company of the 2nd battalion 12th Regiment.

On 26th September the Ballinrobe detachment left its station at an early hour, and proceeded, by march route, to Cong, under Major J. P. Soote and Captain Ruttledge and Lieutenant O'Donnell, from whence it proceeded, by steamer, to Galway, arriving there shortly before the head-quarters party from Castlebar.

The head-quarters, consisting of "B," "F," "G," and "H" companies, under Colonel Cosby and officers,* left Castlebar and pro-

* Majors B. C. Firman and S. J. Butler; Captain W. Joyce; Lieutenants C. H. Cox, T. R. Ward, and G. Freer. Captain and Adjutant Mills,

ceeded by rail, picking up the detachment at Athlone, under Lieutenants MacAdam and Bomford, and detaching the remainder of "A" company, under Lieutenant Crofton, *en route* to Galway, at Athenry, whence it proceeded to Gort.

On arriving at Galway, quarters were taken over in the Shambles and Castle barracks, both extremely small and uncomfortable, and the former condemned for many years as unfit for habitation. There were no married quarters in the Shambles Barracks, and no space either in it or in the Castle to drill upon, and the battalion was forced to hire a field a short distance out of the town for drill and pastimes. The Renmore Barracks (88th Regimental District) were about one mile out of town on the Oranmore road. The musketry range was on the sea-beach, and was limited to seven hundred yards in extent, but was a fairly good one. Previous to the battalion leaving Castlebar, it was with the greatest regret it saw its popular and hard-working paymaster (Captain J. Reid) leave for the Soudan. Captain Reid had served in the battalion upwards of thirty years, and in his various capacities, as sergeant-major, adjutant, and paymaster, had won the respect and esteem of all ranks.

During November of this year orders were received calling for volunteers for service with a force proceeding to Bechuanaland, South Africa, to quell a rising there. Captain and Adjutant Mills and Lieutenants Wemyss and MacAdam volunteered their services, which were accepted.

To fill the vacancies thus caused, the following militia officers were attached to the regiment for duty, viz.: Captain H. Trafford, 3rd battalion, West Yorkshire regiment, and Lieutenant H. Murray and Lieutenant J. M. Tottie, 4th battalion West Yorkshire regiment.

Captain Mills, on proceeding to South Africa, resigned the post of adjutant, and was succeeded by Lieutenant C. Heigham, whose place at the depôt was taken by Lieutenant G. W. Swaine.

A draft of seventy non-commissioned officers and men and four boys left Galway, under the command of Major Price, on 26th December, embarking at Portsmouth on the 27th in H.M.S. *Jumna* for conveyance to join the 2nd battalion in India.

1884.—Second Battalion.

The Amritsar wing joined head-quarters at Sialkote early in January, but as it left a garrison behind it for the fort, the addition was not very formidable. One hundred men, with eight officers, was the force that augmented head-quarters for the drill season.

For a short time the 2nd battalion was left in sole occupation of Sialkote, the other regiments, natives and artillery included, having been taken away for manœuvres &c. By April, however, Sialkote was itself again, and military life was all astir.

On 1st April the annual inspection took place. Lieutenant-General Sir Michael Biddulph, K.C.B., was the inspecting officer, and in the two days in which he saw the regiment, found time to get through a fair amount of work, and left well satisfied with all he saw. In this year "khaki" was extinguished as a parade dress, much to the delight of all, as it was hated by the men, and though perhaps it had considerable merits for its service work, yet its unsuitability for parade and its unsightly appearance at close quarters more than counter-balanced its merits.

On 20th April the Amritsar detachment was furnished again, "D," "G," and "H" companies, under Colonel Dixon, leaving head-quarters for that station.

1885.—First Battalion.

On 19th January the Oughterard detachment, under the command of Lieutenant Walker, was withdrawn, and rejoined head-quarters at Galway by march route.

The barracks at Galway having become overcrowded, instructions were received from the General Officer Commanding for the battalion to furnish a strong detachment at Castlebar.

Consequently, on 26th March, "B," "G," and "H" companies, with the following officers, Major Firman, Captain Trafford, Lieutenants Ward and Freer, proceeded by special train and relieved a detachment of the King's (Liverpool Regiment.)

During April the new valise equipment, 1882 pattern, was received, and found favour with the non-commissioned officers and men, who preferred it to the old on account of its being easier to put on and carry, and more comfortable to wear.

Difficulties with Russia in respect of the Afghan frontier having arisen, the authorities decided that a case of "imminent national danger" within the meaning of the Army Act, was established, and by Special Army Circular, dated 20th April, 1885, a large number of non-commissioned officers and men of the 1st Class Army Reserve were called up for service with their respective regiments. In the West Yorkshire Regiment, all who had been transferred to the reserve between 1st May, 1883, and 21st April, 1885, were ordered to report themselves at the regimental depôt at York by a given date. Some three hundred non-commissioned officers and men of the regiment were thus recalled to their colours. There not being sufficient accommodation for them with the home battalion at Galway, these non-commissioned officers aud men were retained for duty at York, and borne on the strength of the depôt.

On 23rd July "C" company relieved "A" company at Gort, under the command of Captain Ruttledge, Lieutenant Walker remaining at Gort for duty.

The musketry drill of the Castlebar detachment took place at Athlone during this summer.

The battalion was inspected at Galway on 11th August by Major-General Lord Clarina, who expressed his satisfaction at the state of the battalion. The numbers on parade at head-quarters were very small, viz., fourteen officers and two hundred and sixty non-commissioned officers and men. The major-general also inspected the detachments, that at Castlebar on the 8th, and Gort on the 12th.

On Lieutenants Wemyss and MacAdam returning from Bechuanaland, at the close of the year, the services of the militia officers (Lieutenants Tottie and Murray, 4th battalion), were dispensed with. Captain H. Trafford (3rd battalion), also with the permission of the commander-in-chief, was allowed to cease doing duty with the battalion on relief by Captain Daubeny, 4th West Riding regiment.

A draft of one hundred and sixteen non-commissioned officers and men left Galway for Cork, *en route* to India, to join the 2nd battalion at Sialkote.

The difficulties between England and Russia having been diplomatically overcome, a case of "imminent national danger" no longer existed, and the reserve men called up for duty with the colours, were retransferred to the Army Reserve. By a General Order, dated 24th August, they were permitted to proceed on furlough for forty-two days, with pay and allowance, and men within certain limits of service and unmarried were permitted to remain with the colours, if they so desired it.

On 31st August a disturbance took place in the town of Galway between the soldiers and civilians. Galway was always a place where such occurrences were not uncommon, and men walking about alone were not unfrequently badly mauled. The disturbance was much commented on by the local papers, which put a completely different aspect on the case to the actual facts.

Many of the windows in the Shambles Barracks were broken by the volleys of stones hurled by the mob. One stone found in the mess room was preserved as a memento for several years in the officers' mess.

It was in this year that the death roll of officers began, which within two years amounted to no less than nine, and obtained for the regiment the name of the "Unlucky Regiment." Strange to say, the deaths were mainly among the junior officers. Lieutenant McCausland, on 3rd April, was the first on this roll.

1885—Second Battalion.

The annual inspection took place on 29th, 30th, and 31st January. Brigadier-General C. J. East was inspecting officer. The convalescent depôt this year was at Kasauli, where festivities of all descriptions, including excellent theatricals and dances, enlivened the lives of those unfortunate (?) enough to go there.

On 27th October the battalion received orders to march to the Camp of Exercise at Umballa, and subsequently to Mooltan, there to be stationed. A gratifying order was published in Brigade Orders on the departure of the battalion. (See *Appendix Volume.*)

On 28th October the battalion left Sialkote camp at 6 a.m., with

Major A. J. Price — Lieutenant W. Hudson — Lieutenant and Quarter-Master R. Scott — Lieutenant L. N. Younghusband — Major F. W. Harington — Colonel E. W. Saunders — Lieutenant and Adjutant W. Fry — Lieutenant H. G. Vialls — Lieutenant J. C. Yale

Lieutenant F. C. Muspratt — Lieutenant H. E. Watts — Lieutenant W de S. Cayley — Lieutenant F. G. Trevor — Lieutenant P. J. Roberts

Officers of the 2nd Battalion at Delhi Manœuvres, 1885.

a marching-out strength of ten officers and two hundred and sixty-seven non-commissioned officers and men, including the following officers: Colonel Saunders, commanding; Major Harington, Captain Vowell, Lieutenant and Adjutant Fry, Lieutenants Cayley, Hudson, Muspratt, Younghusband, and F. G. Trevor, and Quarter-Master Scott. A depôt, composed of "F" company and attached men (chiefly the English draft), was left behind at Sialkote, with Majors Lemon and Price, Lieutenants Lester, A. N. Roberts, Robin, and H. Trevor. Taking the Jummoo road, and halting at Ghuenki, Dhuski, Gujránwala, Kamooki, Shahdera, Chubal, and Ghurenda, the head-quarters detachment reached Amritsar on 6th November, where it was joined by four companies, with six officers (Major Noyes, Captain Thurston, Lieutenants Vialls, R. Roberts, Watts, and Ridge) and one hundred and twenty-five men. The great fair at Dewali was in full swing. Marching from Amritsar on 7th November, by way of Pundiala, Reya, Girana, Kartapur, Jullundur, Phugwara, Philloor, Loodhiana, Doraka, Khana-ki-Serai, Bara, Ughana, and Moghal-ki-Sarai, the battalion reached Umballa on the 24th, where it was met by the bands of the 5th Fusiliers and 22nd and 88th regiments. It spent a very pleasant ten days at Umballa, and then marched away back by Moghal-ki-Sarai, and thence through Rajpura, reaching Delhi on 7th December. The remainder of the year was spent in manœuvres, under most unpropitious circumstances as regards weather, the rainfall being unprecedented for the time of year.

1886.—First Battalion.

At the beginning of the year it was decided to send out additional drafts to battalions in India, to bring up the European army to its full establishment. A draft of one hundred and twenty-five non-commissioned officers and men was despatched from the 1st battalion, on 13th February, to join the 2nd battalion in India.

On 6th March, Colonel T. P. Cosby, having completed four years service in command, retired, and was succeeded by Lieutenant-Colonel R. Whitting.*

---

* Colonel Reginald Whitting. Ensign, 62nd Regiment, 26th December, 1856;

After the departure of the Indian draft, the duty pressed very heavily on the men at head-quarters, Castlebar, and officers' servants and other permanently employed men had to take their turn of guards. On 27th May orders were received for the battalion to move from Galway and out-stations into Dublin.

On 1st June, 1886, Lieutenant-Colonel Whitting, Major Soote, Captains W. R. C. Baird and C. H. Cox, Lieutenants H. O'Donnell, G S. Bomford, and G. F. Phillips, Lieutenant and Adjutant Heigham, and Paymaster Captain Parkinson, with head-quarters and "A," "D," "E," and "F" companies, proceeded by rail from Galway to Dublin, and took over quarters in Ship Street Barracks, vacated by the 1st King's (Liverpool Regiment), on relief by the 2nd Border Regiment.

This move brought the battalion together again, as ever since it left Aldershot in 1882 it had been much broken up in detachments. "G" and "H" companies, under Major Firman and Lieutenant Alexander, from Castlebar, and "C" company, under Captain Burke and Lieutenant Walker, from Gort, arrived in Dublin the same day. "B" company remained at Castlebar until 2nd June, when it joined head-quarters. Captain King (quarter-master) proceeded to Dublin in advance of the battalion, and Lieutenant Crofton remained behind to take over and give over barracks.

An Army Circular of 1st June, 1886, raised the establishment of the battalion from six hundred to seven hundred and fifty rank and file. The actual strength on this date was four hundred and forty-nine men.

On 11th June sudden orders were received to send every available man to Belfast, to strengthen the garrison there, which had been called out in aid of the civil power. A detachment of three hundred non-commissioned officers and men, under command of Major J. P. Soote, with Captain Burke and Lieutenants Crofton, Freer, Alexander,

---

appointed to 8th King's, 1857 ; lieutenant, 8th King's, 23rd May, 1858 ; captain, 15th January, 1861 ; brevet major, 20th February, 1874 ; major, 7th March, 1877 ; lieutenant-colonel, 7th March, 1882. Exchanged to 14th in 1886 ; retired on half-pay in 1888. Served in the Indian Mutiny in 1858, first as staff-adjutant of a detachment in the Jugdespore jungles. Served in the campaign in Oude in 1858-59, including the capture of Sandu (medal). Died, in command of 14th Regimental District, 12th April, 1891.

and Phillips, proceeded by special train to Belfast early on the morning of 12th June. After a most unpleasant stay at Belfast—picquet-duty being incessant night and day, and the men without a change of clothing, owing to the suddenness of the order—the detachment returned to Dublin on 24th June.

The annual inspection took place on 13th, 14th, and 15th July, the inspecting officer being Major-General Young, C.B.

Twenty-five Enfield Martini rifles were issued to the battalion this summer for trial. It was a quick-loading rifle, using solid-drawn cartridges, having two back-sights, sliding wind-gauge, hand-guard, and safety-bolt, and was sighted to two thousand yards. The bayonet was fixed under the rifle, and the detachable quick-loader fitted on the right of the body of the rifle. It had many defects, and was not accepted as a Service pattern. A draft of one hundred and six non-commissioned officers and men, under Major C. S. Gordon, left Dublin on 8th October for Queenstown, where they embarked in H.M.S. *Crocodile* for India, to join the 2nd battalion.

This draft left the battalion again in a very weak state, and owing to a great deal of sickness among both officers and men, the duties pressed heavily on all ranks.

The musketry drill was performed at the Pigeon House Fort, where a party was always detached for the purpose; the ranges did not afford sufficient accommodation for the troops in garrison, and the shooting generally, owing to the haste, was inferior.

1886.—Second Battalion.

The battalion was occupied in the manœuvres at Delhi till 19th January, when a grand review before the Viceroy (Lord Dufferin) took place. On the 21st of the same month the battalion proceeded by rail from Delhi to Umballa, and thence, by route march, *viâ* Loodhiana and Ferozepore, to Mooltan, where it arrived on 24th February. The infantry lines at Mooltan are built in two lines, with *double-storied* married quarters. A small cricket ground, two racquet courts, large bazaar, good shooting, and some fishing added to the comfort of the regiment.

The annual inspection commenced on 6th March. Brigadier-General Purvis was the inspecting officer, and made a thorough inspection, which lasted several days.

The Mooltan brigade was inspected by Major-General Murray, C.B., on 20th March, and after the inspection the general addressed a few words to the regiment He said that the wing at Amritsar had always behaved remarkably well, and trusted the regiment would always keep up the good name which its detachment had won.

In November of this year Colonel Saunders, who had served in the 14th Regiment for thirty-three years, and had commanded it for the last four, was obliged, under the Royal Warrant in force, to retire from the regimental command. He made his final inspection on the 15th, and left Mooltan on the 16th, the whole battalion bidding him farewell at the railway-station.* He was succeeded by Lieutenant-Colonel Lemon.†

1887.—First Battalion.

On 18th February a draft of one sergeant and ninety-three men, under Captain Burke, left Dublin by steamer for Bristol, *en route* to Portsmouth for conveyance in H.M.S. *Serapis* to India, to join the 2nd battalion.

On 5th March sudden orders were received to immediately send a party of one sergeant and twelve rank and file, under a subaltern, to be trained in stable duties and riding, in order to qualify them for regimental transport duty. Lieutenant Phillips was selected for this duty. Subsequently orders were received for forty men to be trained, and transport equipment, as under, was issued to the battalion:

Three horses, two mules, one waggon, and one cart, with the necessary harness, and pack saddles for the mules. The men on transport duties were issued yellow cord breeches and blue putties.

---

* See "Farewell Order" in *Appendix Volume.*

† Colonel Richard Seymour Lemon. Entered the 14th Regiment, as ensign, 27th December, 1857; lieutenant, 8th October, 1858; captain, 15th June, 1870; major, 1st July, 1881; lieutenant-colonel, 7th March, 1886; commanded the battalion from 20th March, 1887. Retired, as colonel, 23rd December, 1889.

By Army Circular (special), dated 20th June, the establishment of the battalion was fixed as under:

Lieut.-Col.	Majors.	Captains.	Lieuts.	2nd Lieuts.	Adjutant.	Qr.-Mstr.	Total Officers.
1	3	6	8	4	1	1	24

Warrant Officers.	Q.-M. S.	S. I. M.	Col.-Sergts.	O. R. C.	Pay-Sergt.	Sergt.-Dr.	Arm.-Sergt.	Sergt.-Pion.	Sergt.-Cook.	Sergeants.	Corporals.	Drummers.	Privates.	Total all Ranks.	
11	1	1	8	1	1	1	1	1	1	24	40	16	690	812	5 Horses.

The actual strength in rank and file at the time of the publication of this order was five hundred and fifty-four.

This being the year of Jubilee of Her Most Gracious Majesty's reign, her pardon was extended to the following classes of military offenders:

(*a*) Deserters. (*b*) Fraudulently enlisted, as defined by section XIII. Army Act. (*c*) Absented themselves without leave from the regular, auxiliary, or reserve forces. (*d*) Improperly enlisted into the regular forces while serving in the army reserve. Provided they committed the offence before the date of the proclamation and surrendered within two months of that date at home, or within four months abroad.

In addition to the foregoing, all military prisoners undergoing sentence for desertion and fraudulent enlistment were released on 21st June, the unexpired portions of their sentence being remitted.

The Queen's Jubilee was celebrated in Dublin on 28th June, when a grand review of troops was held in the Phœnix Park. The battalion, under command of Major C. A. Morris, was in the 2nd infantry brigade, which was commanded by Colonel R. Whitting (lieutenant-colonel commanding the battalion). H.R.H. Prince Albert Victor of Wales represented Her Majesty, and with His Excellency the Lord Lieutenant of Ireland (the Marquis of Londonderry) inspected the troops on parade, which were under the command of H.S.H. Prince Edward of Saxe Weimar, general commanding the forces in Ireland.

The annual inspection by Major-General the Honble. J. C. Dormer, C.B., commanding Dublin District, took place on 28th July, at the conclusion of which the major-general expressed his entire satisfaction with the state of the regiment.

On 9th September a draft* proceeded, by rail, from Dublin to Queenstown, for embarkation in H.M.S. *Crocodile*, for conveyance to India to join the 2nd battalion 14th Regiment. On 5th November a large detachment was sent to the Curragh Camp, under command of Major Morris, consisting of "C," "D," and "E" companies.

1887.—Second Battalion.

On 6th January, General Purvis gave up the command of the Mooltan brigade, his successor being Brigadier-General Galbraith.

A most successful assault-at-arms was given by the battalion on 3rd, 4th, and 6th January. This was shortly followed by a gala week. The annual inspection took place on 12th, 14th, and 15th February.† General Galbraith was the inspecting officer, and made a most searching inspection. On 6th April, "A" company, under Major C. S. Gordon, with Lieutenant Towsey, proceeded to Dalhousie; strength, five sergeants and ninety-five rank and file.

This detachment rejoined on 21st October. General Galbraith left Mooltan on 29th March, but before leaving he made a final inspection of the battalion, and highly complimented it on the physique of the men and their clean appearance. There were several deaths this year among the officers of the battalion: Lieutenant F. G. Trevor was accidentally killed by a fall from a horse, Lieutenant R. J. Roberts died of fever in October, and Captain J. H. Burke almost immediately afterwards; Lieutenant Jenkinson also died during November, 1887.

1888.—First Battalion.

On 6th January a further draft of two sergeants and one hundred rank and file proceeded to Queenstown, to embark in H.M.S. *Serapis*, to join the 2nd battalion in India.

---

* Draft to India: Captain Swaine, 2nd Lieutenants Barrington and Lang, one sergeant, two corporals, and one hundred privates.

† *Our Journal* states that, of seven hundred and twenty-nine non-commissioned officers and men who landed at Bombay on 1st October, 1878, only ninety-nine (nominal roll given) were serving with the battalion in India on 1st January, 1887.

On 6th March Colonel R. Whitting completed six years as a regimental lieutenant-colonel, and was placed on half-pay; he was succeeded by Lieutenant-Colonel F. W. Harington.*

The position of 2nd lieutenant-colonel having been abolished, no successor was appointed to Lieutenant-Colonel Harington, but Major C. A. Morris becoming senior of his rank, took up the duties hitherto performed by the 2nd lieutenant-colonel, and relinquished the command of a company.

During March the detachment at the Curragh Camp was still further increased to nearly two hundred men.

During this winter a large inlying picquet of fifty privates, with a complement of officers and non-commissioned officers, was constantly held in readiness in the event of disturbances, which, however, did not occur.

A new method of carrying the great coat and cape was brought out for marching order. The coat was now placed under the flap of the valise, and the cape, rolled, was secured by straps underneath it.

On 20th June the Curragh detachment was moved back to Dublin, and took over quarters in Richmond Barracks, vacated by the Coldstream Guards.

The annual inspection took place on 12th July. Major-General J. Davis, C.B., was the inspecting officer, and addressed the men on parade, congratulating them on their good behaviour, which he described as "most marked."

The brown glove for officers was brought into wear during this summer.

Towards the latter part of August the companies quartered in Richmond Barracks were moved into Ship Street Barracks, while

---

* Colonel Frederick William Harington, half-pay. He entered the regiment as ensign, 28th May, 1858; became lieutenant, 3rd October, 1862; captain, 8th May, 1872; major, 1st July, 1881; lieutenant-colonel, 7th March, 1886; commanded a battalion from 7th March, 1888, to 20th March, 1890; colonel, 7th March, 1890; retired on half-pay, 7th March, 1892. Served in New Zealand in the war of 1865-66, including the actions of Kohewa, Gate Pah, and Tauranga (medal), and in Afghanistan in 1880, including the affair at Mazeena and the Kama expedition (medal.) Since these sheets were sent to press, Colonel Harington has succeeded to the command of the 14th Regimental District.

"F," "G," and "H" companies were moved from Ship Street into what had hitherto been known as "The Richmond Bridewell." This was a prison, which had been vacated for some time and handed over to the War Department. When the battalion took it over it was in a most wretched state of dirt and decay. The cooking had to be done in the open in camp kettles ; it was merely a prison in a state of dilapidation. This building, on being appropriated, was named Wellington Barracks, and means were rapidly taken to render it habitable and improve it generally.

On 7th September a draft of two sergeants and one hundred and two rank and file, with one woman and two children, under 2nd Lieutenant Moore, proceeded to Queenstown, for conveyance to India in H.M.S. *Euphrates*, to join the 2nd battalion.

In October orders were received for the battalion to move from Dublin to Fermoy. Accordingly, on 22nd of that month, the battalion* proceeded by rail to Kingstown, and embarked on board H.M.S. *Assistance* for conveyance to Queenstown. On arrival there the following day, it at once disembarked and proceeded by rail to Fermoy, and took over quarters in the Old Barracks, in relief of the Bedfordshire Regiment (1st battalion 16th.) On arrival at Fermoy, "F" company, under Captain Cox, proceeded by march route to Mitchelstown on detachment. Lieutenant Lowing (quarter-master) preceded the battalion to take over barracks, Captain Schuyler being left behind in Dublin to hand over those vacated.

The battalion, which had proceeded in two divisions, was met at the station by the band of the 1st battalion 6th Royal Warwickshire Regiment. This corps gave the battalion a most hearty reception, and the two regiments were on excellent terms during the whole of their stay in Fermoy. The Field Battery, R.A., stationed at Fermoy, was "Q 2," afterwards the 49th. They also fraternized with the regiment, and two pictures preserved in the officers' mess, presented by this battery, are a memento of the harmony and good feeling

---

* Lieutenant-Colonel Harington ; Major Morris ; Captain Cox ; Lieutenants O'Donnell and Trevor ; 2nd Lieutenants Hutchison and Drew ; and Captain and Adjutant Heigham were the officers with the battalion.

prevailing in the Old Barracks, Fermoy, during this period. The musketry practice was carried out on the hill of Corrin, about three and a half miles distant, a bleak and exposed position, by no means conducive to the best results.

1888.—Second Battalion.

Another death among the officers of the 2nd battalion occurred this year, that of Lieutenant J. G. Ridge.

The battalion still remained at Mooltan, gymnastics, theatricals, cricket matches, etc., enlivening the time.

On 21st April, "H" company, under the command of Captain Vialls, proceeded to the standing camp, Dalhousie. After a stay of three years in Mooltan, orders were received to proceed to Meean Meer. The battalion had made many friends, and it was with regret that the final orders were received. Several thefts of arms took place at Mooltan, which was a most unusual thing, for though such thefts are of frequent occurrence near the frontier, yet in a place like Mooltan it seemed almost impossible for the thieves to escape unnoticed. These arms were subsequently found in bales of merchandize packed on camels entering the Peshawur Bazaar. The rifles etc. had been stripped, to facilitate the stowage.

On 1st November the battalion, under command of Lieutenant-Colonel R. S. Lemon, proceeded by route march to Meean Meer, arriving at that station on the 22nd, where it was joined by "H" company from Dalhousie and the draft from England.

The strength of the battalion on the march was seventeen officers, twenty-three sergeants, and five hundred and forty-six rank and file.

During this month news of the death of another officer of the regiment was received, that of Lieutenant Crofton, who had served for eight years in the 1st battalion. His death brought up the total to nine within two years, a most unprecedented record in times of peace, and where no epidemic prevailed.

The troops assembled at Meean Meer for the camp of exercise consisted of three cavalry regiments, three batteries of artillery, and six infantry battalions.

The manœuvres lasted until the end of January.

1889.—First Battalion.

On 26th March, "A" and "H" companies, under Major Socte and Lieutenant Bomford, proceeded to Fort Carlisle, Cork Harbour, where they remained till they were relieved by a detachment of the Royal Warwickshire Regiment on 22nd August.

On 18th May, "F" company, under Captain Cox, rejoined head-quarters from Mitchelstown, its place being taken by "G" company, under Captain Grant-Dalton.

The battalion was inspected on 29th July and following days by Major-General H. J. Davies, the strength of the battalion then being twenty-four officers and six hundred and eighty-six non-commissioned officers and men, of which, however, four officers and one hundred and fifty-two non-commissioned officers and men were on detachment. The inspecting officer, on conclusion, was pleased to express his thorough satisfaction with the condition of the battalion.

On 3rd December, Captain Heigham's time as adjutant expired, and he was succeeded by Captain E. S. Schuyler.

On 6th December a draft of one hundred and three non-commissioned officers and men, under Major Price and 2nd Lieutenant Gardiner, left Fermoy for Queenstown, to embark to join the 2nd battalion in India.

1889.—Second Battalion.

The earlier months of the year were enlivened by athletic sports, in which the battalion took part. In the Lahore divisional assault-at-arms the battalion came well to the front, and the football team gained fresh laurels. The shooting and cricket teams were also very successful.

On 8th April, "E" company, under Lieutenants Towsey and Lang, and the convalescents, under Lieutenant Pearce, proceeded to the hills. On 18th April the head-quarters, with "D" and "F" companies and the band, left Meean Meer by train for Pathankote, *en route* to Dalhousie, for the hot weather. An amusing incident occurred by the way. On arriving at Pathankote, where tents were

Lieutenant W. S. Cayley | Lieutenant A. J. Stephen | Captain H. Midwood | Major C. D. Ferrier | Captain Grant-Dalton | Lieutenant F. W. Towsey | Captain and Adjutant Fry

Captain Vialls | Colonel Lemon | Captain Kitchener | Major Noyes

Captain Yale | Captain Swaine | Major C. S. Gordon | Lieutenant G. G. Lang

Officers of the 2nd Battalion, 1889.

to be issued for the camp and march, these much-needed articles were found not to be forthcoming. Side-walls, poles and lanterns were duly issued, but no tents. By dint of much telegraphing, the deficiency was made good next day, after the troops had passed the night in the open air. The standing camp was reached, after a fatiguing march of four days, through some of the grandest mountain scenery in the world.

The following officers were in camp with the head-quarters at Dalhousie: Colonel Lemon, Majors Gordon and Ferrier, Captains Fry and Swaine, Lieutenant Stephen, and 2nd Lieutenants Fisher and Bell. At Ranikhet were Lieutenants Towsey and Lang. The convalescents were under Lieutenant Pearce. The rest of the battalion remained at Meean Meer, in charge of Captain Thurston. After a very pleasant stay in the hills, the head-quarters at Dalhousie camp, and the detachments at Ranikhet returned to Meean Meer on 28th October.

The battalion left Meean Meer, by route march, for Umballa on 16th November, and arrived there on 5th December.

Some changes had taken place at the depôt at York during the year. A small sum had been laid out in planting creeping shrubs around the mess premises and otherwise improving the rather bare appearance of the barracks. The camp at Strensall was much enlarged, and huts were erected by the Royal Engineer Department regardless of expense. The volunteer battalions of the regiment displayed great activity, and the 3rd Militia Battalion marched in from Leeds during embodiment, to take part in some field manœuvres near York. The column, consisting of three hundred and fifty men, with two guns, advanced on York, and, reinforced by a squadron of the 10th Royal Hussars, proceeded to attack the York garrison on the Knavesmire. After a very interesting field-day, the volunteers were entertained by the garrison in the barracks. Next day they marched to Selby, and the day after returned by way of Pontefract (where further manœuvres took place) to Leeds.

The volunteer camp at Blackpool, in which the 1st and 2nd Volunteer Battalions of the Prince of Wales's Own (West Yorkshire) Regiment took part, proved a great success.

1890.—First Battalion.

The 1st battalion remained at Fermoy during this year. The Mitchelstown detachment was relieved on 17th March by "B" company, under Captain Ward. The battalion was inspected by Major-General N. F. Davis on 22nd July, when eighteen officers and five hundred and twenty-two non-commissioned officers and men were on parade.

The detachment at Mitchelstown was inspected on 11th July, and returned to Fermoy on 15th August, on relief by "C" company, under Major Penno. The detachment was recalled on 15th October, and the barracks at Mitchelstown were handed over. The departure of the military was much regretted by the inhabitants of Mitchelstown, with whom the 14th detachment were on good terms.

The general commanding the troops in Ireland, Lord Wolseley, inspected the 1st battalion at Fermoy on 2nd December, 1890, and made a minute inspection of the barracks. At the conclusion of the parade he expressed himself as very pleased with the steadiness of the men and their general appearance. There were twelve officers and four hundred and forty-nine non-commissioned officers and men on parade.

1890.—Second Battalion.

On 11th March, 1890, the 2nd battalion left Umballa for Dugshai, by route march, where they were to remain during the hot weather. The marching-out strength was fifteen officers and six hundred and six rank and file.

The head-quarters, with "A," "D," "E," "F," and "H" companies, returned to Umballa, under Major Noyes, on 11th November.

1891.—First Battalion.

The advent of the new year was marked by the departure for India of two strong drafts for the 2nd battalion, leaving the 1st battalion in a terribly weak state. The first of these drafts, consisting of one sergeant and one hundred and fifteen rank and file, under Captain

Old Barracks, Fermoy, 1890

Yale, with Lieutenants Daly and Colan, left Fermoy on the 2nd January. The second, consisting of one sergeant and one hundred and twelve rank and file, under Lieutenant Towsey, left Fermoy on the 21st February.

Orders were shortly afterwards received for the 1st battalion to move to England, and, after handing over the regimental transport to the 2nd battalion Worcester Regiment (late 36th Foot) at the Curragh, the battalion bade farewell to Fermoy, on relief by the 1st battalion Seaforth Highlanders (late 72nd Foot). Moving in two special trains, the battalion arrived at Cork on the morning of 17th April, and the baggage, which had been sent on ahead, being quickly put on board, left Queenstown in H.M.S. *Assistance* about 2 p.m., and disembarked at Langtree Dock, Liverpool, on 20th April. The following officers were with the battalion: Colonel Harington; Major Morris; Captains Adye, Ward, O'Donnell, and Cayley; Lieutenants Bomford, Alexander, Phillips, and Drew; 2nd Lieutenants Ryall, Tew, and Hall. On arrival at Liverpool, a detachment of thirty-seven non-commissioned officers and men ("F" company), under Captain Adye, proceeded to Castletown, Isle of Man, there to be stationed. The head-quarters, consisting of twenty-three officers and four hundred and seventy-three non-commissioned officers and men, proceeded by special train to Preston, where the barracks at Fleetwood were taken over from the 1st battalion Royal Sussex Regiment (late 35th Foot). Immediately on arrival, a detachment of thirty men, under Major Morris, Captain Mills, and Lieutenant Berney, were sent to Fleetwood, for duty there.

The major-general commanding made his marching-in inspection of the battalion, in review order, on 30th April, 1891, when there were present on parade seventeen officers and two hundred and six non-commissioned officers and men.

The Lee-Metford magazine rifle was issued to the 1st battalion in the spring of 1891, and although in the hands of well-trained marksmen it proved to be a highly accurate weapon, yet, on the whole, it was not received with favour, probably on account of the sighting, which was quite different from that of the Martini Henry rifle in use. The results of the musketry training were poor.

The new valise equipment, 1888 pattern, was also received, and was much approved by all ranks. The valise was much smaller than that in use, and more easily put on and off; the coat or cape could be more easily got on; the pouches held more ammunition; and the appearance altogether smarter.

A new-pattern serge frock was also issued this June, *without facings*, and with white shoulder-straps. It met with such disapproval on account of its unsightly appearance, the readiness with which the white shoulder-straps got dirty, and the impossibility of making the red collar and cuffs look clean after hard work, that it was condemned, and the white facings restored the year after.

The major-general made his inspection on three separate days, 4th July, 14th August, and 24th September, 1891.

Much to the surprise of all ranks, the battalion was ordered to furnish another draft in the autumn for the 2nd battalion, making three drafts, numbering over three hundred and twenty men, within a year. In accordance with this order, a draft, consisting of one sergeant, eighty-six rank and file, and two women, under command of Captain Vialls, left Preston for Portsmouth, *en route* to India, on 29th October.

The musketry instruction this year, with the exception of the band and recruits, who had fired before the move of the battalion from Fermoy, was carried out at Fleetwood, where Major C. A. Morris was commandant. "E" company, under 2nd Lieutenant Tew, proceeded to Fleetwood to relieve "D" company, under Lieutenant Carey, which rejoined head-quarters on 4th November.

1891.—Second Battalion.

At the assault-of-arms for the Sirhind District, held at Umballa, 19th February, 1891, the 2nd battalion won the following open competitions: Tug of war and bayonet exercise. The polo tournament, for Lord Airlie's Cup, was also won this year by a regimental team, composed of the following officers: Captain H. G. Vialls, Captain J. C. Yale, Lieutenant A. G. Stephen, and Lieutenant G. E. Lang.

2nd Lieutenant J. S. Bartrum — Lieutenant T. H. Berney — Lieutenant G. F. Phillips — Lieutenant C. B. Tew — Lieutenant and Quarter-Master W. Lowing

2nd Lieutenant C. Ryall — 2nd Lieutenant W. M. Hall — Captain T. R. Ward — Lieutenant H. B. Trevor — Lieutenant D. H. Alexander — Lieutenant G. Barry Drew — Captain H. O'Donnell

Captain W. S. Cayley — Captain and Adjutant E. S. Schuyler — Major C. A. Morris — Colonel F. W. Harrington *(commanding)* — Major A. Ruttledge — Major G. Grant-Dalton — Captain C. G. Adye

Captain C. H. Lester — Lieutenant W. S. Carey

Officers of the 1st Battalion, at Fermoy, 1891.

On 10th March the head-quarters of the battalion, with five companies, under Major Noyes, numbering fifteen officers and five hundred and nine non-commissioned officers and men, left Umballa, by march route, for Dugshai, where they arrived on 15th March.

On 20th March, 1891, Colonel R. S. Lemon, having completed his term of service in command of a battalion, was placed on half-pay. He was succeeded by Colonel A. W. Noyes.*

On 27th October the battalion again changed quarters, leaving Dugshai for Lucknow, *en route* for Sitapur and Benares. On 13th December it arrived, with a strength of twenty officers and nine hundred and nine non-commissioned officers and men, at Lucknow, where it took part in the manœuvres.

1892.—First Battalion.

The battalion was inspected by Major-General J. H. Hall, at Preston, on 8th March, 1892, and a medal was presented to Private C. Brown, of the band.

Colonel Harington, having served six years as lieutenant-colonel, was retired under the royal warrant, on 7th March, and was succeeded by Lieutenant-Colonel C. S. Gordon, from the 2nd battalion. A draft of fifty-one non-commissioned officers and men, under Lieutenant Pearce, left for India on 10th March.

On 25th May the battalion proceeded, by march route, to the new camp at Chipping (twelve miles distant), "D" company, under Captain Mills, having preceded it to pitch the camp. The battalion remained under canvas until 19th July, when it returned by march to Preston. During its stay in the camp the battalion was exercised in field training, musketry, and field works. The weather was exceedingly unfavourable and the ground swampy, but in spite of these drawbacks there was little sickness. The battalion was inspected by the major-general on 15th July, when twelve officers and two hundred and ninety-eight non-commissioned officers and men were present on parade. Field firing was practised on the occasion.

---

* Now lieutenant-colonel 2nd battalion, commanding at Sitapur.

The annual inspection took place on 29th September, and the battalion was complimented on its appearance.

During the summer the detachments had been relieved, "F" company, in the Isle of Man, by "H" company, under Captain Cayley and Lieutenant Berney, and the company at Fleetwood by "G" company, under Major Grant-Dalton and Lieutenant Ryall. "B" company, under Captain Ward, furnished the occupation party at Preston and Fleetwood while the battalion was away at Chipping. Musketry for casuals and recruits took place at Fleetwood.

In October, orders were received to proceed to Aldershot. The detachments were recalled on 8th-9th November, and one hundred men were sent off as an advance-party, under Major Mills and 2nd Lieutenant Ingles, to take over the Salamanca Barracks (old C.I.B.) from the 48th Regiment. The battalion left Preston four hundred and eighteen strong, in two trains, the officers accompanying it being as hereunder:

Lieutenant-Colonel Gordon.
Major Ruttledge.
Captains Critchley-Salmonson, Ward, Bomford, and Alexander.
Lieutenants Berney, Barry-Drew, and Tew.
2nd Lieutenant Spry Pellew.
Captain and Adjutant O'Donnell.

On arrival at Aldershot the battalion was brought into the First Army Corps, and, with the 2nd Royal Welsh Fusiliers and 1st Cameronians (26th Foot), formed the Second Infantry Brigade, under command of Major-General Utterson.

The barracks, when taken over, were found to be old, dirty, and infested with vermin.

The battalion was inspected by Major-General Utterson on 16th November. On 6th December, a draft consisting of Major Adye, Captain Cayley, Lieutenant Lang, two sergeants, two corporals, one hundred and forty privates, two women, and three children, proceeded to India, to join the 2nd battalion.

The battalion was inspected by Lieutenant-General Sir Evelyn Wood, V.C., commanding the Aldershot division, early in December.

1892.—Secon Battalion.

After the conclusion of the 1891 manœuvres at Lucknow, the battalion, under command of Lieutenant-Colonel A. W. Noyes, returned to its old quarters at Sitapur, Bengal. A wing, under command of Major A. J. Price, was detached to Benares.

The head-quarters of the battalion remained at Sitapur through the year.

2nd-Lieutenant C. Mansel-Jones — 2nd-Lieutenant A. C. Daly — Lieutenant E. H. S. Cullen — Captain H. G. Vialls — Lieutenant A. J. Stephen — Lieutenant F. W. Towsey

Captain and Quarter-Master R. Scott — Lieutenant G. F. Gardiner — Captain and Adjutant H. E. Watts — Lieutenant-Colonel A. W. Noyes — Major C. S. Gordon — 2nd-Lieutenant E. P. C. Purchas — Major A. J. Price — Captain J. C. Yale

Lieutenant R. F. Lush — Lieutenant G. G. Lang — Captain and Paymaster G. W. Swaine

Officers of the 2nd Battalion, in India, January, 1892

NUMBER OF RECRUITS OBTAINED FROM YORKSHIRE ANNUALLY

SINCE THE

INTRODUCTION OF THE TERRITORIAL SYSTEM, 1ST JULY, 1881.

Year	Number
1881	98
1882	170
1883	185
1884	197
1885	218
1886	183
1887	150
1888	186
1889	176
1890	157
1891	121
1892	175
Total	2,016

FURNISHED ANNUALLY BY THE THIRD AND FOURTH BATTALIONS.

Year	Third Battalion.	Fourth Battalion.	Total
1881	12	4	16
1882	21	18	39
1883	24	24	48
1884	31	23	54
1885	51	22	73
1886	40	29	69
1887	23	16	39
1888	41	27	68
1889	43	41	84
1890	47	49	96
1891	35	36	71
1892	75	48	123
	443	337	780

NOTES ON THE COSTUME AND EQUIPMENTS.

NO evidence appears to exist regarding the very early costume of the Regiment. Although raised in 1685, it unfortunately was absent from the camp at Hounslow Heath, June 30th, 1686, on which date an interesting account was taken, giving full details of the uniforms of the various regiments then present. (*Vide Antiquarian Repertory*, i., 230.)

It may be presumed that its uniform was red. [Of the twelve infantry regiments whose dress is described, one of them—the Earl of Bath's, afterwards the 10th Foot—had blue coats.] With regard to the facings, they may possibly have been buff; but, in the absence of any information, this cannot be adopted with certainty.

"Facings" really meant the lining of the coat—in those days a very easy-fitting frock, single-breasted, without any collar, but with voluminous skirts; the sleeves, made long, were turned up, thereby exposing the lining, which also occasionally shewed when a few of the coat buttons were unloosed, or the motion of walking waved back the skirts.

Loose knee-breeches, with strong stockings and shoes, and slouched felt hats—not as yet cocked, or turned up in the brim,—completed what may have been the earliest costume.

In the absence of exact information, we must pass on to the year 1742, when a work was produced, entitled *The Cloathing of His Majesty's Troops, etc.*—(a copy of this rare book is in the British Museum)—exhibiting coloured drawings or plates of a private soldier of every regiment in existence at that date.

That of the 14th regiment is carefully and accurately reproduced in the illustration, p. 26. The coat, without any collar as yet, is still an easy-fitting garment, with skirts so large that, when the corners were hooked back, they still formed a good covering for the limbs; the chest had a lapel on each side, of buff cloth, with white metal buttons; the cuffs, also of buff, are now permanent institutions of their kind, and not the sleeves temporarily turned back, as formerly. The waistcoat long, and furnished with sleeves, of red cloth; the breeches of the same hue; the belts broad, of solid buff leather. The arms consisted of the heavy musket then in vogue, with a sword, or hanger, and bayonet suspended in a frog from the waist-belt.

For some reason the men wore no lace round the button-holes, or upon the coat-seams; out of some fifty line regiments, only seven were without this lace, which played an important and effective part in the soldier's appearance; it was supposed to give strength to the button-holes and seams, and undoubtedly served, from its great variety of patterns, to be—together with the varied hue of the regimental facing—the main feature in distinguishing one regiment from another at a time when regimental numbers, though allotted, were not placed upon any of the equipments.

The next information is to be found in an oil painting of a grenadier of the regiment, now at Windsor Castle, one of a complete series executed for the King by David Morier, an artist of considerable celebrity, about 1750 or 1752. The illustration, p. 34, represents the figure in question. The coat remains much the same as that last described, in 1742, but the button-holes, seams, coat-edging, and waistcoat trimmed with the new regimental lace, of which patterns had to be lodged at the office of the Board of General Officers, London. This lace, the pattern of which was peculiar to the regiment, and very distinctive, was about three-quarters of an inch wide, of white worsted, with a red stripe down one edge and a yellow one down the other, the centre occupied with a zigzag blue stripe, the cuff-lace being rather wider in pattern. This is the first appearance of partly-coloured lace on the men's uniform, though it had been used in other regiments for some time; it never was worn by sergeants, who were distinguished in their turn by wearing white tape lace, and a red and buff striped sash round the waist. Perhaps the most remarkable feature is the high mitre-shaped grenadier cap; all grenadiers wore this head-dress; originally the tallest and strongest men were selected, the high caps being for the purpose of giving them still greater height.

Evelyn, in his diary, 1678, mentions the new sort of soldiers, called grenadiers; they had furred caps, with sloping crowns, like janizaries, which made them look very fierce.

According to the warrant of 14th September, 1743, the front of the grenadier caps was to be of the same colour as the facing of the regiment, with the king's cipher (embroidered) and crown over it; the little flap (that in front) to be red, with the "White Horse" (the badge of the House of Hanover) and motto. This must not be taken as a special grant to the 14th only; it was universally used by the grenadiers of every regiment.

The back part of the cap was to be red, the turn-up also of the colour of the facing, the number of the regiment in figures on the middle part behind. This is the first mention of the regimental number on the equipment, and, until the authorization of the number on the buttons some eighteen years later, the backs of the caps of the fifty or sixty grenadiers was absolutely the only place in which it was exhibited. It was, however, a leading feature on all regimental colours made after the date of this warrant. The brass match-case may be observed fastened to the front of the shoulder-belt; it was a cylinder, some four inches long, pierced with holes to allow air for the ignited slow match inside; not that anything of the kind was required now, for the throwing of hand grenades had been discontinued for many years, but as an ornament it lingered on, a special mark of distinction for the grenadier company for more than half a century yet.

Unless the evidence of family portraits should be forthcoming, there is no exact information as to the officer's uniform in 1751. A wide-skirted scarlet coat, with buff facings, like that of the men, would be worn, edged all around, however, with narrow silver lace; sword suspended from a waistbelt under the waistcoat; a crimson silk sash over the left shoulder, and a silver aiguillette on the right shoulder. The battalion officers, like the men, had three-cornered cocked hats, but laced with silver, displaying the black cockade of the reigning house of Hanover on the left side. Grenadier officers wore a cap in shape like the men, but handsomely embroidered in gold and silver, and, in common with all commissioned officers, a silver gorget, fastened round the neck, of rather larger shape than that depicted as worn by the officers in 1826. (See illustration, p. 277.)

The earliest recorded inspection of the regiment took place at Stirling, 27th April, 1750, and Lieutenant-General Churchill, the inspecting officer, reports, amongst other things, that "the sergeants "have an odd custom of pulling off their hats when under arms at a "review, which I have ordered to be altered." The officers took off their hats when saluting, and it appears the sergeants followed their example.

A most picturesque appearance must have been presented by the regiment on parade, dressed and equipped according to the regulations of 1751. On the right the grenadier company, three deep, distinguished by its caps; then the battalion companies, nine in number, also in red, with buff facings, their three-cornered hats edged with white lace, formed according to the seniority of their captains—the colonel's company on the right,* lieutenant-colonel's on the left, the major's again next that of the colonel's, and so on, the junior captains' companies being in the centre. The officers in front, espontoon in hand, the latter an ornamental kind of half-pike, whilst the colonel and lieutenant-colonel were on foot in front of all, just in advance of the colours, also with espontoons. The pioneers, with aprons and buff pointed caps, were on the right of the grenadiers, and the drummers formed half on the right of the colonel's company and half on the left of the lieutenant-colonel's company; fifers may, or may not, have been present; they were introduced in 1747, but at first did not become general, or perhaps did not take to the instru-

---

* The colonel, lieutenant-colonel, and major were captains of companies, that of the colonel being commanded by an officer termed the captain-lieutenant; this arrangement, traceable from the time of the civil war, when regiments were first formed, held good in the British army for fifty years afterwards.

ment, for we gather from the inspection return of the 20th regiment in 1750, that "three drummers were absent, learning to play on the "fife."

The drummers would doubtless present a fine appearance, dressed in buff coats, with red facings, and profusely laced with the regimental lace before described; their conical caps, faced with buff, rather lower than those of the grenadiers, and ornamented with trophies of drums and flags, in coloured worsted crewel work.

A few years afterwards the uniform of the officers must have been changed, judging from a remark made by the inspecting officer at Winchester, 26th October, 1765: "The officers' uniforms, faced buff, "with silver button-holes embroidered." The edging of silver lace on the coat seems to have disappeared.

At this date a considerable alteration was taking place in the uniform of the infantry; the red waistcoats and breeches were abolished, and, in the case of the 14th, they were ordered to be buff, and by a royal warrant, dated 14th October, 1765, black bearskin caps were introduced, the regiment permitted, as a special privilege, to have caps fronted with red, with the motto "*Nec aspera terrent*" and the "White Horse" in white metal; the drummers' caps were, at the same time, ordered to be of "white" bearskin, fronted in the same manner.

It being considered desirable to have some additional form of distinction, or difference, between regiments other than that afforded by the colours of the facings, and the varied patterns of the regimental lace, a warrant was issued, September 21st, 1767, requiring that the "number" should appear on the buttons, hitherto quite plain.

A MS. work in the Prince Consort's library, Aldershot, gives a coloured sketch of a grenadier of every regiment, dated 1768, immediately after the period of the before-mentioned change.

The coat appears a more closely fitting garment than before, only just meeting across the chest, leaving the buff waistcoat exposed, the turned-back chest lapels very much narrower, serving little more than to show off the lace looped button-holes; a turn-down collar, or cape as it was called, appears for the first time, fastened down at each end with a button and loop of lace to the top button of the lapel.

The private soldier's lace became now a matter of regulation, the *Official Army List* of 1769 giving, for the first time, particulars of the various regimental laces, under the heading "Succession of Colonels;" that of the 14th being described as white, with a black and red worm. There were various methods of wearing the lace loops, some were square-headed, some pointed, others of the bastion or "flowered" shape; according to this MS., those used by the 14th were square-headed, and set on at equal distances, and were so worn until quite the end of the century, when they were set on by pairs, two and two and so continued as long as loops were worn on the coat of the private soldier, namely, until 1855.

The black bearskin caps in the MS. show the regulation plate worn alike by the grenadiers of every line regiment, namely, black japanned metal, thereon the king's crest (lion, with crown over), surmounted by a scroll or label, bearing the Hanoverian motto, "*Nec aspera terrent;*" whether the specially granted red front of 1765, with the "White Horse," had disappeared, or whether in this case the artist makes an error, must remain a difficult question to clear up;* the regimental tradition, however, is very strong, that the drummers continued to wear the white bearskin caps.

The royal warrant of 19th December, 1768, is very explicit, and full of detail on every point connected with the uniform of the infantry, both men and officers; the costume of the latter was as follows:—scarlet coats lapelled to the waist with buff cloth, lapels three inches wide, fastened back with silver buttons (having the regimental number), and placed at equal distances; the cape, or collar, turned down and fastened by one button-hole to the top button of the lapel. Small round buff cuffs, three and a half inches deep, having four buttons by pairs, cross pockets, in line with the waist, with four buttons, also in pairs; skirts lined and turned back buff.

The silver embroidered button-holes of 1765 had disappeared; in

---

* Since the year 1814 the following appears in the *Official Annual Army List*, under the heading "14th regiment:" "permitted to bear the 'White Horse' and "motto on a red ground on the front of their caps."

fact the regiment, as far as the officers were concerned, was what may be termed a non-laced one, and so continued until 1829; that is to say, instead of silver laced, or silver embroidered button-holes, perfectly plain ones would be used. The silver aiguillette of 1751 was superseded by silver epaulettes; officers of the grenadier company wore an epaulette of silver lace on each shoulder, battalion officers one on the right shoulder only; buff waistcoat and breeches, black linen gaiters with black buttons; crimson sash, tied round the waist (until recently worn over the shoulder); a silver gorget, with the royal cipher engraved upon it, fastened to the neck with buff-silk rosettes and ribbons; hats laced with silver, and the usual black cockade.

Grenadier officers wore the black bearskin cap like the men; they carried fusils, and had white shoulder belts and pouches.

Battalion officers carried the espontoon. Sergeants had buttons of white metal, and narrow loops of plain white tape, hats laced with silver, and crimson and buff sashes; they carried swords and halberts, the latter a light ornamental kind of battle-axe, with a long shaft.

The uniform of 1768 remained unchanged until about 1790. Cross-belts of a lighter make were, however, introduced before the American War of Independence, that is, the bayonet belt removed from the waist to the shoulder, and a brass breastplate, as an ornament, affixed to the latter. The officers' swords were also ordered to be worn suspended from a white shoulder belt, over the right shoulder, with an oval silver breastplate.

A light company was added to every infantry regiment in 1771, jackets, short gaiters, and a leather cap being worn; the officers and sergeants carried fusils and pouches, and in 1784 fusils were ordered to be carried by sergeants of the grenadier company, it being found desirable, in view of their peculiar duties, that they should discontinue the use of the halberts, so as to be able to take some part in the skirmishing and light infantry movements then being developed. The grenadier and light companies received the name of flank companies, and were frequently detached for light operations.

In April, 1786, battalion officers were ordered to discontinue the use of espontoons. Attention must have been directed to this weapon, for at the inspection of the regiment in Jamaica, January

23rd, 1784, the inspecting officer remarks that "the officers were "armed with swords and espontoons." The same order directed officers to wear gaiters like the men on all duties, except on the march, when boots might be worn. Regiments having adopted various modes of dressing the hair, His Majesty was pleased to direct that officers and men in general, when under arms or on duty (the grenadier and light infantry companies, when they wear their caps, excepted), were for the future to wear the hair clubbed. The non-commissioned officers and men to have a small piece of black polished leather, by way of ornament, upon the club. The whole to wear black leather stocks.

On 10th December, 1791, an order was issued that the effective field officers of all regiments should, in future, wear two epaulettes (so far they had only worn one); the officers of the flank companies, who already wore two, were, as a distinguishing mark, to have a grenade, or bugle horn, embroidered upon each.

The regiment arriving home from the West Indies this year was inspected at Portsmouth, on 21st June, by Major-General Hyde, who remarked: "The new clothing not having time to be made since the "arrival of the regiment, they appeared in linen waistcoats and long "trouzer-breeches of the West Indies; the bearskin caps not yet "delivered."

Early in 1792 the halberts, so long carried by the sergeants, were laid aside, and pikes substituted; they had a plain spear-head, with a steel cross-bar just below, the sergeants of the flank companies, however, retained their fusils. Very shortly afterwards the officers of the grenadier and light companies were ordered to discontinue the use of fusils, together with the cross pouch-belt, the sword to be their only weapon.

The illustration, p. 52, represents a private soldier of the battalion companies in 1790, the details taken from Dayes' series of coloured engravings of regiments of the army.* The loops of the regimental lace, shewing the black and red worm, are very apparent.

---

* Copy in the British Museum, and also in the Prince Consort's library, Aldershot.

In the royal library, Windsor Castle, is preserved a series of sketches by De Loutherbourg, R.A. They were evidently intended as studies for some large battle piece, and amongst them is one giving details of the uniform of Colonel Welbore Ellis Doyle, who commanded the 14th at Famars. Unfortunately the artist's work does not extend to the face or head, but still it is sufficient to enable a good idea to be formed of an officer's uniform of that period. The coat is cut very much like that of the private soldier just described, but the stand-up collar perfectly plain, and devoid of any button or ornament; the buff lapels narrow, with ten silver buttons, placed two and two, and narrow imitation buff silk button-holes; a silver gorget worn round the neck, of the accompanying pattern, fastened with two rosettes of buff silk; silver breastplate upon the white shoulder belt for the sword, of the design as in the illustration; the buff cuff is narrow, and has four silver buttons, placed two and two, small in size, rather convex in shape, with the number 14 in the centre.

OFFICER'S SILVER GORGET, 1793.

OFFICER'S SILVER BREASTPLATE, 1793.

This is the earliest indication of the regimental custom of wearing the buttons by twos, or in pairs, which lasted until 1855.

Towards the end of the century, following the Prussian fashion, the coats for all ranks were fastened down to the waist, completely hiding the waistcoat. In the case of the officers, the lapels were to be continued to the waist, so as to button over occasionally (making a double-breasted coat in fact), or to fasten close up the front with hooks and eyes, in which case the buff lapels would be exposed. The jacket for the rank and file was single-breasted, having ten buttons and loops of regimental lace arranged in pairs across the chest.

The warrant of 1796 directed that officers' swords were to have a brass guard, and the sword-knot to be crimson and gold, in stripes; the gorget (hitherto silver in the 14th) was for the future to be gilt, with the king's cipher and crown engraved upon it.

Another warrant, dated April, 1799, ordered that officers and men of infantry (except the flank companies) should wear their hair queued, to be tied a little below the upper part of the collar of the coat, and to be ten inches in length, including one inch of hair to appear below the binding.

The cocked hat worn by the men was discontinued in 1800; a cylindrical shako took its place, ornamented with an oblong brass plate in front, thereon the king's crest; a red and white worsted tuft was fixed in front, rising from a black cockade. The officers, however, retained their cocked hats, wearing them across, even with the shoulders, ultimately changing the position, and wearing them—during the early stages of the war in the Peninsula—fore and aft.

Chevrons for non-commissioned officers were introduced in 1802. Sergeant-majors to have four, sergeants three, and corporals two on the right arm; the first, of silver lace; the second, plain white tape lace; and the third, of the regimental lace, with the red and black worm. The staff-sergeants always wore silver lace, and continued to do so until 1855, notwithstanding the fact that the officers were ordered to wear gold lace in 1831.

To the joy of the army at large, the troublesome queue was abolished in 1808, at the same time the hair was ordered to be cut short in the neck, and a small sponge added to the rest of the soldier's necessaries, for the purpose of frequently washing his head.

December, 1811. Infantry officers were authorized to wear a cap similar to that worn by the men, and a regimental jacket to button across (double-breasted, in fact), a grey overcoat, also grey pantaloons or overalls, as the private soldiers. This was the service dress used by officers during the later Peninsular and Waterloo campaigns.

1814. The general costume of the regiment was as follows:—

officers, long-tailed scarlet coats for parades, levees, etc., buff lapels buttoned back by ten silver buttons, and plain narrow buff silk button-holes in pairs, the silver button now worn being slightly convex, with the numeral 14 within a neat circular device, as example herewith. The collar of buff, with one button and silk button-hole; cuffs with four, also in pairs, cross pockets (in line with the waist) the same; skirts turned back buff, the skirt ornament (placed where the points of the turnbacks met) being—as may be gathered from old lacemen's books—silver embroidered stars, on scarlet, 14 in the centre. White breeches, with black leggings, worn for home, and grey trousers for active service.

OFFICER'S SILVER BUTTON, 1800-14.

The coloured illustration, p. 104, represents an officer in the service dress of 1814 and 1815.

The long straight sword, black leather scabbard, gilt mounted, with crimson and gold sword knot, was worn, according to regulations, suspended in a frog from a white leather shoulder-belt; in the centre of the latter the regimental silver breastplate, as pattern annexed—the plate of burnished silver, the wreath and figure 14 in raised silver-work; it probably succeeded the older oval plate about 1808.

OFFICER'S SILVER BREASTPLATE, 1800-14.

Officers of the light company carried the curved light infantry sabre, suspended by slings from the shoulder-belt (on service, however, this weapon was sometimes used by other officers), and a crimson sash twice round the waist, tied on the left side.

Officer's rank distinguished by the silver epaulette, according to the instructions laid down in the general order of February, 1810. Field-officers wore two, a colonel having a crown and star in gold on the strap; a lieutenant-colonel, a crown; major, a star; captains and subaltern officers (including the quarter-master) wore a single silver epaulette on the right shoulder. Officers of the flank companies, two silver shoulder wings, with gold grenades or bugles thereon respectively. The adjutant wore, in addition to his epaulette, an epaulette strap, without fringe, on his left shoulder.

Paymaster and surgeon wore the regimental coat, single-breasted, with silver buttons, in pairs, with long imitation red silk button-holes, or, more properly, cords across the chest; no epaulettes or sash, the sword suspended by a plain waistbelt under the coat.

Private soldiers had single-breasted red cloth jackets, laced across the chest with square-headed loops of the regimental lace (with red and black worm) four inches long, set on in pairs. The same lace round the high buff collar (shewing a white frill in front), also round the buff shoulder straps, terminating in small white shoulder tufts;* in the flank companies, with a wing of red cloth trimmed with stripes of regimental lace and edged with an overhanging fringe or wing of white worsted; gaiters and breeches, but grey trousers, for service.

Sergeants dressed like privates—in finer cloth, however—having the chevrons of their rank on the arm, which, together with their coat lace, was of fine white tape; sash, crimson with a buff stripe. They carried a straight brass-hilted sword in a shoulder-belt, their other weapon being the pike.

The head dress for officers was a light cylindrical shako of felt, with black leather peak, a black cockade, and small red and white tuft on the left side (green or white for light infantry or grenadiers respectively), a gilt oval plate in front, surmounted with a crown, thereon the cipher G.R., with the regimental number; across the front, a festoon of crimson and gold cord, with tassels on the right side. That for the men of similar make, but the cords and tassels of white worsted. On service, cap covers were worn, of black japanned material.

The late Lord Albemarle informed the writer of these notes that all ranks wore cap covers at Waterloo. His memory was vivid as regards his own case, for he had only received his outfit some few days before the battle, and was distressed by the unusual weight of the head-dress with its waterproof cover; so, without more ado, he

---

* *Vide* coloured illustration, p. 104.

cut off the heavy gold bullion tassel and festoon, and placed them in his pocket, feeling certain that their loss would not be noticed—nor was it!

He also stated that many of the staff officers wore covers to their cocked hats on the 18th June, not probably having had time to change them after the heavy rain of the preceding night; and he well remembered that, whilst the regiment was in square, a mounted staff officer stayed some time conversing with Colonel Tidy; his departure was hailed with delight by the junior ensigns in the centre of the square, they having the impression that the light reflected from the glazed cover might serve to attract the fire of the enemy's artillery.

Further, that the battalion (the 3rd) had no band; and, to the best of his recollection, the drummers were dressed in red, like the private soldiers, but their jackets more profusely laced; the drum-major—a fine man—was, however, dressed in white, with a cocked hat and large silver epaulettes, which had the effect of causing him to be taken for an officer of high rank when in Paris. No bearskin caps were worn, either by the grenadier company or by the drummers.

It may be as well to mention, at this point, that such caps do not appear to have been worn by the—so-called—marching regiments of foot since the great increase of the army, about 1804, though worn on home service by the grenadier companies of the Foot Guards, and also by those of some of the embodied militia regiments.

The 2nd and 3rd battalions of the 14th having been reduced in 1816-17, there only remained the 1st battalion (now the regiment), stationed in India, where bearskin caps were not worn at all, and during its long sojourn there the memory of the special white bearskin caps authorized in 1765 may, to some extent, have faded away.

In 1816 the neat and serviceable felt cap was laid aside, and a broad-topped heavy shako introduced, its shape copied from the head-dress worn by the foreign troops met with in the occupation of France after Waterloo. It was eleven inches in diameter at the top, and seven and a half deep, with brass chin scales—which, when not required, could be fastened to the black cockade in front—ornamented with an upright red and white feather, twelve inches high, and a brass plate in front, with the regimental number. The light company had a green feather. The officers' shako had silver lace, two inches wide, round the top, and a three-quarter-inch silver lace round the bottom. On the lace immediately below the high feather appeared the black cockade, an oval boss of black cord. Silver chin scales generally fastened up to the cockade with black ribbon.

1819. On June 22nd an order was issued by the Adjutant-General that the 14th and other regiments with buff facings should, in future, wear white waistcoats, breeches, and white linings to their coats, instead of buff waistcoats, breeches, and coat linings. Waistcoats, as an article of dress, had not been seen for a quarter of a century, but the word " waistcoats " in the order refers to the undress jackets of the men.*

In the following year the short-tailed coats, or jackets, were abolished for all ranks, and in 1822 the breeches and leggings. The same year a circular was issued, calling attention to the fact that the " gorget " formed part of the officer's equipment. This ancient ornament had evidently fallen into disuse. It was required to be worn on certain duties, and no doubt was so worn, until finally abolished in 1830.

In 1826 the private soldier's coat altered in shape. The chest loops of regimental lace, worn by twos across the chest, were made broader at the top, tapering downwards to the bottom, and the lace taken off the skirts. They were at this time wearing a brass breastplate of the pattern annexed.

PRIVATE SOLDIER'S BREASTPLATE, ABOUT 1820.

In this year the officer's costume was imposing. The coloured illustration, p. 144, represents the parade dress of one of the battalion officers. The shako, as

* Instead of the breeches and leggings, dark grey—and, shortly after, bluish grey—trousers were substituted, though it is doubtful if the service companies in India had worn breeches for some time.

before described, but half-an-inch higher, the broad silver lace of the "Austrian wave" pattern; a small silver star plate had recently been introduced, with a gold girdle and crown, on the former "Corunna; Waterloo," in the centre XIV, surmounted by the "White Horse" (*vide* figure annexed). This is probably the first appearance of this badge upon any of the officer's equipments, but it does not as yet appear to be authorized as a regimental badge, the regulations simply stating that it may appear on the "bearskin caps of the grenadiers and drummers." The feather is red and white —all white for grenadiers* and green for the officers of the light company. It is probable that the latter may have used a silver bugle horn in front, instead of the star plate.

OFFICER'S SHAKO STAR, 1820-29.

The coats long-tailed, and lined in the skirts with white; the skirt ornaments now as pattern herewith—silver embroidered sprays on scarlet; 14 in the centre, also on scarlet. The buff lapels on the chest cut rather broad at the top, making what was known as the cuirass breast; the collar of the so-called Prussian shape, and fastened up the front, precluding the possibility of wearing the shirt collar and large black neckcloth so conspicuous only a few years before; the flap of the cross pocket, not shewn in the illustration, edged or piped with buff. Epaulettes under the same regulations as described in 1814, but larger; the strap of two and a quarter inches silver vellum lace, on scarlet cloth; the large crescent of the pattern termed "scaley," with a raised silver embroidered rose at each end, the fringe some three and a quarter inches deep. It may as well be mentioned that, at this time, almost every regiment had its particular pattern of epaulette. The silver wings

OFFICER'S SKIRT ORNAMENT, 1820-29.

* No bearskin caps in use by the 14th, as the regiment was serving in India.

worn by the officers of the flank companies were handsome, consisting of silver-plated curb chains on scarlet cloth, edged all round with three-eighths silver vellum lace; a gilt grenade or bugle in the centre, on the top of the shoulder; a silver fringe on the outer edge, hanging over the shoulder. The skirt ornaments of these officers were, respectively, silver embroidered grenades or bugles; the gorget gilt, with "G.R." engraved on it. The dress trousers were very full, cossack shape, of light blueish-grey cloth. The silver breastplate upon the shoulder-belt would probably be a new one, bearing the four battle honours now granted (*vide* illustration); and the sword, a new regulation weapon, worn in a frog. Light infantry officers wore silver whistles and chains.

OFFICER'S SILVER BREASTPLATE, 1827-31.

A blue great-coat, otherwise frock coat, was authorised for undress, the crimson sash worn with it, and the sword suspended in a frog from a black waistbelt. As an undress covering for the head, the shako was worn without feather, covered with oilskin; in many regiments a light shako of oilskin was used. The paymaster, quarter-master, surgeon, and assistant-surgeon wore single-breasted coats, as described in 1814, with regimental collars and cuffs, and black leather sword belt under the coat; no epaulettes or sash; all wore cocked hats—the paymaster and quarter-master with silver loop and tassels, the former without a feather; surgeons, plain black silk loop, also without feather.

In December, 1828, a change was ordered to be made in the infantry shakos, that which was then in use in the Prussian Army being closely copied. The silver lace was all stripped off, the height reduced to six inches, and the time-honoured Hanoverian cockade disappeared—never since this date worn on the caps of infantry officers.

The only ornament in front of this head-dress was a new universal gilt shako plate, or star, with crown over. In the centre regiments were allowed to place such regimental devices as they pleased. The 14th used the old silver star as described in 1826, with the addition of "Java" upon the girdle, and a laurel wreath in gilt metal just inside the girdle (see illustration). Also, following the Prussian example, the officers had gold cap lines, having a heavily-braided festoon in front, terminating in two tassels, looped up to one of the coat buttons.

OFFICER'S GILT SHAKO PLATE, 1829-45.

A month afterwards, the feather was ordered to be white for the whole, light infantry excepted; still, however, remaining twelve inches high.

A large amount of lace was worn upon the coats of many infantry regiments. Some, by no means the least distinguished, wore none at all excepting the epaulette and the small skirt ornament, as was the case with the 14th.

To such an extent had the practice of wearing a superabundance of lace grown, that the authorities determined to introduce a universal pattern coat; hence the warrant of February, 1829, authorizing the well-known double-breasted coatee, which remained, with scarcely any alterations, the dress of the officers until the Crimean war.

The coatee adopted by the officers of the 14th had two rows of silver buttons, by pairs, down the front; a buff collar, Prussian shape; on each side two loops of silver lace; buff cuffs with a scarlet slash, thereon four silver loops and buttons by pairs; white turnbacks to the skirts, the extremities ornamented with a silver embroidered star, 14 in the centre on red; scarlet-slashed pockets in the skirts, placed obliquely, with silver lace loops, also in pairs; these pockets, also the coat in front, edged with a narrow piping of buff cloth, not mentioned in the regulations, but kept up as a regimental custom.

Mention is made of lace, which now—for the first time for half a century or more—appeared on the officers' coats. There being no such thing as a regimental lace, one had to be fixed upon. Accordingly, the pattern termed "check and vellum," well known as the officers' regimental lace until 1881, was chosen.

Large silver epaulettes were worn, "on both shoulders," by all ranks of officers (for the first time in the case of captains and subalterns of infantry) excepting the grenadiers and light infantry, who wore large curb chain wings, strictly according to the regulations then introduced. The epaulette strap was of silver vellum pattern, striped with buff silk. The epaulette fringe varied a little in length, according to rank; field officers distinguished by gold crowns, and stars on the strap.

The new, so-called, Oxford mixture was now authorized instead of the old blueish-grey for trousers, and a blue forage cap, with a broad, stiff top, was ordered to be worn; it was ornamented with a welt, and a broad band of buff cloth. A plain scarlet jacket was ordered to be adopted by officers at certain stations. Excepting in tropical climates, officers sat down to mess in the full dress coatee, with epaulettes.

1830. The white—formerly buff—fatigue jacket abolished for the rank and file, and a red one introduced. Fusils also replaced the pikes, so long carried by the sergeants.

The recently-introduced cap lines evidently found little favour. They were abolished, along with the gorget, by the warrant of 1830, upon the accession of William IV.; the tall feathers reduced to eight inches; a green ball tuft for the light infantry; musicians to be dressed in white; and lastly (what was of importance to the 14th), the officers of the regular army ordered to wear gold lace.

With the exception that all silver lace and appointments became of gold, no change took place in the uniform of the officers; the breastplate would continue the same, excepting that the silver wreath, crown, number, and scrolls would be fixed upon a burnished gilt plate.

1832. Field officers ordered to wear brass scabbards. A red seam down the trousers, authorised January 1833. Next year a new forage cap adopted, of dark blue cloth, with a black oak-leaf band, 14 in gold in the centre.

The regiment arriving home from India, necessitated the providing of the grenadier companies with bearskin caps, and it is strongly believed, upon the testimony of old officers, that the regimental custom, dating from 1765, was kept up, by giving the drummers white bearskin caps—a privilege shared, it is believed, by no other regiment.

The officer's undress was now very handsome: a single-breasted blue frock coat, with gilt buttons; shoulder straps edged with gold lace, and terminating in gilt metal crescents; the sword carried in the frog of a black waistbelt, over the crimson sash.

By royal warrant, dated October 10th, 1836, the white lace with the black and red worm, so long worn by the rank and file, was abolished, plain white tape taking its place. Coloured lace, however, was still worn by the drummers, the pattern being buff with a worm stripe of blue and red at each edge; besides the ordinary parts of their coats being laced, after the manner of those worn by the privates, the back, side, and arm seams were covered with this lace; five or six chevrons of it also ornamented the coat sleeves, with large white wings, similar to those worn by the grenadiers, upon their shoulders. The sergeants also were directed to wear double-breasted coatees, without any lace across the chest, and white epaulettes or wings.

OFFICER'S GILT BREASTPLATE, 1810-55.

The coloured illustration, p. 152, represents the costume of an officer of the battalion companies in 1840. The skako star remains as before, but a new gilt breastplate has been introduced (*vide* illustration herewith); the plate gilt, all the ornaments in raised silver. The "Royal Tiger," with the word "India," had been granted to the regiment in 1838, and was now for the first time placed upon its appointments. A new gilt button (*vide* annexed illustration), embodying these distinc-

OFFICER'S GILT BUTTON, 1840-55.

tions, had also been introduced. The skirt ornament now, and up to the discontinuance of the coatee, with its coat tails, was a gold embroidered star, with XIV on a raised boss in the centre, as annexed. Officers of the grenadier and light companies wore the same coatee exactly, but with large gilt curb chain wings over the shoulder, edged with bullion fringe. The light company officers had a gilt whistle and chain upon the shoulder belt, and a green ball tuft in the shako. It was the regulation for officers and men of the grenadier company to wear bearskin caps on home service, and in certain temperate climates, but the regiment being just now stationed in the West Indies, it is very doubtful if this head-dress was used at all.

OFFICER'S GOLD-EMBROIDERED SKIRT ORNAMENT, 1844-55.

In 1844 a new shako for the infantry was authorized, sometimes called the "Albert" hat, six and three-quarter inches high, one-quarter inch less in diameter at the top than the bottom, completely altering the appearance of the head-dress. At the same time grenadier bearskin caps were ordered to be discontinued for those still wearing them. The officers of the regiment lost the large and handsome plate they had been using since 1829, a smaller one altogether, of gilt metal (see illustration), being substituted. It consisted of a universal gilt star, crown over, four and a half inches in diameter, in the centre XIV within a girdle, thereon "Buckinghamshire;" the newly-granted honour, the "Royal Tiger," underneath; and above the girdle, the "Horse," with its motto—its one solitary appearance upon the appointments of any grade in the regiment at this time. Red and white ball tufts were used by the

OFFICER'S GILT SHAKO PLATE, 1844-55.

battalion companies; grenadiers and light infantry, white or green ones respectively. The men still continued to wear their old cap ornament—a small circular brass plate with the regimental number and crown over it.

1845. Sergeants lost their crimson and buff sashes; plain crimson ones substituted.

1848. The undress uniform of the officers was altered very considerably in appearance by the discontinuance of the blue frock coat, with shoulder scales, and the adoption of a plain shell jacket of scarlet cloth, collar and pointed cuff of regimental facing, gold twisted shoulder cords, and small buttons in front, in pairs. A black patent leather sling sword belt was to be worn with the shell jacket, and a great coat, of grey cloth, took the place of the cloak. Officers' forage caps at this time had XIV instead of Arabian numerals, and continued the custom up to the period of the Crimean war.

1850. A plain shoulder belt, without breastplate, to carry the pouch, was authorized; the bayonet being hung in a frog from a waistbelt fastened with a Union locket. This change, like all other authorizations and alterations noted from time to time, did not take place directly, and the regiment was only just fitted with them before the Crimean war.

The band was now dressed in white double-breasted coatees, having red collars, cuffs, and turnbacks; also red shoulder straps, with a red circular tuft at the extremity; black leather waist belts, and, together with the drummers, red ball tufts in their shakos. The drummers' costume still the same as that described in 1836.

The coloured plate, p. 166, gives the winter Crimean appearance of a sergeant and a private soldier on trench duty.

1855. The coat tails of the whole army disappeared. Frock coats, or tunics took the place of the old coatee. The first issue was double-breasted, the buttons of brass,* now placed, in all regiments, at equal distances; no lace used, excepting round the buttons of the cuffs and skirts; the coat edging piped all round with white cloth; dark blue trousers, with a red welt introduced; the shako made

* Hitherto the buttons of the soldiers had been of white metal.

smaller and lighter; officers' and sergeants' sashes worn over the left and right shoulders respectively, instead of round the waist.

The bandsmen also dressed in these tunics, but made of white cloth, having red facings and wings; the buglers and drummers the same tunics, but ornamented with the peculiar regimental drummers' lace; the idea being, it was understood, to strengthen the appearance of the "music."

The year following, with the next issue of clothing, the tunic was single-breasted, and the drummers' coats changed back to red. Officer's rank was now distinguished by the amount of gold lace worn, and by crowns and stars on the collar.

A captain had an edging of the regimental lace (as before, check and vellum pattern) round the top of the collar and the top of the cuff; four loops of lace, diamond shape, round the buttons on the cuff slash; the same loops round the skirt buttons; a crown and star on the collar. Lieutenants and ensigns, a crown and star respectively. Field-officers had additional lace on the buttons of the collar, round the cuff and cuff slash, and on the skirts behind; colonels, a crown and star on the collar; lieutenant-colonels and majors, crowns or stars respectively.

A double-breasted blue frock coat adopted for undress, with gilt regimental buttons and a plain stand-up collar; the crimson silk sash worn over the left shoulder; and the sword carried in a white sling waistbelt, fastened with a gilt Union locket, bearing the regimental number. The officer's buttons made larger in size than before, but pretty much of the same pattern, excepting that the number was expressed in Arabian numerals.* The officer's shako of black beaver was smaller at the top, and altogether lighter; a small gilt star with crown over—the regimental number, 14, in the centre, its only ornament. Lieutenant-colonels and majors distinguished by two rows and one row of gold lace round the top respectively.

The coloured illustration, p. 178, though dated in 1858, very correctly represents the costume above described.

In 1858 an order was issued doing away with the flank companies,

* This button continued in use till 1881.

and the white and green ball tufts, worn by the grenadiers and the light company, disappeared. It was not, however, until 1862 that this change became thoroughly effectual, for in that year companies were ordered to be distinguished by a letter, and to stand on parade according to the seniority of their captains. In this latter year a lighter kind of shako, of ribbed blue cloth, was introduced.

The coloured illustration, p. 194, represents a private soldier in the dress of 1864. In 1866 the regimental drummer's lace was abolished, and a universal drummer's lace introduced—plain white, with small red crowns.

The officer's blue frock coat discontinued, April, 1867, and replaced by a blue patrol jacket; steel scabbards also took the place of those of black leather so long used by the officers.

1868. The slashed cuff on the tunic discontinued; pointed cuffs introduced. For levees, etc., officers were authorised to wear a gold and crimson sash, gold-laced trousers, and sword belt; the shako ornamented with gold cord, and the star replaced by a universal pattern garter and crown, the number inside, and surrounded by a wreath of laurel in high relief.

About 1873 white clothing, so long used by the band, discontinued, and soon after the shell jackets worn by the men abolished. Loose scarlet frocks adopted for undress.

1874. The regimental badge of the "Tiger" ordered to be worn upon the collar of the men's tunics; glengarries taken into use, replacing the old Kilmarnock forage cap, which had not long survived its companion, the shell jacket; and also, as a foreshadowing of coming changes, the old regimental button, worn by the rank and file, ordered to be replaced by a universal pattern army button.

The following Horse Guards letter was received, dated July 18th, 1873:—

"Her Majesty the Queen has been graciously pleased to approve "of the 14th Buckinghamshire Regiment of Foot wearing on the "second, or Regimental, Colour the White Horse and motto '*Nec* "*aspera terrent*,' as formerly worn in the Grenadier Corps."

This was followed next year by authority to wear this honourable badge in silver over the number on the forage caps of the officers, a distinction shared by only one other regiment (the 8th King's), though it has been pointed out as having been used, but without special authority, on the officers' shako plates from 1820 to 1855.

1880. The helmet introduced; also a round undress forage cap for officers, with circular peak over the forehead; badges of rank removed from the collar and displayed on gold cord shoulder straps.

The coloured illustration, p. 234, represents the last of the old buff facings so long and honourably worn by the regiment, for the following year, 1881, was signalized by the introduction of the territorial system, when the regiment lost its number and became the Prince of Wales's Own West Yorkshire Regiment, and, in accordance with the new regulations, adopted the white facings universal for all English corps* which had not the distinction of being Royal regiments; the officers at the same time exchanging their regimental gold lace for one of the universal rose pattern.

S. M. MILNE.

---

* *Vide* illustration, p. 264. The collar badge of the plume is, perhaps, rendered too large, and the magazine on the left side was not used on home service.

NOTES ON THE COLOURS.

THE regiment being raised by Sir Edward Hales, Bart., in 1685, would, as a matter of course, receive its colours very soon after its formation. At first composed of ten companies, one of grenadiers was added the year after—but it is doubtful if the latter company ever carried a colour, certainly every other company did. The colonel, lieutenant-colonel, and major, besides being field officers, were also effective captains of companies.

In the royal library at Windsor Castle is to be found a set of coloured drawings of the ensigns of Sir Edward Hales' regiment, taken in the year 1688, just before the abdication of King James II., in all probability executed to his own royal order.

It was customary at this time for the colonel's flag, carried by his company, to be of a perfectly plain colour throughout, and the lieutenant-colonel's, major's and captain's colours, to be, as a rule, of the same hue, but bearing the device of England—namely, the red cross of St. George—in a white field, which latter heraldic injunction was fulfilled by placing a narrow edging, or fimbriation, of white on the borders of the red cross, thus separating it from the hue of the ensign itself.

From the evidence above quoted, the colonel's flag was plain throughout, of a full red shade, with a gilt spear-head, and red and silver cords and tassels. That of the lieutenant-colonel's company, full red, and bearing the red cross of St. George, edged white, and three yellow or golden sun rays spreading from each angle of the cross towards the outer corners. The major's, the same, but differenced with a white flame or pile, wavy, issuing from the upper corner. The first captain's as that of the lieutenant-colonel's, but bearing a silver ball in the centre of the cross, *vide* frontispiece.

Unfortunately, no other examples are given; consequently it is a little doubtful how the other captains' colours were distinguished; that there would be some mark of distinction between the standards of the various companies is certain, judging from analogous cases; the probability being that the next captain's colour would have two such silver balls on the cross, the next three, and so on.

No special meaning can be attached to the fact that the colours bore the sun rays, they were used upon those of three other infantry regiments in James' army. Nor is there also any attempt to place any part of the colonel's coat armour in the colours of his regiment, a practice exceedingly common at the time.

Sir Edward Hales lost the colonelcy upon the accession of William III., and was succeeded by Colonel William Beveridge; but it is impossible to say, in the absence of information, whether the new colonel provided a set of colours or was content with the old ones.

The union with Scotland, in 1707, very much altered the appearance of the national colours, by the introduction of the white saltire cross of St. Andrew upon a blue field, forming the so-called "Union" colour.

By this time, pikemen having been discontinued, infantry soldiers consisted of musketeers only, armed with muskets and bayonets, and the number of colours, or flags, reduced to two per regiment. Those surviving being the two senior ones, namely, those of the colonel and the lieutenant-colonel; all the rest—major's and captain's—had fallen into disuse.

From the few examples which have been met with, it appears it had become customary to introduce a small "Union" into the upper corner of the first, or colonel's, flag, the field being fitted up in a variety of ways, generally displaying some part of his armorial bearings. The second colour, that of the lieutenant-colonel, was the "Union" throughout, sometimes with, sometimes without, a badge or distinction in the centre.

This continued until 1743, when the royal warrant, dated September 14th, completely altered the existing state of things, abolishing the colonel's and lieutenant-colonel's colours altogether, and prohibiting the use of any private armorial bearings.

In their stead two colours for each regiment were authorized: one the king's, the other the regimental; the first the "Union" throughout, the second to be of the colour of the regimental facing, with a small "Union" in the upper corner, in the centre of both the regimental number, in Roman characters, within a wreath of roses and thistles on one stalk. Regimental numbers were already in existence, but this was the first time they appeared upon any part of the regimental equipment.

The earliest recorded inspection of the regiment, then known as Herbert's, took place at Stirling, April 27th, 1750, the colours being reported upon by Lieutenant-General Churchill as "in bad condition," whilst, soon after, at Berwick, November 12th, they were reported upon as "good;" consequently it may be assumed that, during the interval, a stand of colours, strictly in accordance with the new regulations, and probably the earliest of their kind, had been given out to the regiment.

In 1765 the regiment was inspected at Winchester, October 26th, and the colours reported upon as "good, being new in 1764." These colours, if at all corresponding to similar sets given out to other regiments at the time, would have the number REGT. XIV. embroidered upon a red oval cartouche, or shield, with a floriated yellow silk border, all within a wreath of ordinary—not heraldic—roses and thistles, wildly discursively spread out in accordance with the fashion of the day.

The warrant of December, 1768, for the first time, gave the size of infantry colours, six feet six inches flying, and six feet deep on the pole; length of the pike (spear and ferrule included) nine feet ten inches. These dimensions remained in force until 1857.

The regiment was inspected at Coxheath camp, November 1st, 1779, and the colours mentioned as having been "new in 1778." This stand, again, would be very much like the last mentioned, but the wreath becoming more staid and regular, the number of roses on each side exactly balancing, signs that the wild studied disorder in fashion some ten years before was passing away.

In 1782 the regiment received orders, dated August 31st, to assume the title of "The Fourteenth, or Buckinghamshire, Regiment of Foot;" but nothing came of this for the present—county titles had no place on infantry colours until 1817. Judging from the inspection report, Chatham, June 1st, 1792, the regiment appears to have received a new stand that year. This time the regimental number would be placed upon a red heart, or heater shaped shield, extremely plain in character, with a thin, almost meagre, wreath around it, very similar to those used afterwards. *Vide* illustrations pp. 86 and 124.

The union with Ireland, 1801, introduced the red saltire cross of St. Patrick on a white field into the "Union" flag, thus completing this ensign as it is now known; it was further ordered that the shamrock be given a place in the "Union" wreath. Without any direct evidence it may be considered certain that a new stand was made very soon after the union with Ireland, embodying the changes just noticed.

The inspection returns are not very complete at this period, the earliest after this referring to the 14th is dated 1809, when the regiment had arrived in India, the colours being reported upon as "good."

In the meantime a 2nd battalion was raised, in 1804, which, in turn, would have colours, carried during its services in the Peninsula and Mediterranean; it distinguished itself at Corunna, and was authorized to bear that distinction on its colours January 6th, 1812, and this duly appeared in the following *Annual Army List*. Whether the honour was ever actually placed on the colours it then carried must be a matter of uncertainty, as they have been lost sight of; but another stand was made for the battalion at some date before it was disbanded, in 1817, and the word "Corunna" was embroidered upon the regimental colour only, immediately over the heart-shaped shield. *Vide* illustration, p. 86, representing these colours (now at Claydon). The distinction XIV. REGT. 2ND BATT. is embroidered upon the centre shield, and the wreath is very plain and thin in design, in fact the regulation wreath of the period.

The 3rd battalion, raised in 1813, received new colours in 1814, which were carried at Waterloo; they are represented in the illustration, p. 124, and are of the same plain type as those of the 2nd battalion.

The 1st battalion in India still used its old colours, rapidly falling to pieces in consequence of its arduous services in Bengal, the Isle of France, and Java. At the inspection at Berhampore, May 15th, 1814, the inspecting officer reports, "the colours are worn out, but "Colonel Watson has applied for new ones, which he expects from "Europe." The commanding officer had, however, to wait some years before new ones could be obtained; time after time they were reported upon as "bad," or "fairly worn off the staves."

It was not until 1819 that a new stand arrived, and were presented to the regiment at Meerut, by Lieutenant-Colonel Tidy, C.B., who had commanded the 3rd battalion at Waterloo. These new colours were altogether different in design to any yet received by the 14th regiment. The county title was now used upon all infantry colours, and, in this case, "Buckinghamshire" (changed from Bedfordshire in 1809) was placed upon a red girdle, within this girdle $\overset{14}{\text{REGT.}}$, the number now in Arabian numerals, the whole surrounded with a Union wreath crossed at the bottom and tied with ribbon, the two rosebuds at the top also meeting and crossing.

Three battle honours—"Corunna," "Waterloo," and "Java,"—had been up to this date authorized, and were displayed upon red scrolls, one each side and one below the central devices of each colour, king's and regimental. Only one, "Java," had been won by the 1st battalion (authorized May, 1818); the other two were gained by the disbanded 2nd and 3rd battalions, and, as was customary at the time, ordered to be borne upon the colours of the remaining battalion—now in fact the regiment.

By the year 1833 these colours had fallen into bad order; at the inspection at Athlone, November 1st, the following remark was made, "those in use mere shreds, but the commanding officer has got a "new stand in his possession;" they were retired in 1835; and since then appear to have been lost sight of.

Bhurtpore was duly granted to the regiment in 1827 and appeared on the next stand, presented in Dublin, 18th June, 1835. This stand was very similar to the last one, the wreath a trifle fuller in leaf and flower perhaps; it also shewed the four battle honours on red scrolls, one above and below and one each side, both colours exactly alike in ornamentation. *Vide* illustration, p. 50.

On February 24th, 1836, "Tournay" was at length tardily authorized to be borne upon the colours, and on November 8th, 1838, the "Royal Tiger," superscribed "India," was also granted to the regiment in consideration of its services in India from 1808 to 1831. Neither of these honours were ever placed upon the colours given out 1835 and retired 1853.

It is remarkable that in the metal work of any badge worn by the regiment, for instance—the officers' breast and shako plates and buttons from 1838 to 1855—the tiger was always displayed "cowed," that is, the head down, and the tail curled between the hind legs; whereas, upon the colours the embroiderers—acting presumably under the authority of the inspector of regimental colours—invariably depicted the tiger *passant*, head erect and tail curved high over the back.

In 1844 an order was issued ordering that no device of any kind should appear upon the Queen's colour other than the regimental number in Roman characters, surmounted by the crown.

By the year 1850—*vide* inspection return, at Newport, April 18th—the stand of 1835 was getting worn; Colonel Love, C.B., states, "Queen's colour worn out, from duty in Dublin and other places." This particular colour, from the frequency with which it was employed, and also, possibly from the numerous seams necessary to compose the "Union," always fell to pieces first.

In 1849 an order was issued prohibiting any ceremonial being gone through upon the presentation of colours; most of the stands of colours then given out were taken into use quietly. Had it not been, therefore, for the assistance of an officer* then serving with the regiment, the author of these notes would have been unable to give the date when the Crimean colours were unfurled.

Upon the Limerick parade ground, and on the occasion of the celebration of the Queen's birthday, 1853, the new stand was first used in the ranks of the 14th Regiment. The colours were the same size as before, the Queen's having only the number and crown in the centre; the regimental, on the other hand, possessed a very finely

---

* Captain C. Monck Wilson, to whom the writer is indebted for much interesting and useful information.

embroidered wreath with crown over; in the centre, on red, the regimental number in plain Roman characters, surrounded by a girdle, thereon "Buckinghamshire;" below the wreath, at the bottom of the flag, an embroidered tiger, with "India" on a blue scroll immediately above it, the five battle honours arranged also in blue scrolls, one below the wreath and two on each side.

These colours were carried by the regiment until 1876, when it had the honour of having new ones presented, at Lucknow, by the Prince of Wales, who in turn was pleased to accept the old ones of 1853.†

The new set were totally different, coming under the regulation of 1858, which prescribed the size to be four feet flying and three feet six inches on the pole. Gold fringe surrounded each colour, and the old spear-head replaced by the crest of England (lion and crown). The Queen's bore the simple numeral and crown; the regimental had its central device much as the last, the wreath, however, the regulation pattern of 1858, with three roses at regular distances on each side. One of the principal features was the new badge of the "White Horse," with its motto "*Nec aspera terrent*," which occupied the three corners, worked directly upon the buff silk of the flag. This honour, so rarely granted to infantry regiments, was ordered to be placed upon the colours and appointments by special warrant dated July 18th, 1873, although it had been authorized so long since as 1765 to be worn upon the caps of the grenadiers and drummers. The "Tiger," with "India" superscribed, was placed quite at the bottom, and immediately below the wreath a blue scroll, inscribed "New Zealand" (an honour recently won by the 2nd battalion), on each side of the wreath the six battle honours previously granted, three and three, all upon blue scrolls.

On 6th June, 1876, it was announced in the *Gazette* that the regiment was to be styled "The Fourteenth (Buckinghamshire) Prince of Wales's Own Regiment of Foot," and be permitted to bear the Prince of Wales' plumes on the second colour.

As a consequence, some time after, but before the year 1881, the regimental colour was altered, in order to display the new title and

† They are now at Sandringham.

badges. The words "Prince of Wales's Own" were placed upon the red girdle, the centre being occupied by **XIV.**, surmounted with the "Prince's plumes," and the motto "*Ich dien.*" "Buckinghamshire" was embroidered upon a long blue scroll placed across the bottom of the wreath; below this the "Tiger," with "India" superscribed; a fourth scroll bearing, "New Zealand," was added on the left side, and a similar one, bearing "Afghanistan" (authorized in 1880), on the right side.

A 2nd battalion, raised in 1857, received new colours at the hands of the Countess of Eglington, in Dublin, towards the end of the following year, the pattern being that of 1858. These colours were retired in 1877,* when a new stand was presented at Belfast, bearing the same devices as those described for the first battalion, a small scroll being added under the centre, with **II. BATT.** upon it.

The introduction of the Territorial scheme in 1881 had a great effect upon infantry regiments, especially upon their colours, as all numbers were swept away. The regiment itself received the title of "The Prince of Wales's Own West Yorkshire Regiment;" its time-honoured buff facings gave way to the universal white,† but, fortunately, having a strength of two battalions, it escaped the fate of the single battalion regiments junior to the 25th, retaining its own hard-won devices, badges, and battle honours intact and undisturbed.

S. M. MILNE.

---

* Now in the Chapel, Eton College.

† The next colours the regiment receives will be white, with the red cross of St. George.

APPENDICES.

A.

I. Biographies of Colonels of the Regiment, in order of Succession, from 1685.

II. Biographies of some Eminent 14th Officers.

III. Succession List of Lieutenant-Colonels during the last Fifty Years.

IV. Complete Roll of the Officers of the Regiment, from Monthly Army List, 1st January, 1893.

B.

Volume II.—Miscellaneous.

V. Copies of some Letters received by the Regiment.

VI. Letters from Old 14th Men, Anecdotes, Legends, Etc. relating to the Regiment.

VII. "Ça Ira," the Regimental Quick-step.

VIII. Billy, the Regimental Deer.

IX. Trophies, Relics, Etc. in the Regiment.

X. Regimental Mess Plate, Pictures, Etc.

XI. Athletics.

XII. Masonry.

XIII. Recollections of Dublin Duty.

APPENDIX A.

I.

BIOGRAPHIES OF COLONELS

OF THE REGIMENT,

IN ORDER OF SUCCESSION, FROM 1685.

1. Sir Edward Hales, Bart.

Appointed 22nd June, 1685.

Sir Edward Hales, Bart., son of Sir Edward Hales, Bart., of Woodchurch, near Ashford, Kent, a distinguished loyalist in the reigns of Charles I. and Charles II., whose project, in concert with Sir Roger L'Estrange, to carry off Charles I. from Carisbrooke Castle is related at length in *Clarendon's History*. He succeeded to the baronetcy at his father's death, soon after the Restoration; and when the Court showed a disposition to favour Papacy, he openly changed his religion to Roman Catholic. He raised a company of foot for King James at the time of Monmouth's rebellion, and was appointed colonel of the regiment formed from this and other companies, and which became the 14th Foot. He was sworn of the Privy Council, made a lord of the Admiralty, Deputy-Warden of the Cinque Ports, and Lieutenant of the Tower of London. He was convicted under the Test Act for accepting a military appointment when disqualified by religion from taking the prescribed oath, but moved the case into the Court of King's Bench, and obtained judgment in his favour. Being in King James' confidence, he was employed to make arrangements for the king's flight to France at the Revolution of 1688. On the night of 10th December Sir Edward, with Quarter-Master Edward

Syng, of his regiment,* accompanied the king in a hackney-coach from Whitehall Palace to Horse-Ferry, crossed the Thames in a boat, got down to Faversham in disguise, and went on board a custom-house hoy, designing to cross the Channel to France. They were suspected to be Romish priests, and were seized by the country people. The king was afterwards recognised and persuaded to return to London, but ultimately escaped from Rochester to France. Sir Edward attempted to conceal himself to elude the fury of the populace, who at the time of his apprehension were shewing their detestation of his change of faith by looting his country-house and killing his deer. He was kept a prisoner for eighteen months in the Tower, and on his release proceeded to France. His eldest son died fighting for King James at the Boyne. When in France Sir Edward was created Earl of Tenterden by King James. He died in France in 1695, and was buried in the church of St. Sulpice, Paris.

2. William Beveridge.

Appointed 31st December, 1688.

William Beveridge served under the Prince of Orange in one of the British regiments in the pay of Holland. At the Revolution of 1688 he was promoted to the colonelcy of the regiment, late Sir Edward Hales'. He commanded it for four years, and was killed in a duel by one of his captains on 14th November, 1692. (See foot-note to p. 7.)

3. John Tidcomb.

Appointed 14th November, 1692.

John Tidcomb (or Tidcombe) is described in the records of the University of Oxford as son of Peter Tidcombe, of Calne, Wiltshire, "pauper puer," a "servitor" of Oriel College, where he matriculated, 22nd March, 1661, at the age of nineteen. Of his early military life there are no details; but in 1687 he was third senior among the captains of Lord Huntingdon's Regiment, since the 13th Foot, and

* See King James' own account of the circumstances in Dr. Clark's *Life of James the Second.*

was made lieutenant-colonel of that regiment 31st December, 1688. He served under Major-General Mackay in Scotland, and distinguished himself at Killiecrankie. Afterwards he fought under King William at the Boyne. Returning to Ireland with Lieutenant-General the Earl of Marlborough (afterwards the famous Duke of Marlborough), he was at the capture of Cork and Kingsale and in various skirmishes. He was rewarded with the colonelcy of the 14th Foot, with which he served in Flanders, where he was present at the battle of Landen and the siege of Namur. After the latter the University of Oxford conferred on him the degree of D.C.L., *causa honoris*. He was promoted to the rank of brigadier-general in 1703; major-general, 1st January, 1704; and lieutenant-general, 1st January, 1707. He died at Bath in June, 1713.

4. Jasper Clayton.

Appointed 15th June, 1713.

Jasper Clayton obtained his first commission 24th June, 1695, and served under King William until the Peace of Ryswick. In the reign of Queen Anne he fought under Marlborough, and was appointed colonel of the 11th Foot, with which he served in Spain. The regiment suffered heavily at the battle of Almanza in 1707, and Clayton returned to England the year after to recruit. In 1709 he distinguished himself at the siege of Mons, where he was wounded. He was at the forcing of the French lines in the following year, and was rewarded on 8th December, 1710, with the colonelcy of a newly-raised regiment of foot, with which he served in the disastrous expedition against Quebec in 1711, when his regiment had three officers and one hundred and seventy-one men drowned in the St. Lawrence, or, as it was then called, the Canada River. At the Peace of Utrecht, in 1713, his regiment was disbanded, and on 15th June he was appointed colonel of the 14th Foot. He served in Scotland in 1715-16, and commanded a brigade at the battle of Dumblain. He was made lieutenant-governor of Gibraltar, and commanded the troops during the siege of 1727 by the Spaniards. He became a major-general in 1735, and lieutenant-general, 2nd July, 1739. He served under King George II. at the battle of Dettingen, where he was killed while directing the artillery to play upon the French as they were retreating over a bridge. General Clayton, who was much regretted by the army, was buried with military honours in the private chapel of the Prince of Hesse at Hanau. He was father (?) of Jasper Clayton, who was many years an officer in the 14th Foot, and was lieutenant-colonel of the 46th (then 57th) regiment in Scotland, in 1745.

5. JOSEPH PRICE.

Appointed 22nd June, 1743.

JOSEPH PRICE obtained a commission in 1706, and rose to be captain and lieutenant-colonel in the 1st Foot Guards. In January, 1741, he was appointed colonel of the 57th (since 46th) Foot, then first raised, from which he was transferred to the 14th Foot on the death of Lieutenant-General Jasper Clayton. He became a brigadier-general, 6th June, 1745. He distinguished himself at the battle of Val, or Laffeldt, on 2nd July, 1747, when his brigade held the village of Val; and his conduct was mentioned in the Duke of Cumberland's despatch. He died at Breda in November, 1747.

6. THE HONOURABLE WILLIAM HERBERT.

Appointed 1st December, 1747.

THE HONBLE. WILLIAM HERBERT, fifth son of Thomas, eighth Earl of Pembroke (and father of the first Earl of Carnarvon), entered the army 1st May, 1722, and became captain and lieutenant-colonel, 1st Foot Guards, 15th December, 1738. On 1st December, 1747, he was appointed colonel of the 14th Foot, and in January, 1753, was removed to the Queen's Bays. He was a major-general, a groom of the bedchamber to George II., and M.P. for Wilton, Wilts. He died 31st March, 1757.

7. EDWARD BRADDOCK.

Appointed 17th February, 1753.

EDWARD BRADDOCK, born in 1695, was son of Major-General Edward Braddock, who was regimental lieutenant-colonel of the Coldstream Guards in 1703, and died in 1720. He was appointed ensign in his father's regiment 29th August, 1710, and was lieutenant of the grenadier company in 1716. He became captain and lieutenant-colonel in 1735, major in 1743, and regimental lieutenant-colonel in 1745. He commanded the grenadier companies of the Guards in the Duke of Cumberland's march from London in 1745 (the "march to Finchley"). He commanded the 2nd battalion of his regiment in Admiral Lestock's expedition to Lorient, and afterwards in the Low Countries in 1746-48. In 1753 he was appointed colonel of the 14th Foot, which he joined at Gibraltar. On 29th March, 1754, he was appointed commander-in-chief in North America. Having completed arrangements for several concurrent attacks on the French settlements, he took the field himself with a force of one thousand two hundred, regulars, provincial troops, and Indians, for an attack on the French post of Fort Duquesne, on the present site of Pittsburg, U.S. On 9th July, 1755, after crossing the Monagahela River, he was suddenly attacked by a French ambuscade, which threw his men into confusion. Great slaughter followed, ending in a panic-stricken rout. Braddock, who did all a man could do, and had five horses shot under him, was at length himself struck down by a bullet that passed through his right arm and lodged in the lungs. He was carried from the field in a cart, and died four days later at a place called Great Meadows, fifty miles from the battle-field. "We shall know better how to deal with them next time" are said to have been his last words, during a momentary return of consciousness. Like other unsuccessful commanders, he has been unsparingly condemned; but General Wolfe, who knew him, describes Braddock as "a man of courage and good sense, although not a master of the art of war."

8. Thomas Fowke.

Appointed 12th November, 1755.

Thomas Fowke received his first commission 25th May, 1705. Fifteen years afterwards, on 25th June, 1720, he became lieutenant-colonel 7th Dragoons (now Hussars). In January, 1741, he was appointed colonel of the 54th Foot (since 43rd Light Infantry), then raising, and in August, the same year, was transferred to the 2nd (or Queen's). He became a brigadier-general, 1st January, 1745, and commanded a brigade under Sir John Cope at Preston Pans. He became a major-general in 1747, and lieutenant-general in 1754. He was governor of Gibraltar in 1754, when the island of Minorca was attacked by the French, and the unfortunate Admiral Byng was shot for failing to relieve it. Fowke having disobeyed the orders of the Secretary at War to send a reinforcement to Minorca, was tried by a general court-martial, and was sentenced to be suspended for nine months; but George II., despite his fifty years' service, ordered him to be dismissed the army. At his death a pension was granted to his widow.

9. Charles Jefferies.

Appointed 7th September, 1756.

Charles Jefferies was appointed lieutenant-colonel of Cholmondeley's Regiment (34th Foot) 17th February, 1746. Owing to the carelessness in the military entries, there is some doubt as to his previous career. Cannon assumes—probably rightly—that he was the Charles Jeffreys, or Jefferies, who succeeded Colonel Robert Moore as lieutenant-colonel of Price's (14th) Foot, and who is stated in the *London Gazette* to have died in February, 1746. He signalized himself in command of the 34th at the defence of Port Mahon, Minorca, in 1756, particularly in repulsing an assault on the place, on which occasion he was taken prisoner. Meanwhile he had been appointed, on 2nd January, 1756, to be colonel-commandant of a battalion of the newly-raised 62nd Royal Americans (since 60th King's Royal Rifles). In September, 1756, he was transferred to the colonelcy of the 14th Foot. He became a major-general 27th June, 1759, and died in 1765.

10. The Honourable William Keppel.

Appointed 31st May, 1765.

The Honble. William Keppel was fourth son of William Anne, second Earl of Albemarle, K.G., and great uncle of the sixth earl, who, as the Honble. George Keppel, served as an ensign in the regiment at Waterloo. (See "*Biographies of 14th Officers*," App. A.) He became captain and lieutenant-colonel, 1st Foot Guards, 28th April, 1750, and gentleman of the horse to King George II. in December, 1752. On 21st July, 1760, he became regimental major with the rank of colonel, and on 17th December, 1761, was appointed colonel of the 56th Foot. He accompanied his regiment on the expedition against the Havana, in which he held the local rank of major-general. When the city capitulated, in August, 1762, he occupied the fort of La Punta, and afterwards, as commander-in-chief, delivered up the city to the Spaniards in accordance with the Treaty of Peace of 1763. He became a major-general 10th July, 1762; was transferred to the colonelcy, 14th Foot, 31st May, 1765; became a lieutenant-general in 1772, and was appointed commander-in-chief in Ireland in 1773. On 18th October, 1775, he was transferred to the 12th Dragoons. He sat in Parliament for the borough of Windsor for several years. He died 1st March, 1782.

11. Robert Cunninghame.

Appointed 18th October, 1775.

Robert Cunninghame served some years in the 35th Foot, in which he became captain in December, 1752. He was afterwards for many years adjutant-general in Ireland. He attained the rank of lieutenant-colonel in 1757, became colonel in 1762, and colonel, 58th Foot, in 1767; a major-general in 1772, and was transferred to the colonelcy of the 14th Foot, 18th October, 1775. He became lieutenant-general in 1777; was transferred to the colonelcy, 5th Royal Irish Dragoons, in 1787, and attained the rank of general in 1783. He died in 1787.

Whilst General Cunninghame was colonel, the 14th received its first county-title of the "Bedfordshire" Regiment.

12. John Douglas.

Appointed 4th April, 1787.

John Douglas, many years an officer in the Scots Greys, with which he served in Flanders in 1747-48, and in Germany in 1759-63, distinguishing himself on several occasions. He became lieutenant-colonel of the Greys in 1770, aide-de-camp to the King in 1775, a major-general in 1779, and in April of the same year was appointed colonel of the 21st Light Dragoons (the second of the regiments that have borne that number), which was at that time formed out of the *light* troops of the Greys and other cavalry regiments, and was disbanded at the close of the American War in 1783. He became colonel of the 14th Foot and a lieutenant-general in 1787, and on 27th August, 1789, was transferred to the colonelcy of the 5th Dragoon Guards, which he held until his decease in 1790.

13. George (4th) Earl of Waldegrave.

Appointed 27th August, 1789.

George (4th) Earl of Waldegrave, eldest son of the third earl, was, whilst Viscount Chewton, appointed to an ensigncy in the 3rd Foot Guards 10th May, 1768, became lieutenant and captain in 1773, and captain-lieutenant and lieutenant-colonel in the Coldstream Guards in 1778. In the following year he raised, and was appointed lieutenant-colonel-commandant of, the 87th Foot (the second regiment so numbered in succession), which, after service in the West Indies, was disbanded at the peace of 1783. He attained the rank of colonel in 1782; succeeded his father as Earl of Waldegrave in 1784, and was master of the horse to the Queen and aide-de-camp to the King. On 27th August, 1789, he was appointed colonel of the 14th Foot, and died six weeks later.

14. George Hotham.

Appointed 18th November, 1789.

George Hotham, a younger brother of Admiral Lord Hotham, born in 1741, was appointed ensign in the 1st Foot Guards 14th May, 1759, and became lieutenant and captain in 1765, captain and lieutenant-colonel in 1775, and aide-de-camp to the King, with the rank of colonel, in 1781. On 18th November, 1789, he was appointed colonel of the 14th Foot. He became a major-general in 1790; lieutenant-general in 1797; general in 1802. Died, 1806.

General Sir Harry Calvert, Bart., G.C.B.,
Commander-in-Chief of the 14th Regiment from 1806 to 1826.

15. Sir Harry Calvert, Bart., G.C.B., G.C.H.

Appointed 8th February, 1806.

Sir Harry Calvert, Bart., G.C.B., G.C.H., was appointed 2nd lieutenant, 23rd Royal Welsh Fusiliers, in April, 1778. After spending some months at the Royal Military Academy, Woolwich, he joined his regiment in America, and served with it at the siege and capture of Charleston, and afterwards in the campaigns in North Carolina and Virginia, under Lord Cornwallis, up to the surrender at York Town, 17th October, 1781. He remained a prisoner of war until the peace of 1783, when he proceeded with his regiment to New York, and returned home with it in 1784. After travelling on the Continent, he got his company in October, 1786, and in 1790 exchanged to the Coldstream Guards. He was aide-de-camp to the Duke of York in Flanders in 1793-94, and brought home the despatches announcing the capture of Valenciennes. He returned home from the Continent early in 1795, and in May that year was sent on a confidential mission to Berlin. In 1796 he was made deputy adjutant-general of the forces. He obtained the brevet rank of colonel in 1797, and in 1799 was appointed lieutenant-colonel of the 63rd Foot. On 9th January, 1799, he was appointed adjutant-general of forces, a post he held, under the Duke of York, during the remainder of the war and for some years after. He devoted his best energies and unwearied attention to the carrying out of the many important measures introduced during that eventful period for the improvement of every branch of the interior economy of the army, the bettering of the recruiting service, the establishment of regimental schools, the improvement of general hospitals, etc. His courteous manners and kindly consideration for the feelings of all with whom he came in contact were long gratefully remembered—nowhere more so than in the department over which he so long presided. He was rewarded with a baronetcy in October, 1818, and in 1820 relinquished the post of adjutant-general. He had in the meantime attained the rank of major-general in 1803, and lieutenant-general in 1810. He was appointed colonel of the 5th West India Regiment in 1800, and was transferred to the colonelcy of the 14th

Foot in 1806. He was a commander of Chelsea, a G.C.B. and G.C.H., and in 1826 attained the rank of general. He died suddenly of a fit of apoplexy while staying with his family at Claydon Hall, in Buckinghamshire, on 3rd September, 1826.

16. THOMAS, LORD LYNEDOCH, G.C.B., G.C.M.G.

Appointed 6th September, 1826.

THOMAS GRAHAM, laird of Balgowan in Perthshire, was born 19th October, 1798. He first joined the army, as a volunteer, with Lord Mulgrave's force at Toulon, in 1793, when over forty years of age. Returning home from Toulon, he raised the 90th (or Perthshire) Volunteers, of which regiment he was appointed lieutenant-colonel-commandant in 1794, and colonel in 1803. He was British military commissioner with the Austrian army shut up in Mantua in 1796, and distinguished himself at the capture of Minorca in 1798; commanded a brigade at Messina in 1799; captured Malta in 1800, and joined the army in Egypt after the fall of Alexandria. He sat for Perth in several Parliaments, and made a maiden speech in 1806 in favour of "short service" enlistments. He was with Sir John Moore in Sweden and at Corunna, and commanded a brigade, of which the 2nd battalion 14th Regiment formed part, in the Walcheren expedition. In 1811 he was sent with an expeditionary force to defend Cadiz against the French, whom he defeated at the memorable battle of Barossa, 5th March, 1811. Joining Wellington's army, with the local rank of lieutenant-general, he commanded the first division and an army-corps during the sieges of Ciudad Rodrigo and Badajos and the advance on Salamanca. Returning from sick-leave, in 1813, he commanded the left of the British at the battle of Vittoria. He commanded at the siege of San Sebastian, and the force that crossed the Bidassoa and established the British army on the soil of France, 7th October, 1813. When again on sick-leave, he was appointed to command the army in North Holland in the winter of 1813-14, where he failed in a desperate attempt to carry Bergon-op-Zoom by storm in March, 1814. He was raised to the peerage, under the title of Lord Lynedoch of Balgowan, in 1815. He became a full general in 1821, and was colonel of the 14th Foot from 1826 until transferred to the 1st Royal Regiment in 1834. He was the founder of the United Service Club. He died on 18th December, 1843, at the age of ninety-five, having retained his mental powers and much of his bodily activity to the last.

17. The Honourable Sir Charles Colville, G.C.B., G.C.H.

Appointed 12th December, 1834.

The Honble. Sir Charles Colville, G.C.B., G.C.H., second son of the ninth Lord Colville, was born in 1770. Ensign, 28th Foot, 13th June, 1787; lieutenant the same year; captain of an independent company, 1791; captain, 13th Foot, 26th May, 1791; major, 1795; lieutenant-colonel, 1796. Served with the 13th Foot in Jamaica in 1792-93; in San Domingo, 1794-95; in Ireland, 1798; at Ferrol in 1800; in the campaign in Egypt in 1801, and afterwards at Gibraltar and Bermuda. He commanded a brigade at the capture of Martinique in 1809, became a major-general in 1810, commanded a brigade at Fuentes d'Honor, and the fourth division at Badajos (twice wounded). He commanded a brigade at the battles of Vittoria and the Pyrenees; the sixth division at the Pass of Maya; the third division at Nivelle and the battles on the Nive; and the fifth division at the investment of Bayonne. He commanded the reserve at Waterloo, and captured Cambray during the march on Paris. He afterwards commanded the troops at Bombay. He became a lieutenant-general in 1819, and general in 1837. He was colonel of the 14th Foot from 12th December, 1834, to 25th March, 1835. He died in 1837.

18. The Honourable Sir Alexander Hope, G.C.B.

Appointed 25th March, 1835.

The Honble. Sir Alexander Hope, G.C.B., second son of the second Earl of Hopetoun, and brother of Sir John Hope, who was afterwards Lord Niddry and Earl of Hopetoun. Appointed ensign 63rd Regiment, in 1786; became lieutenant, 64th, in 1788; and raised an independent company in 1791, which was drafted into the 17th

Foot. In 1793 he exchanged from the latter regiment to the 1st Foot Guards, as lieutenant and captain, and was one of the officers selected for the light companies when light companies, four in number, were first formed in the 1st Guards, in 1793. He was brigade-major of the Guards at Lincelles; became major in the 81st Regiment and lieutenant-colonel in the 2nd battalion 90th Regiment, from which he exchanged with Colonel Alexander Ross to the 14th Foot. He commanded the regiment at Gueldermalsen, 8th January, 1795, where he received a dangerous wound in the left shoulder, which deprived him of the use of one arm and induced partial paralysis of the whole side. He remained supernumerary lieutenant-colonel of the regiment until promoted, but did not rejoin. He was appointed lieutenant-governor of Tynemouth in 1797, and of Edinburgh Castle in 1798; became a brigadier-general in 1807 and major-general in 1808. After some years at the Horse Guards, as deputy quarter-master-general, he was appointed governor of the Royal Military College, Sandhurst. He was sent on a special military mission to Sweden in 1813, to report on the forces available for co-operation in Germany. He returned to the Royal Military College, where the Riding School and Infirmary were built and the grounds planted during his governorship. He exchanged back to Edinburgh Castle in 1819, and was appointed lieutenant-governor of Chelsea Hospital in 1826. He became a lieutenant-general in 1813, and general in 1830. He was appointed colonel of the 14th Foot, 25th March, 1835. He was a G.C.B., D.C.L. Oxford, and sat in Parliament for forty years. He died at Chelsea Hospital, 19th May, 1837.

19. SIR JAMES WATSON, K.C.B.

Appointed 24th May, 1837.

SIR JAMES WATSON, K.C.B., son of Major James Watson, of the Royal Invalids, was born in 1772, and was appointed ensign, 64th Foot, 24th June, 1783. He was placed on half-pay immediately afterwards. In 1787 he was brought on full pay in the 14th Foot, in which he became lieutenant 18th April, 1792; captain, 11th March, 1795; major, 9th December, 1802, and lieutenant-colonel,

15th May, 1808. He served with the regiment in the campaigns in Flanders and the retreat to Bremen in 1794-95; at St. Lucia, Trinidad, etc., in 1796-97; went with the 1st battalion to India in 1807; commanded it at the capture of the Isle of France in 1810 and the reduction of Java in 1811, including the attack of Fort Cornelis. Commanded the regiment at the storming of D'Joejocarta in 1812, and commanded the expedition against Sambas, in Borneo, in 1813 (repeatedly mentioned in despatches, C.B. and gold medal for Java). Commanded the regiment in the Nepaul War and at the capture of the Fort of Hatrass. Commanded a brigade in Bundelkund during the Pindaree and Mahratta wars, and was engaged at the capture of Dhomone, Mundhla (where he led the storming party), Gurrahkote, and the fortress of Asseerghur (medal). He became a major-general in 1821, and commanded a division in India from 1830 to 1837, bringing up his total service in India to over twenty-seven years. During the temporary absence of Lord William Bentinck he acted as commander-in-chief in India. He became a lieutenant-general and was appointed colonel of his old corps in 1837; was made K.C.B. in 1839; and attained the rank of general, 11th November, 1851. He died at his seat, Wendover, Bucks, 12th August, 1862, aged ninety.

20. Sir William Wood, K.C.B., K.H.

Appointed 13th August, 1862.

Sir William Wood, K.C.B., K.H., was appointed ensign, 14th Foot, 27th January, 1797, in which he became lieutenant 27th December, 1799; captain, 3rd December, 1802; and major, 14th May, 1807. He served with the regiment six years in the West Indies; was present with the 2nd battalion at Corunna (medal) and at Walcheren. He commanded the battalion when sent back to Walcheren to bring off the troops remaining there at the end of the year, and afterwards proceeded with it to Malta. He was appointed lieutenant-colonel, 85th Light Infantry, 8th April, 1813, and served with that regiment in the South of France; proceeded with it to America, and commanded it at the battle of Bladensburg, where he received four severe wounds, had his horse shot under him, and was taken prisoner. He became brevet colonel 22nd July, 1830; major-general, 23rd November, 1841; lieutenant-general, 11th November, 1851; and general, 31st August, 1854. He was appointed colonel of the 3rd West India Regiment in 1849, and transferred to the 14th Foot on 13th August, 1862. He commanded the troops in the West Indies in 1851-55. He died 8th August, 1870.

21. Maurice Barlow, C.B.

Appointed 9th August, 1870.

Maurice Barlow, C.B., was appointed ensign, 85th Light Infantry, 21st July, 1814; became lieutenant 23rd March, 1815; and was placed on half-pay of the regiment in 1818. He purchased a company in the 3rd Buffs 20th December, 1821, and a majority 12th June, 1828; exchanged to 14th, as major, 25th June, 1830, and became brevet lieutenant-colonel 23rd November, 1841. On 25th December, 1847, he purchased a lieutenant-colonelcy in the 44th Foot, but reverted to his old regiment the same day by a three-cornered exchange with Colonel M. Everard, C.B., K.H., who retired on half-pay of the Buffs. He went out in command of the 14th Regiment to Malta; became brevet-colonel, 20th January, 1854; arrived with the regiment in the Crimea 10th January, 1855; served as a general in the trenches during the siege of Sevastopol from 11th June, 1855, and was in command of a brigade when the place fell (medal and clasp, C.B., Knight of the Legion of Honour, 5th Class of the Medijie, and Sardinian and Turkish medals). After serving twenty-seven years in the regiment, he exchanged to half-pay, unattached, in 1857. He became a major-general 26th October, 1858; lieutenant-general, 24th April, 1866. He was appointed colonel, 3rd West India Regiment, in 1863, and transferred to the 14th Regiment 9th August, 1870. He died 20th November, 1880.

22. James Webber Smith, C.B.

Appointed 13th April, 1875.

James Webber Smith, C.B., son of General James Webber Smith, C.B., Royal Artillery, entered the army as ensign, 48th

Foot, 11th July, 1826. In the course of his nine years' service as an ensign in that regiment he made the campaign against the rajah of Coorg in April, 1834, and was wounded. Became lieutenant 25th December, 1835; captain, 7th September, 1838; Major, 11th December 1849; exchanged from 48th to 95th, July, 1851; and purchased the lieutenant-colonelcy of the 95th, 24th December, 1852. Commanded that regiment in the Eastern campaign of 1854, and was severely wounded (C.B., medal and clasp, 5th Class of the Medijie, and Turkish medal). Exchanged to unattached in 1855, and was appointed aide-de-camp to the Queen, with the rank of colonel, July, 1855. Became major-general 20th April, 1866; lieutenant-general, 15th May, 1874; general, 1st October, 1877. Appointed colonel, 14th Foot, 13th April, 1875. Died 31st December, 1878, aged sixty-nine.

23. Sir Alfred Hastings Horsford, G.C.B.

Appointed 5th January, 1879.

Sir Alfred Hastings Horsford, G.C.B., son of General George Horsford, an officer of long service in the West Indies, was educated at Sandhurst, and on 12th July, 1833, was appointed 2nd lieutenant in the Rifle Brigade, in which corps he became 1st lieutenant 23rd April, 1839; captain, 5th August, 1842; major, 26th December, 1851; and lieutenant-colonel, 28th May, 1853. He served with the 1st battalion Rifle Brigade in the Kaffir War of 1846-47; returned to the Cape with the battalion in 1852, and commanded it in the Kaffir War of 1852-53 (medal); accompanied the battalion to the East in 1854, and served with it in Bulgaria and the Crimea, including the battles of the Alma, Inkerman, and Balaklava, and the early part of the siege of Sevastopol (C.B., Knight of the Legion of Honour, British medal and clasps, and Sardinian and Turkish medals). He became brevet colonel 28th November, 1854; was appointed one of the lieutenant-colonels of the present 3rd battalion Rifle Brigade when formed at Portsmouth in 1855; took a wing of the battalion out to Calcutta, where it landed in October, 1857. He commanded the 3rd Rifle Brigade as part of Walpole's brigade at the battle of

Cawnpore and in the advance on Lucknow. Commanded a brigade from February, 1858, at the siege and capture of Lucknow and in the campaign in Oude and the Trans-Gogra. When Lord Clyde returned to Lucknow, after the final defeat of the rebels at the Raptee, 30th December, 1858, Horsford was left with his brigade to watch the passes into Nepaul. He returned home soon after; was adjutant-general at the Horse Guards, 1860-66; brigadier-general at Aldershot, 1866-69; major-general on the staff at Malta, 1870-72; major-general commanding South-Eastern District, 1872-74; military secretary at the Horse Guards, 1874-80. In 1874 he was sent to represent Great Britain at the International Congress on the Usages of War, held at Brussels. He became a K.C.B. in 1860; a major-general, 1st January, 1868; lieutenant-general, 1874; G.C.B., 1875; general, 1st October, 1877. He was transferred from the colonelcy of the 79th Highlanders to that of the 14th Regiment 5th January, 1879, and 1st November, 1880, from the 14th to colonel-commandant of a battalion of the Rifle Brigade. He died 13th September, 1885, aged sixty-seven.

24. ALFRED THOMAS HEYLAND, C.B.

Appointed 21st November, 1880.

ALFRED THOMAS HEYLAND, C.B., was appointed ensign in the 95th Foot 4th April, 1833; lieutenant, 18th July, 1836; captain, 13th December, 1842; brevet major, 20th June, 1854; major, 1st December, 1854; brevet lieutenant-colonel, 12th December, 1854; and regimental lieutenant-colonel, in succession of Colonel James Webber Smith, C.B. *(vide supra)*, 5th June, 1855. He served with the 95th in the Eastern campaign of 1854-55, including the battle of the Alma* (severely wounded; arm amputated) and the siege and fall of Sevastopol (mentioned in despatches, brevet lieutenant-colonel, C.B., medal and clasps, 4th Class of the Medijie, Sardinian and Turkish medals). At the close of the Crimean War he was appointed to command a depôt battalion in Ireland. In August, 1857, he proceeded to India

* He was senior captain at the Alma.

as junior lieutenant-colonel, 56th Regiment, and served there until 1867, the last five years—from June, 1862 to May, 1867—in command of a brigade of the Bombay army. He became brevet colonel 20th March, 1858; major-general, 6th March, 1868; lieutenant-general, 1st August, 1877; colonel, 14th Regiment (now P.W.O. West Yorkshire), 21st November, 1881; general, 12th April, 1881.

THIRD AND FOURTH BATTALIONS,

FROM 1881.

Third Battalion (late 2nd West Yorkshire Militia).

H. VAN STRAUBENZEE, appointed honorary colonel from lieutenant-colonel commandant, 24th September, 1873. Formerly in the 14th Light Dragoons.

Fourth Battalion (late 4th West Yorkshire Militia).

G. P. FAWKES, appointed honorary colonel, 22nd December, 1888. Formerly captain, 83rd Foot.

BIOGRAPHIES
OF
SOME EMINENT 14TH OFFICERS,
ALPHABETICALLY ARRANGED.

GENERAL SIR JAMES EDWARD ALEXANDER, C.B., K.C.L.S., F.R.S.E., etc.
Soldier.—Traveller.—Author.

GENERAL SIR JAMES EDWARD ALEXANDER, C.B., K.C.L.S, F.R.S.E., etc., eldest son of Edward Alexander, Esq., of Powis, Clackmannonshire, born in 1803; educated at the Royal Military College, Sandhurst; and in 1820 obtained a Madras cadetship. He served for a short time in the Madras Light Cavalry, and was appointed by Sir Thomas Munro adjutant of the Body Guard. On 20th January, 1825, he obtained a cornetcy in H.M. 13th Light Dragoons; became a lieutenant on 26th November, 1825, and exchanged from half-pay to the 16th Lancers in 1827. In the meantime he had served in Burmah during the latter part of the First Burmese War, and was aide-de-camp to Colonel Macdonal Kinneir, the British Envoy in Persia, during the war between Persia and Russia. He entered the Senior Department, Royal Military College, Sandhurst, on 28th March, 1827, when on half-pay, and passed his examination and received a first certificate, 9th December, 1828. He obtained an unattached company, 18th January, 1830, and was transferred to the 42nd Highlanders, 9th March, 1832. He visited the seat of war in the Balkans during the war between Russia and Turkey, and Portugal during the Miguelite War. He also visited South America, and made two voyages of exploration up the Essequibo. He was aide-de-camp to Sir Benjamin Durban during the Kaffir War of 1835. He was afterwards at the head of a Government Exploring Expedition into Namaqualand and North Damara Land in 1836-38, for which service he was knighted. On 11th September, 1840, Sir James

Alexander, who had exchanged from the 42nd Regiment to half-pay, unattached, exchanged to full-pay in the 14th. He served as aide-de-camp and private secretary to Sir Benjamin Durban, in Canada, and subsequently as aide-de-camp to Sir William Rowan. During his fourteen years staff service in North America, he spent much time in exploring the forests of New Brunswick in concert with the Boundary Commission. He joined the 14th Regiment in the Crimea at the beginning of June, 1855, and as major and brevet lieutenant-colonel served with it in the trenches before Sevastopol, including the assault of 18th June, 1855. Was present at the battle of Tchernaya, and was in command of the regiment at the fall of the city. After the return of the regiment to Malta, he was appointed to a depôt battalion, but on the 30th March, 1858, was reappointed to the regiment a few weeks after the date of the order for forming a second battalion. He raised and commanded the present 2nd battalion; took it out to New Zealand and commanded a military force there in 1860-62.

Sir James Alexander, who had the first-class of the Persian Order of the Sun and Lion, and had received seven war medals, was awarded a distinguished service pension in February, 1864, and was made a C.B. in 1872. He was a deputy lieutenant of Stirlingshire, and a fellow of various learned societies. He became a full general in 1882. He died 2nd April, 1885.

Sir James Alexander was an able and prolific writer. Of eighteen works published by him, which are enumerated in the British Museum Catalogue of Printed Books, those of most interest—regimentally—are, "*Passages in the life of a Soldier*" (London, 1857), the second volume of which gives his Crimean experiences, "*Bush Fighting*," and "*Incidents in the Maori War*," which describe the campaigns in New Zealand of 1860-61.

It was chiefly through the exertions of Sir James Alexander that the famous obelisk, known as Cleopatra's Needle, now on the Thames Embankment, was rescued from destruction and brought over to England.

LIEUTENANT-GENERAL MAURICE BARLOW, C.B.

See "*Biographies of Colonels*," App. A.

LIEUTENANT-GENERAL MONTAGU BURROWS.

LIEUTENANT-GENERAL MONTAGU BURROWS, youngest son of the Rev. John Burrows, rector of St. Clements Danes, London, and Christ Church, Southwark, was born in 1775, and educated at Merchant Taylors' School. He was appointed ensign in the 29th Foot, 15th February, 1793; lieutenant of an independent company (Captain Bell's), 15th February, 1793; captain of an independent company (*vice* Butler), 29th March, 1794; captain in the 102nd (General French's) Regiment of Foot, 25th August, 1794; and captain in the 14th Foot, 2nd September, 1795. Soon after entering the service he embarked with a detachment doing duty as marines in the Newfoundland fleet, and was afterwards brigade-major to General Campbell, of Monzie, in Scotland. After his appointment to the 14th Regiment, in which he served over twenty-five years, he was on the staff of Sir George Prevost, in the West Indies. He came home as a major, having obtained his majority 29th April, 1802. He served with the 1st battalion in Ireland, in 1806, and in India in 1807-8. In the latter year an exceedingly severe illness compelled him to invalid home from Bengal. On the passage he was all-but shipwrecked in in the *Diana*, Indiaman, during a hurricane. He was major of the 2nd battalion during the Scheldt expedition, and as lieutenant-colonel took out the battalion, in 1810, to Malta, where—to its intense regret—it remained while the Peninsular War was raging. Frequent efforts were made to obtain an order for the battalion to proceed to the Peninsula, but in vain. Colonel Burrows got leave to repair to Lord Wellington's head-quarters to try the effect of a personal application, which proved equally fruitless. He overtook Lord Wellington near Salamanca, and was present in one of the artillery combats which preceded the great battle near that place on 22nd July, 1812, but was sent back to Malta. He commanded the battalion at Malta during the visitation of the plague, in 1813; at Genoa, in 1814; at Marseilles, in 1815, where, after the departure of Sir Hudson Lowe, he commanded a division of the Army of Occupation, having its head-quarters at Marseilles. He commanded the battalion in the Ionian Islands, in 1816; at Malta, in 1817, when he returned home with it.

When the battalion was reduced he exchanged to the 64th Regiment, at Gibraltar. He became a major-general and brevet general, 10th January, 1837. He died 23rd February, 1848, and was buried at Alverstoke, Hants. He left no papers bearing on the history of the 14th Regiment.

[Particulars furnished by Captain Montagu Burrows, R.N., Chichele-Professor of Modern History at the University of Oxford, the eldest surviving son of Lieutenant-General Burrows.]

Lieutenant-General Sir Kenneth Douglas, Bart.

(FORMERLY KENNETH MACKENZIE.)

Originator of Light Infantry Drill in the British Army.

Lieutenant-General Sir Kenneth Douglas, Bart., entered the army on 26th August, 1779, at the age of thirteen, as ensign in the 33rd Foot, with the depôt of which he served in Guernsey, until the peace of 1783, when he was put on half-pay as a lieutenant. He was brought on full-pay in the 14th Foot soon after, and served with the regiment in Jamaica until its return home, in 1791. As a lieutenant he commanded the light company of the 14th Foot, forming part of a flank battalion at the sieges of Valenciennes and Dunkirk, at the latter of which he was severely wounded by a grape shot in the shoulder. He was promoted to a company in the 90th Foot on 13th May, 1794, and soon after succeeded to a majority and a lieutenant-colonelcy in that regiment, with which he served in the expedition to Isle Dieu and at Gibraltar. He commanded a battalion of grenadiers and light companies in Portugal, in 1797, which he drilled as light infantry. He was deputy adjutant-general at the reduction of Minorca, in 1798, and afterwards held the same post for two years in the Mediterranean, and at the same time held command of the 90th Foot during the absence of Lieutenant-Colonel Rowland Hill—afterwards General Lord Hill. Mackenzie devised and introduced the system of breaking up a battalion into skirmishers, supports, and reserve, which was seen and approved by Sir John Moore, and subsequently introduced by him at Shorncliffe (See Moorsom's *Historical Records, 52nd Light Infantry*). Mackenzie served with the 90th Foot in Egypt, and on the 13th March, 1801, when the regiment was prominently engaged with French cavalry before Alexandria, and brought it out of action when Colonel Hill was disabled. He was promoted to a lieutenant-colonelcy in the 44th Foot. When it was decided to form a regiment of light infantry, experimentally, General Moore's regiment, the 52nd, was selected for the purpose, and Mackenzie was transferred to the regiment to instruct it in his system of light infantry drill and manœuvre, under Moore's direction. A violent concussion of the brain—caused by a fall from his horse while at Shorncliffe—disabled Mackenzie for several years, which he spent on half-pay. He was a colonel on the staff under General Graham for a short time, at Cadiz, in 1810. In 1811 he was made a major-general, and appointed to the Kentish District, all the light infantry in England being placed under him for instruction. He served in North Holland in 1813-14, and was selected by the Prince of Orange to command at Antwerp, in which post he was continued by the Duke of Wellington until the withdrawal of the British troops from the Netherlands. After Waterloo, Mackenzie succeeded to the family baronetcy in 1831, and assumed the name of Douglas. He died a lieutenant-general and colonel of the 58th Regiment.

MAJOR-GENERAL WELBORE ELLIS DOYLE.

Colonel, 53rd Foot.

MAJOR-GENERAL WELBORE ELLIS DOYLE, a younger son of William Doyle, Esq., Master in Chancery, in Ireland, and brother of General Sir John Milley Doyle, G.C.B., K.C., was born in 1758, and on 12th December, 1770, was appointed ensign in the 55th Foot, in which he became lieutenant, 17th February, 1773, and captain-lieutenant, 5th November, 1777. In 1778, Lord Rawdon, afterwards Earl of Moira and Marquis of Hastings, undertook to form a regiment at Philadelphia out of the Irish who were constantly deserting from the ranks of the American line. The corps was called the "Volunteers of Ireland," and Captain-Lieutenant Welbore Ellis Doyle was appointed acting lieutenant-colonel. Like many other officers appointed to loyal provincial corps, he appears to have sold his commission in the 55th Foot, as his name disappeared from the

Army List in 1779. The Volunteers of Ireland were a difficult lot to manage, and their officers, although selected for special fitness for their posts, were powerless to prevent the men deserting almost as fast as they joined. Lord Rawdon is said to have adopted a singular expedient. He solemnly handed over a deserter caught in the act to be adjudged by his comrades. The officers withdrew, and a few minutes later the deserter was seen hanging from a neighbouring tree. His comrades had lost no time in lynching him, and the example is said to have proved most effectual. With all their faults, the "Volunteers of Ireland" fought splendidly. At the battle of Camden, under Lord Cornwallis, half the number killed and wounded belonged to the regiment, and in the second action near the same place, on 25th April, 1781, otherwise known as Hobkirk's Hill, the proportion of casualties was still larger. In 1782, the "Volunteers of Ireland," which had been previously regarded as a local corps, in recognition of their services, were made a line regiment, with the title of the 105th (King's Irish) Regiment of Foot; and Lord Rawdon and Welbore Ellis Doyle were appointed respectively colonel and lieutenant-colonel by commissions dated 21st March, 1782. The regiment was disbanded at the peace, in 1783, and the officers placed on half-pay. On 13th March, 1789, Lieutenant-Colonel Doyle exchanged to full-pay in the 14th Foot, which he joined on its return home from Jamaica, and commanded at the battle of Famars, 23rd May, 1793, as described in another place. Later in the same year, Doyle was appointed adjutant-general of a force placed under the command of the Earl of Moira (Lord Rawdon), to co-operate with the French Royalists in La Vendée. After some desultory operations in the Channel, the troops landed at Ostend in the following summer, and marched to join the Duke of York's army at Mechlin. A redistribution of brigades then took place, and Lord Moira and Doyle returned home. The latter was employed for a time on the staff at Southampton; became a major-general in 1795, and colonel of the 53rd Foot in 1796. Soon after he was appointed to the command of the troops in Ceylon, where he died in 1798, in the prime of life.

The late Sir Francis Doyle, Regius Professor of Poetry at the

University of Oxford, has left the following amusing sketch of his grandfather:—

"My grandfather, Welbore Ellis Doyle, though some years younger "than Sir John, was his senior in the service. Sir John had at first "been intended for the Bar, and this, of course, threw him back as a "soldier. He was a man of a different type from his elder brother— "scarcely as amiable, and with less of what is commonly called talent; "but he possessed immense force of character, and a gift of "dominating others, which, as it is one of the greatest of human "powers, so it is one of the most difficult to understand or explain. "He died in the vigour of life as Governor of Ceylon. The mountain "centres of that island were still under the rule of the kings of Kandy, "so that all Europeans were confined to the unhealthy districts of the "coast; and this proved fatal to him. Had it not been for his "untimely death, his extraordinary personal qualities must have "secured him the highest distinction; as it is, he has left a mark "upon the British army, which, though trifling in itself, ought yet, "among his brother soldiers, to keep his name alive. To him it is "owing that the revolutionary tune, 'Ça ira,' has remained to the "14th Regiment since 1793 as their chosen quick march. At the "battle of Famars, in that year, the French attacked so fiercely that "his regiment wavered for a moment. The revolutionary fever, in "truth, blazed forth as a new element in war, and everywhere the "discipline learnt under average drill sergeants was at a loss how to "meet it. He, however, was not at a loss; for, dashing to the front, "he called out in a loud voice, 'Come along, my lads, let's break "these scoundrels to their own damned tune. Drummers, strike up "Ça ira!' The effect was irresistible, and the enemy found them-"selves running away (it was an Irish exploit, and a bull is excusable) "before they could look around. Again at the siege of Valenciennes "a redoubt had to be stormed, and he was selected to storm it. He "called his men together, and addressed them thus: 'My lads, the "general in command has done us a great honour. We have been "selected to perform an important, and, I will not disguise from you, "a dangerous duty. We have to carry yonder redoubt, said to be "mined underneath; we must carry it therefore in such a fashion that

"the enemy may not have time, as he retires, to blow us and it up
"together. I want a hundred of you to follow me there—volunteers,
"ground your arms!' The whole regiment grounded their arms, as
"if by the action of a single will. 'Very good,' continued the
"colonel, 'then I'll take the hundred next for duty.' And with that
"hundred next for duty the redoubt was so rapidly stormed that the
"enemy had no time to explode their mine. His last recorded
"interview with his favourite regiment is not less remarkable, and
"it shows how long a noble strength of mind retains its power over
"those who submit to it, not as slaves but as freemen, because it is
"noble. The 14th Regiment, some years after he had left it, being
"under orders for India, mutinied on the beach of Southampton,*
"where my grandfather was then military governor, and positively
"refused to embark. I have heard from the description of an eye-
"witness how the general (he was then a general) received the news.
"Seizing the excited messenger by the collar, he swung him out of
"the saddle, and, jumping on his horse, galloped down to the scene
"of action. His staff followed him, the eye-witness I speak of being
"one of them. We learned from him that the moment the
"mutineers caught sight of their old colonel, a kind of thrill trembled
"through their ranks, as if an electric current was traversing them.
"He then drew up his horse close in front of the disorganised crowd,
"and quietly gave the following order: 'Grenadiers, recover arms;
"shoulder arms; to the right wheel, march!' The regiment obeyed
"at once, and passed on into the boats without saying another word.
"Then, having enforced discipline, as he was bound to do, the
"general—not one of those officers who govern by mere sternness,
"without any sympathy for their men—inquired into their grievances,
"and insisted that those of which they complained justly should be
"redressed before the transports sailed. My grandfather died before
"his time, and this was a real loss to the British army. Whether I
"should have been better off if he had refrained from joining the
"majority, like his brother John, till past eighty-four, is quite another

* This must have been on embarkation for the West Indies. There is no record of the alleged occurrence.

"question. I fancy, with all his fine qualities, he was a bit of a "Tartar, and how a short-sighted, blundering, unmethodical grand-"son might have fared at his hands is a problem which perhaps it "is just as well should have been left unpresented by destiny, and "unsolved. Indeed, it may be doubted whether any necessity for its "solution would ever have arisen."

Sir Francis Doyle's poem on Famars is given in the *Appendix Volume*.

MAJOR-GENERAL RICHARD GOODALL ELRINGTON, C.B., K.H.

MAJOR-GENERAL RICHARD GOODALL ELRINGTON, C.B., K.H., entered the 14th Foot as ensign, 27th July, 1791; served with it at Famars, Valenciennes, and Dunkirk, where he was severely wounded (see p. 61). He was mentioned by Lord Cathcart in his despatch for distinguished conduct at Gueldermalsen, 8th January, 1795. He was promoted to a company in the 115th Foot, and subsequently transferred to the 2nd West India Regiment, with which he served in the West Indies, and was wounded at St. Vincent. In 1800 he was appointed to the 47th Foot, and served with that regiment in South America in 1806-7, and in the Persian Gulf in 1809. He was appointed lieutenant-colonel of the 47th in 1813, and commanded the regiment for over thirty years, during which time he commanded brigades in the Pindaree and Mahratta Wars, in 1817-18; in Arabia, in 1819-20; and in the First Burmese War in 1825-26. He died a major-general in 1845.

MAJOR-GENERAL MATHIAS EVERARD, C.B., K.H.

MAJOR-GENERAL MATHIAS EVERARD, C.B., K.H., third son of Thomas Everard, Esq., of Randilestown, county Meath, was appointed ensign in the 2nd, or Queen's, at Gibraltar, 28th September, 1804, and became lieutenant, 21st March, 1805. In December, 1805, the company of the Queen's to which he belonged and two others of the regiment, and two of the 54th Foot had a singular series of adventures. Returning home after some years service at Gibraltar, they were captured by a French naval squadron of six sail of the line and some frigates, under Admiral Guillaumont, bound for Mauritius (Isle of

France). The troops were put on board the *La Volontaire*, frigate, and were carried about at sea for about three months, when *La Volontaire* put in to Table Bay for water, in ignorance of the recapture of the Cape of Good Hope by Sir David Baird. The frigate had to strike to Fort Amsterdam and the other batteries, and the troops were put on shore. Everard was temporarily attached to the 54th companies, which were sent to South America with other reinforcements, and served as mounted infantry with the force under Sir Samuel Auchmuty. Whilst so employed, Everard led the forlorn hope at the storming of Monte Video, 3rd February, 1807, when twenty-two out of thirty-two men with him were killed and wounded. For this service he was promoted to a company in the 2nd battalion 14th Foot; was presented with a sword by the Patriotic Fund, and received the freedom of the City of Dublin.

He served with the 2nd battalion 14th Foot in the campaign in Spain in 1808-9, and battle of Corunna, and in the Walcheren expedition. When in the latter he was thanked in General Orders for his conduct at the siege of Flushing, 12th August, 1809, upon which occasion the flank companies of the 2nd battalion 14th Regiment, one of which he commanded, supported by the rest of the battalion, in conjunction with some companies of the King's German Legion, stormed one of the enemy's batteries and effected a lodgment within musket-shot of the walls. Everard commanded one of the companies sent to Tarifa, and was temporarily attached to the 47th, for the defence of that place, in 1810.

After the arrival of his battalion at Malta, he was transferred to the 1st battalion in India, in which he long commanded the light company (No. 10). He commanded a battalion at the siege of Hatrass, in 1817; commanded a provisional battalion of flank companies against the Pindarees, in 1818-19; became regimental major in 1821; and commanded the regiment at the storming of Bhurtpore, in 1825 (C.B., and brevet of lieutenant-colonel). He became regimental lieutenant-colonel 12th July, 1831, commanding it for sixteen years at home, in the West Indies, and North America. He retired 25th December, 1847, and became a major-general in 1851. He subsequently received a distinguished service pension. He died at Southsea, 20th April, 1857.

Dawsons.Ph.Sc.

Lieut. Colonel W. Hewett.
Served in the 14th Regt. at Waterloo.
The last Surviving Officer. Died 26th Oct 1891.

Lieutenant-Colonel Wm. Hewett,
(late unattached.)

The last survivor of the British Army that fought at Waterloo.

Lieutenant-Colonel Wm. Hewett, third son of General the Right Honble. Sir George Hewett, Bart., G.C.B., Commander-in-Chief in India, was born in 1795, educated at the Royal Military College, Sandhurst, and on 19th December, 1811, appointed to an ensigncy in the 22nd Foot. He became a lieutenant in the Bourbon Regiment, 16th July, 1812; lieutenant, 33rd Foot, 23rd July, 1812; captain, half-pay, 92nd Foot, 24th November, 1814; and was brought on full-pay in the 3rd battalion 14th Foot, 15th April, 1815, in which, when not twenty, he served as a captain at Waterloo and at the occupation of Paris. He became captain, 33rd Foot, 24th October, 1816; captain, Rifle Brigade, 14th October, 1823; major, half-pay, unattached, 10th September, 1825; major, Rifle Brigade, 8th June, 1826; lieutenant-colonel, unattached, 19th August, 1828. Retired from the service in June 1836. Died at Southampton, 26th October, 1891.

General the Honble. Sir Alexander Hope, G.C.B., D.C.I., etc.

See "*Biographies of Colonels*," App. A.

Honble. George Thomas Keppel,

Afterwards 6th Earl of Albemarle.

Honble. George Thomas Keppel, son of the fourth earl, was born in 1799, educated at Westminster School, and on 4th April, 1815, was appointed ensign in the 14th Foot. He served with the 3rd battalion in the Waterloo campaign, and afterwards with the 2nd battalion in the Ionian Islands. When the latter battalion was disbanded, he was appointed to the 22nd Foot in Mauritius. He returned home with that regiment in 1819, and was for a time equerry to H.R.H. the Duke of Sussex. He was promoted to a lieutenantcy in the 24th Foot in 1821, was transferred to the 20th Foot in India and served as aide-de-camp to the Governor-General the Marquis of Hastings. With the aid of a scanty knowledge of Persian, picked up during the long voyage out round the Cape, he travelled home from India overland through Persia and Russia. He afterwards published an account of the journey. For a while he was aide-de-camp to the Marquis Wellesley, when lord-lieutenant of Ireland. He attained a company in the 62nd Foot in 1825, and an unattached majority in 1827. He was not brought on full-pay again, but ultimately obtained the rank of full general (on half-pay of his former regimental commission) in 1874. He succeeded to the title on the death of his brother, the fifth earl, in 1851.

Few men were longer known or more generally popular in London society than the late Lord Albemarle. He was the author of several books, in one of which, entitled "*Fifty years of my life*," he recorded his interesting recollections of the Waterloo campaign, extracts from which have been given in this work.

On the seventy-first anniversary of the great battle (18th June, 1886), a silver tankard, subscribed for by the officers of the 1st battalion of the regiment, was forwarded to the veteran earl. A telegraphic answer was received from him the same evening.

'Lord Albemarle, to Colonel Whitting, 14th Regiment, Dublin.

"Lord Albemarle's warmest thanks to his old regiment for the beautiful "letter and tankard, which has been greatly admired by the Prince of "Wales, the Commander-in-Chief, and the Duke of Connaught."

Lord Albemarle, who was one of the last survivors of Waterloo, died in London, 21st February, 1891, aged ninety-two. The following paragraphs appeared in a society paper at the time :

"The seventy-fifth anniversary of Waterloo saw the last of the 18th "June receptions in Portman Square, and the children who are "accustomed to play there will miss this summer the bent figure of "the kindly nonagenarian who took so much interest in their sports. "The junior ensign of Wellington's army retained his memory to "within a few hours of his death, and in the autumn Lord Albemarle "listened with interest to an account of a visit paid to the battle-field "by his daughter, Lady Augusta Noel. For some years the leveés of "the 'junior ensign' replaced the Waterloo banquets of the "commander-in-chief, and they were always attended by the Duke of "Cambridge. Lord Albemarle always maintained his relations with

Dawsons Ph.Sc.

Earl of Albemarle,
Served in the 14th Regt at Waterloo.
from a portrait presented to Officers 1st Battn 1890.

" the 'Old and bold Fourteenth," and invariably received congratula-
" tions from the officers on each anniversary. He delighted to talk
" about his Waterloo reminiscences, and his favourite walk was from
" Portman Square to Apsley House, where he listened to the Duke of
" Wellington's last speech, when he broke down while proposing the
" health of a foreign guest. George Keppel came of a race of soldiers
" and sailors, and his study in London was filled with relics of his
" ancestors, as well as souvenirs of his own career. Perhaps the one
" great sorrow of his life was the sale, two or three years since, of the
" magnificent portraits by Romney, Kneller, and Reynolds, which
" adorned the ancestral home of the Keppels—Quiddenham Hall—
" where Lord Albemarle passed painlessly away on Friday. He had
" been at the time of his death sixty-one years on the retired list; but
" strangely enough he never received any decoration beyond the
" Waterloo medal. His son, Lord Bury, who succeeds him, will be
" sixty next year.

" The death of Lord Albemarle reduces the number of Waterloo
" officers who are still living to two, namely, General George Which-
" cote, and Colonel William Hewett. Curiously enough they all
" belonged, like Lord Albemarle, to the 14th Regiment of Foot,*
" which at the time of the great battle was irreverently dubbed the
" 'Peasants,' on account of its consisting chiefly of raw lads from
" Buckinghamshire. They distinguished themselves in so marked a
" manner at Waterloo as to be described as the most gallant infantry
" regiment of the British line. Lord Albemarle, who was just sixteen,
" carried the colours.

" The late Lord Albemarle, who made his *début* as a soldier in
" 1815, may be said to have finally appeared in that character as
" recently as the opening day of last year's Military Exhibition at
" Chelsea, when, in the capacity of a Waterloo veteran, he was
" supported to the dais and formally presented to the assembled
" royalties amid the plaudits of the spectators, who thus witnessed the
" practical end of seventy-five years' 'soldiering.' "

---

* This, of course, is wrong. General Whichcote belonged to the 52nd Regiment.—Ed.

Major-General Stringer Lawrence.

"*Father*" *of the Indian Army.*

Major-General Stringer Lawrence was a native of Hereford, where he was born in 1695. On 22nd December, 1727, Major-General Jasper Clayton nominated him to a vacant ensigncy in his regiment at Gibraltar. Lawrence's name cannot be found in the books of the Admiralty or Ordnance, and the inference is that most likely he served at the previous defence in the ranks of the 14th or some other regiment there, the muster-rolls of which no longer exist.

He served in the regiment as a subaltern officer for over twenty years, at Gibraltar, in Flanders, and at Culloden, obtaining his lieutenantcy on 11th March, 1736. He appears to have left the regiment in Scotland after Culloden.

In January 1748, Lawrence, then described as "a soldier of great "experience," was appointed to command "all the company's troops "in the East Indies," with the rank of major, a salary of £800 a year, and a seat in Council. He received the King's brevet of "major in "the East Indies only" on 9th February the same year. One of his first acts was the formation of the Madras European Regiment, afterwards famous as the Madras Fusiliers (now the 1st Royal Dublin Fusiliers), out of the independent companies which the East India Company had long maintained for the defence of their posts and factories. Lord Clive first smelt powder under Lawrence at the capture of Devicote, in 1749.

Lawrence held the chief command of the Company's troops during the wars in the Carnatic, amidst which the foundations of our Indian empire were laid, down to 1759, when he retired with the rank of major-general in India, which he held until his death. He died in London, 10th January, 1775. He was buried in the little parish church of Dunchidiock, in Devonshire, and a tall column—a well-known landmark in the West country—was erected to his memory on the neighbouring hills by his friend, Sir Robert Palk, who had been Governor of Madras.

On a monument erected to his memory in Westminster Abbey by the late East India Company, is inscribed the word "Trichinopoly," and on the shield of a figure of Fame, "For discipline "established—fortresses protected—settlements extended—French "and Indian armies defeated—and peace restored in the Carnatic."

Sir John Malcolm has said of Lawrence *(Life of Clive*, vol. iii.) "He neither was nor pretended to be a statesman, but he was an "excellent officer. He possessed no dazzling abilities, and his acts "never displayed the brilliancy which men admire as the accompani- "ment of genius; but, nevertheless, he was a rare and remarkable "man. We trace in all his operations that sound practical knowledge "of his profession, which, directed by a clear judgment and firmness, "secured to him an uninterrupted career of success, under circumstan- "ces of great danger and difficulty. As one of the chief causes of this "success, may be mentioned the absence of that common but petty "jealousy which renders men afraid lest they should detract from "their own merit by advancing that of others. . . . . Lawrence "early discovered and fully employed the talents of those under his "orders, and we find him, on all occasions, more forward to proclaim "their deeds than to blazon forth his own. . . . . . To this "quality, the best test of a high and liberal spirit, England is "principally indebted for all the benefits she derived from the "successes of Clive."

Lieutenant-General Sir Jasper Nicolls, k.c.b., *Colonel of the 5th Fusiliers.*

Lieutenant-General Sir Jasper Nicolls, k.c.b., entered the army as ensign in the 45th Foot, 24th May, 1793, and served with part of that regiment in the West Indies for nearly six years, during which time he was employed as acting adjutant, paymaster, and deputy judge-advocate. In 1802, being then a captain in the regiment, he went to Bombay as aide-de-camp to his relative, General Oliver Nicolls, commander-in-chief in that presidency. In 1803 he joined the force in the field, under command of Sir Arthur Wellesley, immediately after the battle of Assaye, as a volunteer. He was attached to the 78th Highlanders, which was short of officers, and commanded a company of that regiment at the battle of Arguam and the siege and capture of Gawilghur. Rejoining the 45th as major, he served with it in the Hanover expedition in 1805, and in South America in 1807. At Buenos Ayres, when the attack on the city failed, he distinguished himself by retaining possession of the Presidencia, and capturing two guns brought against it by the Spaniards. For this service he found himself, on landing at Cork, gazetted to the lieutenant-colonelcy of the York Rangers.

He was transferred to the 14th, and remained on the strength of that regiment as a lieutenant-colonel until his promotion to major-general, 12th August, 1819. He commanded the 2nd battalion of the regiment at Corunna (mentioned in despatches, and gold medal), and in the Walcheren expedition, and at the siege of Flushing (mentioned in despatches). After holding the appointments of assistant adjutant-general, under Sir H. Calvert at the Horse Guards, and deputy adjutant-general in Ireland, he was made quarter master-general of the King's troops in India.

During the early part of the first Nepaul war he was selected by Lord Moira, afterwards Marquis of Hastings, to take command of a force consisting exclusively of Sepoys, with which he attacked and captured Almorah, the capital of Kumaon, which province—together with that of Ghurwal—was annexed to the British dominions (mentioned in despatches, and C.B.) He commanded a brigade in the Pindaree war of 1817-18. He returned home on promotion, but in 1825 was reappointed to the staff in Bengal, and commanded the second division of the army at the storming of Bhurtpore. He returned home in 1831. In 1838 he was appointed Commander-in-Chief at Madras, and was Commander-in-Chief in India from 1839 to January, 1843. He became a lieutenant-general 10th January, 1837. He was colonel commanding of the 93rd, 38th, and 5th Fusiliers. He died at his seat near Reading, 4th May, 1849.

Colonel Francis Skelly Tidy, C.B.,

24th Foot.

Colonel Francis Skelly Tidy, C.B., son of the Rev. T. Holmes Tidy, M.A., entered the army as a volunteer in the 43rd Foot, with which he served five months, and in 1792 was appointed to an ensigncy in the 41st. He embarked with the latter corps for the

West Indies, and during the occupation of the French Islands was stationed at Point à Pitre, Guadaloupe, where a mortality of ten to thirteen men daily reduced the regiment to ninety-six men. When Guadaloupe, after being defended inch by inch, was retaken by the French, the 41st had only two officers and twenty men fit for duty. Tidy was taken prisoner and kept for fifteen months on board a hulk, when he was sent home to France, and was allowed to visit England on parole. He was appointed adjutant of the 43rd, and embarked for the West Indies as a private person, being still a prisoner until exchanged. On arrival in the West Indies, Tidy was appointed to a company in the 1st West India Regiment, with which he served against the brigands in St. Lucia. Transferred to the 1st Royals, he was one-and-a-quarter years an assistant quarter-master-general in Scotland, and in September, 1802, joined the 2nd battalion of the Royals at Gibraltar. In May, 1803, he embarked a third time for the West Indies, his company being one of three which embarked with sealed orders at two hours' notice. These companies were afterwards joined by the rest of the battalion, and took part in the capture of St. Lucia, 22nd June, 1803. He was appointed Colonial Secretary of the island. This post he subsequently resigned, and was sent on detachment to Dominica and employed as major of brigade and aide-de-camp to Sir Wm. Myers and to Sir George Prevost, commanding the troops in the West. He was promoted to a majority in the re-formed 8th West India Regiment, and appointed to the 2nd battalion 14th Regiment on 10th September, 1807. He served as an assistant adjutant-general at Corunna and in the Walcheren expedition, and as major of the 2nd battalion 14th Regiment in Malta, during the visitation of the plague in 1813, and afterwards at Genoa and Sarzana. He became a brevet lieutenant-colonel, 4th January, 1813, and was sent home to the new 3rd battalion 14th Regiment, which, as major and brevet lieutenant-colonel, he commanded in the Waterloo campaign (mentioned in despatches, C.B., and Waterloo medal). After disbanding the 3rd battalion, he served with the 2nd battalion in the Ionian Islands. Sir Thomas Maitland, the Lord High Commissioner ("King Tom"), who used to say that Colonel Tidy was one of the few officers who understood him, made him

Resident at Zante. Tidy took the 2nd battalion home and disbanded it, and subsequently, when still a major and brevet lieutenant-colonel, commanded the regiment in India. He was appointed lieutenant-colonel, 44th Foot, 25th November, 1825. He served as deputy adjutant-general to Sir Archibald Campbell in the first Burmese war, and in 1826 was one of the officers deputed to arrange terms of peace with the Burmese. With his rare gift of conciliating all sorts and conditions of men, Tidy won the friendship and confidence of the Burmese prime minister, helping, in no small degree, to the success of the negotiations. After a few years as Inspecting Field Officer of the Glasgow Recruiting District, he exchanged to the command of the 24th Foot in Canada, in March, 1833. He died at Kingston, Upper Canada, 9th October, 1835. A monument was erected to his memory at Kingston by the officers and men of the 24th Regiment.

It was well known that Colonel Tidy always expressed a wish to die in the regiment in which he had served the longest—the 14th. In proof of the respect in which his memory was held in his old corps, the following anecdote is told: "Six years after Colonel Tidy's death, "the 14th, which had come to Kingston, lost their sergeant-major— "Maltby—who had been a sergeant in the 3rd battalion under "Colonel Tidy, at Waterloo, and whose last request was to be "buried near his old Colonel. This was done, and at the special "wish of the men, the firing-party of the 14th was so drawn up as to "fire the parting volleys over *both* graves."

Colonel Tidy had two sons in the army, the elder of whom—the late Major-General Thomas Holmes Tidy, some time Deputy Adjutant-General in Jamaica—served for twenty-eight years (1825-53) in the 14th, with which he was present at the storming of Bhurtpore.

Colonel Tidy's daughter, who married Captain Ward of the 91st Grenadiers, was a well-known authoress. From a small volume, entitled, *Reminiscences of an Old Soldier*, by this lady, much of the information relating to her father has been taken.

General Sir William Wood, K.C.B., K.H.

See "*Biographies of Colonels*," App. A.

General Sir James Watson, K.C.B.

See "*Biographies of Colonels*," App. A.

III.

SUCCESSION OF LIEUTENANT-COLONELS

COMMANDING

DURING THE LAST FIFTY YEARS.

Mathias Everard, C.B., K.H.	12th July, 1831
Maurice Barlow, C.B.	24th Dec., 1849
Ralph Budd	27th Jan., 1857

REGIMENT FORMED IN TWO BATTALIONS.

FIRST BATTALION.

Ralph Budd - -	27th Jan., 1857	T. P. Cosby - -	7th March, 1882
John Dwyer - -	13th July, 1867	Reg. Whitting -	7th March, 1886
W. H. Hawley -	8th May, 1872	F. W. Harington -	7th March, 1886
C. E. Grogan - -	15th Aug., 1877	*Commanding Batt.*	7th March, 1888
F. Barry Drew, C.B.	13th Sept., 1879	C. S. Gordon - -	7th March, 1892

SECOND BATTALION.

E.W. D. Bell, V.C.	8th Jan., 1858	D. S. Warren, C.B.	30th March, 1878
Sir J. E. Alexander	30th March, 1858	E. W. Saunders -	1st July, 1881
C. W. Austen - -	10th June, 1862	R. S. Lemon - -	23rd Dec., 1886
W. C. Trevor, C.B.	8th Dec., 1863	*Commanding Batt.*	20th March, 1887
J. S. Thomson -	30th April, 1873	A. W. Noyes - -	20th March, 1891

THIRD AND FOURTH BATTALIONS FROM 1881.

THIRD BATTALION.

G. J. Hay	24th Sept., 1873

FOURTH BATTALION.

W. Pollard - - *Lieutenant-Colonel Commandant.*	24th Aug., 1871	R. F. Meysey-Thompson - -	19th Jan., 1889
G. P. Fawkes - -	3rd Feb., 1887		

IV.

ROLL OF OFFICERS OF THE PRINCE OF WALES' OWN (WEST YORKSHIRE) REGIMENT.

(From the Monthly Army List for January, 1893.)

Honorary Colonels: Heyland, General A. T., C.B., *1st and 2nd Battalions.*
H.R.H. George F. E. A., Duke of York, K.G. *3rd Battalion.*

1st and 2nd Battalions.

Lieutenant-Colonels (2):

2 Noyes, A. W.
1 Gordon, C. S.

Majors (8):

2 Price, A. J.
1 Ruttledge, A.
d. 1 Penno, T. W. L.
1 Grant-Dalton, G.
2 Adye, C. G.
s. Kitchener, F. W.
1 Mills, E. C.
2 St. George, A. W.

Captains (12):

m. *Cox, C. H.*
2 Vowell, H. A.
mc. *Lester, C. M., p.s.c.*
s. *Wemyss, G., p.s.c.*
Fry, W.
m. *Heigham, C. J. M.*
v. *Walker, H.*
2 Swaine, G. W.
2 Vialls, H. G.
d. 2 MacAdam, F. R. P.
2 Yale, J. C.
1 Critchley-Salmonson, H. B. S.
2 *Watts, H. E., Adjt.*
1 Ward, T. R. R.
1 *O'Donnell, H., Adjt.*
2 Cayley, W. de S.
1 Bomford, G. S.
1 Alexander, D. H.
1 Phillips, G. F.
1 Trevor, H. B. C.

Lieutenants (19):

2 Towsey, F. W.
2 Stephen, A. J.
2 Pearce, F. B.
d. 2 Barrington, T. P.
2 Lang, G. G.
d. 1 Carlisle, R. H.
1 Berney, T. H.
2 Lush, R. F.
2 Fisher, A. A.
Moore, C. H. G. (prob.)
Hutchinson, C. R. M. (prob.)
1 Drew, G. B.
Haldane, C. L. (prob.)
Carey, W. S.
2 Gardiner, G. F.
1 Bartrum, J. S.
Cullen, E. H. S. (prob.)
Colan, W. R. B. (prob.)
1 Ryall, C.
2 Daly, A. C.
Harrison, W. C. W. (prob.)
2 Purchas, E. P. C.
Musgrave, W. W. F. C. prob.)

Lieutenants (continued):

2 Mansel-Jones, C.
2 Hall, W. M.
1 Tew, C. C. B.
1 Carlyon, C. W.

2nd Lieutenants (12):

2 Hawes, C. H.
2 Purvis, E. W. M.
2 Wilde, H. N.
1 Spry, L. H.
1 Ingles, A. W.
2 Luard, C. E.
1 Pellew, F. H.
2 Fulton, H. T.
2 Ames, A. G.
1 Paget, J. B.
1 Costello, E. W.
1 Howard, T. N. S. M.

Paymaster:

2 Swaine, G. W., *captain (acting).*

Adjutants:

2 Watts, H. E., *Capt.*
1 O'Donnell, H., *Capt.*

Quarter-Masters.

2 Scott, R., *Hon. Capt.*
Pye, F., Hon. Lieut.
1 Lowing, W. H., *Hon. Lieut.*
Wilson, T., Hon. Lieut.

3rd Battalion.

Lieutenant-Colonel:

ps. Hay, G. J., *Hon. Col.*

Majors:

I'Anson, J., *Hon. Lieut.-Col.*
ps. Bickaby, J., *Hon. Lieut.-Col.*

Captains (8):

ps. Trafford-Rawson, H., *Hon. M.*
Hine-Haycock, R. W. *Hon. M.*
ps. Sagar-Musgrave, A. M.
ps. Benson, J. M.
Boynton, F.
ps. Straker, J.
ps. Stott, J.

Lieutenants and 2nd Lieutenants (12).

Lieutenants:

Bartlett, Sir E. A., *Knt.*
Whiteley, A. S.
Vipan, R. C. H.
ps. Edwards, C. W., *I. of M.*
ps. Saunders, F. A.
Pearson, W. S.
Hodson, M. S.

2nd Lieutenants.

ps. Hambrough, W. D. C.
Massey, W. F. E.
ps. Benwell, C.

Instructor of Musketry:

Edwards, C. W., *Lieut.*

Adjutant:

Cox, C. H., *Captain W. York Regt.*

Quarter-Master:

Pye, F., *Hon. Lieut.*

Medical Officer:

Ramsay, J., M.D., *Surg.-Maj.*

4th Battalion.

Hon. Colonel:

Fawkes, G. P.

Lieutenant-Colonel:

Meysey-Thompson, R. F., *Hon. Col.*

Majors:

ps. Maude, W. W., *Hon. Lieut.-Col.*
Turnor, W. W., *Hon. Lieut.-Col.*

Captains (8):

ps. Gott, W. H., *Hon. M.*
Lee, J. T., *Hon. M.*
Pollard, W., *Hon. M.*
ps. Bennett, A. C., *Hon. M.*
Bulkeley, H. C.
Murray, H. S.
Mahon, W. H.
Tottie, J. B. G.

Lieutenants and 2nd Lieutenants (12).

Lieutenants:

Scaife, G. S. G.
Johnstone-Scott, C.
Swainston Strangwayes, D'A. E.
Swainston Strangwayes, J. P. N.
Scarlett, J. L.

2nd Lieutenants:

Herbert, G. W.
Yorke, H. R.
Jobling, C. E.

Adjutant:

Heigham, C. J. M., *Capt. W. York Regt.*

Quarter-Master:

Wilson, T., *Hon. Lieut.*

Medical Officer:

Walker, J., *Surgeon-Lieutenant-Colonel.*

HISTORICAL RECORDS

OF THE

14th REGIMENT.

APPENDIX B,

OR

VOLUME II.

MISCELLANEOUS.

Devonport:
A. H. SWISS, 111 AND 112 FORE STREET.

APPENDIX B.

(OR VOLUME II. – MISCELLANEOUS.)

V.

SOME LETTERS RECEIVED BY THE 14TH REGIMENT.

Horse Guards Letter to the old Second Battalion.

Horse Guards,
2nd January, 1816.

Sir,

I am directed by His Royal Highness the Commander-in-Chief to acquaint you that the Prince Regent has been pleased, in the name and on the behalf of His Majesty, to approve of the 2nd battalion 14th Regiment of Foot being permitted to inscribe on their colours and appointments the word "CORUNNA," in consequence of the distinguished conduct of the battalion in the action of 16th January, 1809, near that town in Spain.

I have the honour, etc.,
(Signed) WM. WYNYARD.

Lieutenant-General Calvert, or
Officer Commanding 2nd battalion 14th Foot.

Horse Guards Letter to the Regiment.

Horse Guards,
2nd March, 1816.

Sir,

I have the honour to acquaint you that as the 3rd battalion of the 14th Regiment of Foot bore a distinguished part in the memorable battle of Waterloo, and on that occasion received the Prince Regent's gracious permission to bear on its colours and appointments the word "WATERLOO," and as that battalion has been reduced, and the men are to be transferred to the 1st and 2nd battalions, His Royal Highness the Prince Regent has been pleased to approve of the 1st and 2nd battalions of the 14th Regiment being permitted to bear on their colours and appointments, in addition to any badges or devices that may have been heretofore granted to the regiment, the word "WATERLOO," in commemoration of the distinguished gallantry displayed by the 3rd battalion in the action on 18th June, 1815.

I have the honour, etc.

(Signed) H. CALVERT,
Adjutant-General.

The Officer Commanding
1st battalion 14th Foot.
The same to Officer Commanding 2nd battalion 14th Foot.

Letter received by Officer Commanding the Regiment, at Barbadoes.

Sir,

We have the honour to acquaint you, the following memorandum appeared in the *Gazette* of the 11th inst., viz., "His Majesty has been "graciously pleased to permit the 14th Foot to bear on its colours "and appointments, in addition to any other badges or devices which "may heretofore have been granted to the regiment, the word "'TOURNAY,' in commemoration of the distinguished conduct of the "brigade, consisting of the 14th, 37th, and 53rd Regiments, in the "action fought near Tournay on the 22nd May, 1794."

We have the honour, etc.,

(Signed) COX AND CO.

London, 12th March, 1836.

Letter from the Inhabitants of London, Canada.

We, the inhabitants of the town of London, understanding that the head-quarter division of the 14th Regiment will march from this garrison on Monday morning next, cannot permit you, the officers, non-commissioned officers, and men of that distinguished regiment, to take your departure from amongst us without expressing the high sense we entertain of the excellent and very soldierlike conduct of that corps during its stay in this town.

The regiment which you replaced in the garrison of London, the gallant 83rd, left behind them a character which caused in many of the inhabitants the feeling that "we should never see their like again," for which reason, perhaps, the 14th came amongst us with greater disadvantages than any of the regiments that had preceded it; that feeling ceased to exist very soon after the arrival of your regiment in this place, owing to the desire evinced by the officers to create a friendly feeling, and to promote British sports among the inhabitants, which they have done with more success than had yet attended similar attempts.

We have much pleasure, also, in bearing testimony to the conduct of the men, which has been such as to maintain and support the true character of the British soldier.

Accept, then, our hearty wishes for your welfare, and be assured that the future career of the 14th Regiment will be watched with much interest by many of the inhabitants of London.

On behalf of the President and Members of the Board of Police, and the inhabitants of the town of London.

(Signed) EDWARD MATTHEWS,
President of Board of Police.

London, Canada,
September 16th, 1843.

Letter to the Second Battalion on their departure from New Zealand.

Head-Quarters, Auckland,
15th October, 1866.

The Major-General Commanding cannot allow the 2nd battalion 14th Regiment to leave New Zealand for the Australian Colonies without recording his sense of the value of their services in the country during an eventful period, and more especially in the late operations, in which they bore so prominent a part under his own observation. In their gallantry at the assaults on the enemy's stronghold, and in their exemplary endurance of the unusual fatigue of the march through the forest behind Mount Egmont, they exhibited the highest qualities of British soldiers; and again would the Major-General acknowledge their distinguished services. The high opinion of the corps which the Major-General formed from witnessing its valour in the field, has been raised still higher by finding at his recent inspection that its interior economy and discipline are unexceptional. To Lieutenant-Colonel Trevor, C.B., who bravely led and still ably commands them, to brevet Lieutenant-Colonel Dwyer, and to all the officers, non-commissioned officers, and men, the Major-General now bids farewell, with the sincerest wishes for their continued welfare. He is sure that wherever the 2nd battalion 14th Regiment serves in peace or in war, their future career also will reflect credit on themselves and honour on the character of the British army.

Letter to Officer Commanding First Battalion 14th Foot, from Major-General Beatson, Commanding the Allahabad Division.

The half battalion of your regiment having been ordered to Cawnpore, I cannot allow them to leave Allahabad without expressing to you my regret at losing them from this station. The gentlemanly bearing of Captain Furneaux and the other officers, had made them universal favourites at Allahabad, while the conduct of the non-commissioned officers and men, since I have had them under my command, has been most soldierlike.

"Horse Guards, War Office,
"18th July, 1873.

"*Re* WHITE HORSE.

"Her Majesty the Queen has been graciously pleased to approve "of the 14th Buckinghamshire Regiment of Foot wearing on the "second, or regimental, colour, the 'White Horse,' and motto, '*Nec* "*Aspera Terrent*,' as formerly worn in the Grenadier corps.

"(Signed) J. W. ARMSTRONG, D.A.G."

Extract from Brigade Orders by Brigadier-General Sir H. Gough, V.C., K.C.B., Commanding Sialkote Brigade.

The Brigadier-General cannot permit this battalion to leave his command without placing on record his sense of the good conduct and discipline of the regiment, and their general efficiency during the time they have served under his immediate command, and during their three years stay in Sialkote.

Brigadier-General Sir H. Gough wishes the 2nd battalion West Yorkshire Regiment a prosperous and successful career wherever their future might lead them.

27th October, 1885.

VI.

LETTERS, ANECDOTES, LEGENDS, Etc.

RELATING TO THE REGIMENT.

Letter written by Captain (afterwards Major-General) William Turnor from the field of Waterloo.

Mont St. Jean,
Field of Battle, ten miles from Brussels,
19th January, 1815.

Though the papers will give you better information relative to the sanguinary conflict of yesterday, I am unwilling to permit a courier to proceed to England without acquainting you that your friends of the 14th are well. The contest just terminated commenced at 12 o'clock, and lasted without intermission until 9 in the evening. It was the most bloody, as well as the most decisive, battle that has been fought since the French Revolution, and its results will be more important than those of Leipsig. The cannonade was tremendous on both sides. The French fought with desperation, and I am fully convinced that no troops on earth but English could have won the victory. They are, in action, *savagely* courageous. The cavalry of the enemy particularly distinguished themselves, and charged our infantry, when in squares, four, five, and six times, but they were not to be broken. Our infantry had immortalized itself, and its conduct has never been surpassed—indeed, never equalled. We were so fortunate as not to have suffered any very great loss, having been posted on the right of the line to hold in check a very strong body of the Imperial Guard.

The whole day we were exposed to the fire of several batteries of artillery, particularly to that of two pieces brought to bear upon us. The situation was trying in the extreme, but our young soldiers behaved well. They would have been glad to have been led against infantry, but we dared not lose sight of the cavalry. Many regiments, both of cavalry and infantry, are almost annihilated, but it is said that some regiments of dragoons were not as forward as they should have been. One regiment of hussars* is particularly mentioned as having refused to charge. The field of battle this next morning presents a most shocking spectacle, too dreadful to describe. Every effort was made by Buonaparte to turn our right, within two hundred yards of which we were posted. He showed the greatest courage, and led in person many charges, both of cavalry and infantry. Those officers who were in the Peninsula regard the battles there as mere *combats* in comparison with that of yesterday, and they may easily be credited when we reflect that Napoleon was fighting for his crown, and was opposed to the greatest general of the age. The escape of Lord Wellington amounted to a miracle, for he was exposed during the whole day to the hottest fire. We know not the amount of our loss, but it must be great indeed.

Extracts from a Letter from Ensign Keowan.—The Night After Waterloo.

It was a fine moonlight night. . . . . The night was calm, and the cries and lamentations of the wounded were most awful, some with their last breath faintly calling upon the names of those that were nearest their heart, and whom they should never see again; some calling upon their God, and others faltering the name of some beloved friend or relation, and awful were the execrations of the delirious, who were heard cursing the foe by whom they had fallen, and many who bore the semblance of death, when they felt the plunderers stripping them, would lift up their heads and *look* what they could not utter.

I wandered through the field by the light of the burning chateau and the moon to look for something to eat (for the call of nature was pressing) and for some straw to sleep upon. I found the hind leg of some animal (it was either a cow or a horse) with the flesh half

---

* A new German levy—not part of the King's German Legion.

stripped off it. I took it along with me and dressed a bit of it myself, for my servant was busy robbing the dead, "that were past all pain." I also found a bundle of straw, and with this on my back and the leg of the beast in the other hand, dragging it along the ground, I frequently fell over the dead bodies. It was the most romantic and the most awful sight I ever witnessed, or *could* be witnessed, to see by the moon and the flames of the farmyard, that was still burning, the faces of the dying and the dead; from the remembrance of it, I should like to see that part of it over again, it was so sublimely beautiful. We could get no water to drink that was not deeply tinged with blood; such was the wine we drank at our cannibal feast. The extent of the battle was so great, and the slain so thick in every place, that not one drop of crystal water could be obtained within less than three miles of the place.

The place where our regiment bivouaced that night was in the grove or orchard * (oh, how I should like to retrace the spot!); it was among the avenues of the trees, all green, except the red stains. Under a little bank at the root of a tree I slept that night on the bundle of straw, and I gave Captain Adams part of my bed, for if we had not made something by way of a bed, they would have taken us for dead, and have stripped us of everything that was worth taking. Before I slept I had the grace to say something by way of thanks to God for having escaped the honour that death had conferred upon the many who *slept sound* beside me. I was greatly fatigued, and had taken a great-coat off a dragoon who had been killed; it was thick with blood, but it kept the dew off. In this manner we lay down and slept tolerably well, with occasional interruption from the shrieks of the dying and from the agitation of our minds, for the waves will roll high after the storm has ceased, and as much of the fight recurred to me as I had time to dream of, and in my sleep I saw the deep columns and the extended line of the enemy as they appeared that morning before we had routed them; I saw their eagles fly, and the cavalry seemed to sweep over the plain amidst the tempest of the battle.

* Of the Chateau of Huguomont. The extracts are taken from letters published a short time ago in *The Brigade of Guards' Magazine.*

The next morning we felt keenly the calls of hunger, and were chilled by sleeping in our wet clothes, which were not quite dry, although we had sat upon the trees that were felled by the shot, opposite the flames, to dry ourselves. . . . . The trees of the grove were all shattered to pieces; there was scarcely any interval between the marks of the shot.

Colonel Tidy at Waterloo.

The following story used to be good-humouredly told in after years by a late general officer. The officer in question, then a colonel, was present on the field of Waterloo as an amateur. Having picked up a fragment of a *buff* regimental colour, he took it to his friend, Colonel Tidy, asking mildly if he had lost it. Colonel Tidy fairly rode at him, shouting, "Lost a colour, sir! No, sir! the 14th never lose their colours!" and wheeled about without deigning to look at the treasure-trove.

More Anecdotes of Colonel Tidy.

Thirty years ago a belief was current among the men of the regiment that the late Major-General Thomas Holmes Tidy—who was a very popular officer—was born on the field of Waterloo and wrapped in the regimental colour. According to Mrs. Ward, the story originated in the following manner: Colonel Tidy's family joined him on the arrival of the army at Paris, and a son—a younger brother of the Major-General—was born immediately afterwards. The Colonel received the news on returning to his quarters from one of the grand reviews held in the presence of the Allied Sovereigns. Some of his officers who had come with him insisted on seeing the new-comer and drinking his health, but an answer was sent that the baby was not dressed; whereupon Tidy, taking the regimental colour from the corner of the room, ran out, and returned with the infant wrapped in its folds.

When the 2nd battalion was at Cephalonia, a young 14th soldier was sentenced to death by a general court-martial for threatening to shoot a superior officer. Mrs. Ward thus relates the sequel:—

"The parade was ordered; the troops marched to the grave to "the solemn strains of the 'Dead March;' the prisoner stood hand- "bound beside his coffin. It was a sad sight, and the Colonel was "the last man to wish to see such a spectacle, but his duty had "to be done, and he did it.

"The prisoner is blindfolded, he kneels upon his coffin, and the "firing party forms up. Colonel Tidy advances and stands between "the bayonets and the prisoner, giving the command himself.

"'Make Ready!'* and the rattle of the muskets is heard over "the whole parade, which is silent as death.

"'Present!' A line of musket-barrels face the Colonel. The "least slip, and he is a dead man. It is a spectacle never to be "forgotten by any of the audience. The prisoner awaits the death- "dealing word. No more. The prisoner is bidden to rise. A "reprieve, at the earnest request of the Colonel, has been granted, "but the remission was accompanied by an order that the ceremony "was on no account to be omitted."

An Incident at the Taking of Cambray.

The following anecdote is related on the authority of a 14th officer present:—

"The colour-sergeant of the 3rd battalion light company was a "remarkably active man. At the attack of Cambray, after placing "our ladders in the ditch, which alone separated us from the citadel, "the colour-sergeant, deeming it his duty to be foremost man, was "the first on the ladder, and had descended a few steps, when Private "William Smith, a truly religious man and an excellent soldier, "hastened to follow, but so quickly that he trod on the colour-

---

* This was the word then in use.

"sergeant's hand, bearing the whole weight of his body on it with "one foot, which gave the sergeant such pain as to cause him to utter "a violent oath in desiring Smith to lift his foot. The latter quietly "exclaimed, 'For God's sake, sergeant, pray don't swear; think "where you are.' He had scarcely finished speaking, when the "sergeant was released by a shot cutting the ladder in two and pre-"cipitating both men into the ditch, without either sustaining further "injury than a blow received by the sergeant on the breast."

Survivors of Waterloo.

The last of the Waterloo officers to leave the regiment was Captain A. Ormsby. He was an ensign in the battle. He left the regiment in 1838, and died on half-pay in 1851.

The last Waterloo man to leave was Captain and Quarter-Master Samuel Goddard. Promoted to sergeant on the formation of the 3rd battalion, he served with it in Captain W. Turnor's company at Waterloo; was quarter-master-sergeant at the siege of Hatras and in the Pindaree and Mahratta wars; and quarter-master at the capture of Bhurtpore, and in the West Indies and North America. He left the regiment at Limerick in March, 1853, and died a Military Knight of Windsor.

The *Cambrian News* of June, 1886, gives the following particulars of the death of, probably, the last survivor of the Waterloo rank and file. The particulars agree with the record of a soldier who, under the name of "John Marsh" or "March" (the name is spelt both ways in the rolls), joined the 3rd battalion on its formation, fought in it at Waterloo, and purchased his discharge from Captain Everard's (No. 10) company of the regiment at Fort William, Bengal, early in 1821.

"On 4th June, 1886, there died at Aberystwith, in his 92nd year, "Mathew John March, a Waterloo veteran; his wife, who had "attained the age of 76, having been buried on the previous day. "Early in the present century, he was made midshipman of a sloop "of war, through the influence of the Duke of Clarence (afterwards "King William IV.), and served at the bombardment of Cadiz. 'Returning to England, he enlisted in the Royal Westminster "Middlesex Militia, and volunteered into the 14th Foot, with the "3rd battalion of which he fought at Waterloo, where he was slightly "wounded, and narrowly escaped death at the storming of Cambray, "where a shot passed under his arm and killed a comrade behind "him. He joined the 1st battalion, and served in the Mahratta war. "With his prize money, and a legacy that came to him at this time, "he bought his discharge, returned to England, and drove a coach "until the railways took away his avocation. He then retired into "private life at Aberystwith, where he was much respected. On his "way home from India he witnessed the funeral of Napoleon at "St. Helena."

Of two officers of the regiment, Lord Albemarle and Colonel Hewett—the latter believed to be the last survivor of the whole British army present at Waterloo—particulars have been given on a preceding page.

The Bhurtpore Gun.

The famous brass gun, which has so long stood in front of the Royal Artillery Barracks at Woolwich, is one of 187 pieces of ordnance captured at Bhurtpore. It was presented by Lord Combermere to King George IV., in the name of the victorious army, of which the 14th was the senior infantry regiment. It was presented by the King to the corps of Royal Artillery and Royal Engineers in 1828. The following description of it is taken from official sources. It bears the following inscriptions :—

ON THE CHASE :

THE FATHER OF VICTORY.
THE REVIVOR OF RELIGION.
MUHAMMAD, AURANG-ZEB, ALAM-GIR.
THE WARRIOR, THE VICTORIOUS KING.

The third line of the translation gives the three names of Aurang-zeb. Their verbal meaning is :—

MUHAMMAD - - -	Extolled.
AURANG-ZEB - - -	Throne-adoring.
ALAM-GIR - - - -	World-subduing.

"The Revivor of Religion" is a title peculiar to Aurang-zeb. The title of "The Father of Victory" was borne by Shah Alam also; the other titles are common to all the Mogul emperors.

ON THE SWELL:

YEAR OF THE HEJIRA 1087,
THE 20TH OF THE REIGN.

1087 of the Hejira corresponds with A.D. 1677; according to Mohammedan chronologists, exactly the twentieth year of Aurang-zeb's reign.

UNDER THE RIGHT TRUNNION:

THE GUN, THE AID OF ALI.

Ali, the hero-saint of the Indian Mohammedans, is invoked by them in every difficulty, and especially in battle. His titles are "The Victorious Lion of God," "The Remover of Difficulties."

UNDER THE LEFT TRUNNION:

ACCORDING TO THE WEIGHTS OF SHAH JEHAN.
THE BALL - 30 SIRS.
POWDER - - 13 SIRS.

The weights and measures, as established by the Emperor Shah Jehan, are those still used in Hindostan. The *sir* is about 2 lbs. avoirdupois. The verbal meaning of "Shah Jehan" is "King of the World."

This gun is 16 feet 4 inches long, and weighs about 17¾ tons; the calibre is 8 inches. Tradition has always maintained that the precious metals enter into its composition, but recent analysis of metal taken from three places dispel this idea. The metal is variable; it contains from 9 to 15 per cent. of lead, with traces of arsenic and antimony; the other components are 3 to 5 per cent. of tin, and the remainder copper. It is remarkable, however, that the exterior of the breech is of a totally different metal from the body of the gun, being, in fact, brass, containing nearly 37 per cent. of zinc, and has been cast over the body of the gun subsequently to the first completion of it, as one of the ornamental scrolls is partly covered by it.

Anecdote of Colonel Everard.

The following story is repeated by several old 14th men, and is undoubtedly true. In the year 1833 the regiment was at Portsmouth in garrison with the 2nd, or Queen's; 8th, the King's; and a division of Royal Marines. Colonel Everard at that time was very sore because the regiment had not been made "Royal Light Infantry," in recognition of its long services in India. It was said that the Colonel had been asked by the authorities whether he would like the regiment made "Royal" or "Light Infantry," but that he refused one without the other. One day, there was a garrison parade—at that time a less frequent occurrence than now—and the brigade was called to attention, the several corps being addressed by their commanding officers as "Queen's," "King's," and "Royal Marines." Fuel was at once added to the smouldering flame in Colonel Everard's breast, and he addressed his regiment: "Neither 'King's,' nor 'Queen's,' nor 'Royal Marines'—only plain 14th, that have done 25 years in India and never unfixed bayonets—'Attention!'"

Anecdote of Colonel John Dwyer.

Colonel Dwyer used to relate how, when the yellow fever was raging in the West Indies, in 1837, he was a lieutenant commanding a company detached to one of the smaller islands. The epidemic was very fatal, so that the company lost twenty-one men in a single week. Dwyer addressed a letter to head-quarters, grimly asking what he should do when the company was dead? The reply was an order for its removal to a less deadly post.

Anecdote of Colonel Maurice Barlow.

When the regiment was in Canada, in 1845-46, the officers gave a dinner to the leading inhabitants where they were stationed. A merchant from Cork, addressing Colonel Barlow—who was in command—said, rather pompously, "Pray, Colonel, have you many Cork men in your regiment?" "No," said Barlow, "there are a good many wooden ones."

The Regiment in 1831.—Letter from an Old Hand.

November, 1892.

Dear Sir,

If any of my recollections of the 14th are of any service, you are most welcome to make use of them, for I have most pleasant recollections of my old regiment, and it did me a world of good.

I joined the 1st battalion in July, or August, 1831, at Chatham, soon after it returned from India. My parents, however, were greatly opposed to my being in the army, and they purchased my discharge in March, 1832, so that I was only in the service about eight months.

Of course I remember several little incidents, and some stories and gossip obtained from my comrades, many of whom had just come back from India. One was that only ten returned to England who went out with the regiment twenty-three years previously; they were, one officer—Captain L'Estrange, eight men, and one woman. The latter was the wife of the hospital sergeant, she had been—I think—nurse-maid to one of the officer's wives, and was a daughter of one of the soldiers.

At Waterloo there were in the 3rd battalion more than four hundred who were then under twenty years of age, and when the French war ceased, those fit for service were transferred to the 1st battalion, in India; for, in 1825, at the siege of Bhurtpore, the front rank of the grenadier company of the 14th all wore Waterloo medals. Many of these were killed at that memorable siege, but a fair number returned to England in 1831, being too old to volunteer or be transferred to other English regiments in India.

At the siege of Bhurtpore two gongs were taken. One—the most beautiful and valuable—was taken in the inner fortress, and this gong became quite part of the Regiment. It was about two-and-a-half to three feet in diameter, and was said to be made of many valuable metals, and of great value. When it was at Haslar I quite well remember that it was placed in front of the main guard house; there was an hour-glass by the side of it, and it was the duty of the sentinels regularly to turn this hour-glass, and then to slowly strike the hour, day and night—the above was part of the drill. Several people used often to congregate at the entrance gates just before the hours, to see and hear it struck. The gong was deep in tone, and it could easily be heard four miles off: many at a distance thought it was a clock. I cannot help thinking that when the 14th was ordered on foreign service, this gong, being of great weight, was left behind in some English military store.

Lieutenant-Colonel Everard was a first-rate officer, and a strict disciplinarian; he was much liked, as the following slight story will prove. A civilian at Gosport happened to say to some of my comrades, who had returned with the regiment from India, "Your "colonel is a little insignificant looking man!" "Little man," exclaimed my comrade, "you should have seen and heard him at "Bhurtpore."

When I joined, on the brass plate, where the belts in front crossed, was inscribed, "Java," "Corunna," "Bhurtpore," "Waterloo." I belonged to the grenadier company, and we wore black bear-skin caps, but they did not wear them in India. The "White Horse" was on them, but I cannot remember what coloured ground it was on. I certainly do not think that the drummers wore white bear-skin caps, but they may have worn white caps.

You ask about the kit; all I can remember about it is the following: Three shirts, three pairs of stockings, I think (not socks), two flannel (cholera) belts, two pairs of boots, two shoe brushes, and one other—a clothes brush (these were kept in the haversack), razor, soap brush, and box. Our names—thus, T. Webster—and our company had to be on everything. I go back, remember, as far as 1831-32, whereas I expect the John Bignell is later, and so is some one else you mention, for you say the latter joined in 1842, nearly eleven years after my time.

My father was a large farmer in Buckinghamshire, where our family had been for a great many years. One of my brothers belonged to the Yeomanry commanded by the Earl of Orkney. This nobleman was a friend of our family, and he most kindly took the trouble to come to Haslar, for I well remember being sent for by our colonel when we were at drill, and on going to the colonel's quarters finding to my surprise Lord Orkney there arranging and carrying out my father's wishes with regard to my discharge. The colonel very much wished me to remain in the 14th, or go into some cavalry regiment, but not to leave the army. However, I felt it my duty to do as my father wished, although I did not obtain my discharge for a month or two, owing to the riots at Bristol and elsewhere and the consequent necessity to keep all the regiments up to their full strength.

I am, yours faithfully,

THOS. WEBSTER.

Another little incident connected with one of the 14th.

In the summer—I think—of 1864, I and a relative drove to Windsor on a Sunday morning, in order that we might attend the service in St. George's Chapel. On our way, after leaving our carriage at one of the Windsor Inns, we ascended the hundred steps, and passing through the Cloisters we saw an old gentleman in full uniform in front of us; as we passed him I noticed that a Waterloo medal was on his breast. I looked at him with greater attention, and turned round, and although it must have been at least thirty-one years since I had last seen him, I am positive that he was the quarter-master when I was in the 14th, in 1831-32. Then it was stated as a fact, that he had been one of the drummer boys at Waterloo.

I cannot remember his name, although I have tried my best, and I have no reference books I can obtain here that would give me the names and old regiments of the Military Knights of Windsor, this old gentleman being one of them in 1863, 1864, or 1865. I have a slight notion that his name was Goddard.

Letter from an Old Hand.

Dear Sir,

In a small barrack room, accommodating eight men, in the Plymouth Citadel, in 1846, were three *Johns*, privates of the 14th Regiment, whose services I annex :—

NAME.	Year of Enlistment.	Year of Retirement.	Honorary Rank on Retirement.	Years' Service.	Remarks, 1892.
John Mills - -	1842	1877	Captain	35	Still alive
John Bignell -	1846	1884	Major	38	,,
John Moore -	1847	1882	Major	35	,,

Extract from "The Carmarthen Journal," March, 1850.

"MILITARY MOVEMENTS.

"On Saturday morning three companies and a half of the 77th "Regiment of Foot, under the command of Major Egerton, dis-"embarked at Blackpool, and marched from thence to this town, "where the men were immediately distributed in billets, to await the "departure of the 14th Foot. On Sunday morning, at St. David's "Church, the Rev. D. Evans, curate, took the opportunity (at the "close of an eloquent and talented sermon,) of delivering a valedic-"tory address to the troops then assembled in the Church, and who "were to leave on the ensuing day. . . . . We need scarcely "say that this well-deserved tribute to the brave corps which has "lately sojourned amongst us, was listened to with breathless "attention.

"At eleven o'clock on Monday morning the troops defiled from "the barracks, under the command of Major Tidy, and proceeded "to Blackpool, where they embarked on board the *Phœnix*, which "steamed for Bristol at half-past one. A vast crowd accompanied "the men to the place of debarkation, and with hearty and reiterated "cheers bade them adieu. It is to be hoped that the present regi-"ment will be equally as peaceable and orderly as its predecessor."

The Regiment in 1843-58. — Notes by an old Colour-Sergeant of the 14th.

SOLDIERS' FOOD AND DRESS IN 1843.—Two meals a day only were served out at this time, *i.e.*, breakfast at 8 a.m., consisting of a pound of bread and a pint of coffee; and dinner at 1 p.m., consisting of three-quarters of a pound of boiled beef, two pounds of vegetables, and a pint of soup.

[Those were the days of the dear loaf, when household bread sold at double its present price, and more. Many correspondents speak of the infamous character of the bread often issued as soldiers' rations. One old hand, who joined in 1830, relates that he had often seen a ration of bread stick fast when flung against a brick wall.]

Dress in marching order consisted of the large ball-topped shako of the day; a coatee, the front covered with white lace; white trousers, laced-up boots, cross-belt, sixty rounds of ball cartridges in the pouch, the great-coat rolled on top of the knapsack, the mess tin behind. The knapsack fitted with slings and breast strap. In the knapsack was carried, in marching order: a forage cap, one pair of white trousers; one shell jacket, not lined; two linen shirts, two pairs of socks, one pair of cloth trousers; one hold-all, complete; three brushes, one pair of boots—representing in all a weight of fifty pounds.

Clothing then allowed: A coatee, a pair of cloth trousers, a pair of ammunition boots. Everything else had to be paid for by stop-pages out of the soldier's pay.

[Another veteran relates that, in the days of white duck trousers, it was customary, when a man ran short of clean ones, to wash and pipeclay a pair before he went to bed, put them under his blanket, and sleep on them. They were then all right for guard-mounting or parade next morning.]

Barrack Room Stories current in Quebec.—Lieutenant and Adjutant John Spence, whose son died quarter-master of the 28th in the Crimea, had reported to him a man for being dirty at guard-mounting parade. This was esteemed a very bad crime. The Adjutant, after looking at the man, said, "You have dirty hands and dirty face; you have dirty buttons and dirty lace. To the guard room! Right about face, double, and be d——d to you. Orderly sergeant, make out a crime against him."

Two men belonging to the grenadier company told the orderly corporal they wanted to speak to the Colonel, who was then walking up and down in front of the officers' quarters. The corporal, approaching the Colonel, saluted, informing him that two men wished to speak with him. "This is not the place; go to the orderly room." One of the men, speaking for himself and comrade, said, "Please, sir, this is Bhurtpore day." "Yes, I know that; I was there," said the Colonel; "What do you want?" "Leave for the day, sir." "Why not put in a pass?" "But we are confined to barracks, sir." "You want to be released from confinement?" "No, sir; if you will lend us the day we will pay it back." "Very well—you may have it, but mind you don't get drunk. Corporal, see they pay it back." The colonel and both the men had been in the battle, the former being wounded.

* * * *

Brown Bess.—The old flint-locks of the 14th were exchanged at Quebec for the new pattern musket.

* * * *

Fire at Quebec Theatre.—The writer mentions a fire in Quebec theatre, of which the following account appeared in a military paper in July, 1846:—"Lieutenant Thos. Hamilton, 14th Regiment, who "perished in the flames, had attended the exhibition in company with "his betrothed bride and her sister. When the flames broke out he "succeeded in getting the sister in safety from the building, and then "returned to seek his intended, when both perished. Lieutenant "Armstrong, of the same regiment, saved himself by a desperate "effort of personal strength, having sprung up to a window, the "position of which he knew, and through which he ultimately forced "his way out."

* * * *

CRIMEAN MEMORIES.—In December, 1854, the women, children, and invalids were sent home. We embarked for the Crimea in a hired transport. The light company was the first to land.

In January, 1855, I was out on night reconnaissance with Sir Colin Campbell, one very cold and frosty night, when my moustache and beard were frozen together.

Another man, named Healey, was startled by a ball passing through the thigh of his trousers. I warned both against lying in such an exposed position; but, it being a little shaded under the bank, they chanced it, as many did. Captain Trevor, Sergeant Hopkins, and Healey can vouch for the truth of this, although it is not recorded in regimental books.

On the 18th June, 1855, Captain Hammersley led the storming party of 200 of the 14th against a 22-gun battery, called the Crow's Nest, at the end of the Woronzoff road. Casualties reported:—15 killed; 45 wounded.

Left the Crimea when peace was proclaimed, bound for Malta. Alexander Gordon put us in shape, after nearly two years in the field. Captain Trevor was made major. A severe shock of earthquake split the walls of officers' mess, and did other damage.

The Orderly-Room Five Pound Note.

There is a story that "once upon a time" the adjutant of the 14th had in his possession a five pound note, which was to be given to the first man in the regiment who succeeded in *minding his own business for the space of twelve months.*

For a very long time, so the story goes, no one claimed it. At last a claimant came forward, and explained his errand to the adjutant;

"mind your own business," was the reply, and the applicant went sorrowfully away.

There is no five pound note about the orderly-room now, so it may be assumed that, after all, somebody was found in the regiment, in the olden days, who did mind his own business.

The Old "Bomb-Proofs."

In the Crimea the regiment acquired the nickname of the "Bomb-proofs," from the fact that it did duty in the trenches for some time without having a man injured. A correspondent writes: "I have "often heard men of other corps say, when forming up for duty, 'Oh, "we shall have a quiet time to-night, for the old *Bomb-proofs* are "with us.'"

14th Regimental Family Names.

The families of Kenyon and Armstrong have long been represented in the regiment, and the name continues, and it is hoped will long continue regimental ones. The late Quarter-Master-Sergeant Kenyon, of the Cheshire Militia, and late of the 14th, entered the latter regiment about 1830.

The families of Johnston and Forrest are also historic, and have been in the regiment some two or three generations.

Regimental Friendships.

There are, I believe, (says a writer in the *Brigade of Guards Magazine*, 1888) many inter-regimental friendships dating from the Peninsular war, some of which may be traced to the simple fact of the regiments having then served together in the same brigade—such as the 43rd and 52nd—a friendship that seems likely to last for many years, in spite of the attempts to obliterate regimental numbers by substituting so-called territorial titles. From this cause, there is little doubt, they will eventually die out, and that some have already done so is much to be regretted.

An old friendship, for instance, used to exist between the 14th, 37th, and 53rd Regiments (known as the "Fighting Brigade") which dated from the action before Tournay, in 1794, when the brigade formed of these three regiments, and numbering one thousand one hundred and twenty men, left no fewer than five hundred and thirty-three on the field on the 18th May; and when the British force was again attacked, a few days later, the same brigade, in its then weakened condition, is said to have actually won the battle. It was said by a historian of the campaign that, "Perhaps there is not on "record a single instance of greater gallantry and more soldierlike "conduct than was exhibited by those three regiments on that "occasion." It is not to be wondered at that a mutual feeling of respect, admiration, and affection should have existed between them, but although such a feeling did exist, and outlive many generations, it is doubtful whether at this moment the soldiers of the West York, Hampshire, and Shropshire Regiments have any peculiar consideration for each other.

Another ancient friendship, dated from the Waterloo campaign; the 14th and the 51st (now 1st Yorkshire Light Infantry). The two regiments were brigaded together throughout the campaign, and so perfect was the harmony existing between them that it was considered extraordinary to see a man or party of men of one corps amusing themselves apart from the other. Colonel Tidy used to illustrate the fact, in his pithy characteristic way, by saying that, "Whenever you "saw a man of either corps drunk, you saw a man of the other corps "taking him home."

Yet another friendship, of later date, was that subsisting between the 14th and the 89th (now 1st Princess Victoria's). The corps had served together in the West Indies, North America, Ireland, and the Crimea. "The more than brotherly feeling manifested whenever the "two regiments met," writes an old Crimean hand, "was something "grand to see."

VII.

"ÇA IRA."

REGIMENTAL QUICK-STEP.

LAROUSSE, a French authority, gives June-August, 1790, as the probable date of the origin of the tune. It was at first a mere chant of liberty, much in vogue with the workmen preparing the ground for the national fêtes of that year:

> Ah ! Ça ira
> La Liberté va s'etablira ;
> Malgré les Tyrans, tout réussira.

Larousse gives the original words and music in full. (See *Dictionnaire Universelle*, vol. iii.)

Like the "Marseillaise" and the "Carmagnole," the air became popular. It was played by the French regimental bands, and became adapted more and more to the truculent spirit of the time.

> Ah ! Ça ira, Ça ira, Ça ira ; les aristocrats à la Lanterne,*
> Ah ! Ça ira, Ça ira, Ça ira ; malgré les mutins, tout réussira,

was yelled by the crowds escorting victims to the guillotine during the Reign of Terror.

The accepted tradition of the adoption of the air as the 14th Regimental Quick-step is, as stated in the preceding historical narrative, that during the attack of the Duke of York's troops on the French works covering the entrenched camp at Famars, in Flanders, on 23rd May, 1793 (see vol. i., pp. 58, 59), the 14th was at first repulsed by a body of French playing the tune, then a recognised national air of the Republic. Lieutenant-Colonel Welbore Ellis

---

* A slang word for the guillotine.

Doyle, commanding the 14th, rallied his men, and bade the drummers strike up "Ça ira," saying "We'll beat them to their own damned tune." The men responded to their leader; the French were beaten; the battery won; and, in the end, the camp at Famars taken. By express order of the Duke of York, the air was adopted as the 14th Regimental Quick-step, and as such has been played ever since, with two brief exceptions—once at Plymouth, when it was thought that the revolutionary character of the tune might give umbrage to some foreign princes, and again when the 1st battalion tried "God Save the Prince of Wales" for a short while, but afterwards reverted to the old tune.

An old 14th man, who entered the regiment in 1842, and served in it for thirty-three years, states that when he joined, the tradition current in the ranks was somewhat different. It was that at Famars, after fighting all day, the regiment found itself towards evening hemmed in by French troops, and that the colonel directed the band to whistle the tune until they were familiar with the notes. When night came, the 14th, with colours flying and the band playing the "accursed tune," passed safely through their foes and regained their own lines.

Legendary and romantic as this appears, the incidents are, doubtless, true in the main; although they seem to refer to the action in front of Tournay the year after—where the 14th did fight all day, and did cut their way right through the enemy at nightfall—rather than to Famars.

There is no doubt that "Ça ira" was acquired at Famars, as stated by the late Major-General Elrington and by Sir Francis Doyle.

The correspondent above quoted writes: "It used always to be "said in the regiment that no one but a 14th man could march past "to the old tune; and there really was some truth in the saying, for "I have often seen other regiments in brigade with us thrown out of "step at once on the striking up of 'Ça ira.'"

Since the compiler of these notes has been engaged in collecting information about the regiment and corresponding with old 14th soldiers, he has found this question almost invariably asked, "Do they play the dear old tune still in the regiment?" From every grade the question comes. It was one of the first asked by the late Earl of Albemarle, and it has been repeated by the numerous correspondents to whom the compiler is indebted for the notes and legends given on preceding pages.

The short poems on Famars annexed will be read with interest by all 14th men. The first was written by the late Sir Francis Doyle, Regius Professor of Poetry at the University of Oxford, and grandson of Lieutenant-Colonel Welbore Ellis Doyle; the other is by the mother of an officer at present serving with the 1st battalion.

THE XIV. QUICK-STEP.

WHEN first the might of France was set
'Gainst creeds and laws, long years ago,
And the great strife—not ended yet—
Tossed crowns and nations to and fro;
Now buried deep beneath those wars,
That since have made the earth their prey,
Our hard-won triumph at Famars
Was famous in its day.

Here, trained through steadfast work, and drilled
Till as one thought they moved along,
By the old land's old memories filled,
Our English lads were calm and strong.
There—drunk on hope as on new wine,
That in their veins like madness wrought—
With power half devilish, half divine
Each restless Frenchman fought.

Wealth, numbers, skill, they heeded not,
Trampling them down as common things;
Man's spirit was a fire, made hot
To burn away the strength of kings.
Thus armed, as roars before the blast,
At forest trees the prairie flame,
On our firm silence, fiercely fast,
Their howling frenzy came—

Until (why shun the truth to speak?)
The courage rooted in the past,
Struck, as by sudden storms, grew weak,
And wavered like a wavering mast.
Still, kept their time the well-taught feet;
Nor dreamed the soldier yet of flight,
Though deepening shadows of defeat
Fell on him like a blight.

Straight out in front their leader dashed
(A God-given king of men was he),
And from his bright looks on them flashed
One sparkle of heroic glee:
"They hold us cheap," he cried, "too soon,
"We'll break them—frantic as they are—
"Unto their own accursed tune.
"Strike up, then, '*Ça ira!*'"

The drums, exulting, thundered forth
Whilst yet with trumpet tones he spoke,
And in those strong sons of the North
The old Berserker laugh awoke;
Their bayonets glowed with life; their eyes
Shone out to greet that eagle glance,
And, in her rush, a strange surprise
Palsied the steps of France.

Then, like a stream that bursts its banks,
To "*Ça Ira*" from fifes and drums,
Upon their crushed and shattered ranks
The cataract charge of England comes;
Whilst their own conquering music leapt
Forth in wild mirth to feel them run.
Right o'er the ridge that host was swept,
And the grim battle won.

Thus, in the face of heaven and earth,
From their first home those notes he tore,
To live, as by a second birth,
Linked fast with England evermore.
Yes, evermore; that through them still
To coming ages might be shown
Whose arrowy thought and iron will
Had made that prize his own.

Thence, as each panting year rushed by
  With garments rolled in blood—His march
Went sounding onwards, far and nigh,
  Beneath cold rains or suns that parch,
Northward or southward—east or west,
  Where still the heirs of that renown
Behind some other colonel pressed
  To the field, hurrying down.

For him, alas! on Java's shore
  It throbbed unheard through purple skies;
Nor marked he under dark Bhurtpore
  The blood-bought battle-hymn arise.
New Zealand's fern gloom, as they stept,
  Might quiver to that piercing tone;
But him it stirred not where he slept,
  In a far land—alone.

And whilst o'er its old ground the strain
  Smote with high scorn our ancient foe,
Called he upon those drums again?
  Shared he their closing rapture? No!
His grave lay deep in dust before
  They pealed through Belgian corn-crops, when
The baffled Eagle fell, no more
  To tear the hearts of men.

Yes, he died young; and all in vain
  We dream how much he left undone:
Painting upon an idle brain
  The glorious course he should have run.
Forgotten by the reckless years,
  He rests apart—and makes no sign;
Even his proud march no longer cheers
  The 14th of the line.*

Still, if elsewhere of this no trace
  Remain, by some as worthy deed,
O youthful soldiers of his race,
  Against oblivion for it plead.
Thus, if his death-lamp have grown dim,
  Relight it; thus force Time to spare
This leaf of laurel earned by him
  For the old name we bear.

---

* Written under the erroneous impression that the old tune was no longer played in the Regiment.

BEATEN backward in the press,
  Reeled the old Fourteenth ;
And in triumph shrill arose,
The yell of the triumphant foes,
As, where the British Lion flew,
Flaunted "white and red and blue,"
For well the fiery Frenchmen knew
  The fame of the Fourteenth.

Beaten backward in the press,
  Reeled the old Fourteenth ;
Cheerily their colonel spoke,
As the red line round him broke ;
Laughing, waving with his hand,
To the leader of the band—
As again they took their stand,
  The men of the Fourteenth—

"Play the Frenchman's march," he said,
  The Chief of the Fourteenth ;
"Strike it up — strike loud and clear ;
As I stand before you here,
We will prove our mettle soon ;
Ere yon pale sun rides at noon,
We'll beat them to their own brave tune,
  We men of the Fourteenth !"

Joyously the cheer arose
  From the old Fourteenth ;
"Ça ira !" gay and full and strong,
Rang the rallied ranks along ;
English hearts and steel were good ;
Rushing onward like a flood,
Naught their furious charge withstood,
  The charge of the Fourteenth.

And they play "Ça ira" yet
  In the old Fourteenth,
In memory of the glorious day
When they swept their foes away !
In memory of the fight begun
When, beneath the Southern sun,
To the Frenchman's tune they won,
  The men of the Fourteenth.

ÇA IRA.

The March of the 14th P.W.O. Regiment.

"Billy," 1st Battalion.

VIII.

"BILLY," THE REGIMENTAL DEER.

"BILLY," the pet deer, for so long associated with the 1st battalion, lies buried behind the canteen, in the old barracks, Fermoy. A stone placed in the wall, immediately over his grave, bears the following inscription :

BILLY,

SERVED IN THE 14TH REGIMENT,

NOVEMBER, 1878, TO OCTOBER, 1889.

"Billy" was presented to the regiment in November, 1878, when at Ranikhet, by the Rajah of Kashipur. For some years "Billy" used to march out at the head of the regiment, led by two drummer-boys, for which purpose he was provided with a silver collar and chain, which, as well as his skin, may still be seen hung up in the mess of the 1st battalion.

At cricket matches, tennis parties, and other festivities at which the band might officiate, "Billy" would always be well to the fore. In fact, on these occasions he was apt to make himself rather a nuisance, for, in his eagerness to get to the tea-table, or anywhere where the ladies might be congregated and so suggest sugar to his mind, he would rush frantically to the place, regardless of what he might upset in the way. He had a marked aversion to small children and to dogs ; and, though neither vicious nor dangerous in any way, was endued with considerable strength and obstinacy, and it has on more than one occasion taken time and trouble to dislodge him from a cricket or tennis net, in which he might be entangled, or drive him away from the magic tea-table. He received an injury to his back early in October, 1889, and, after some days' intense suffering, the order for his destruction was most reluctantly given.

IX.

TROPHIES.

A REGIMENT which has served in all quarters of the globe, and in so many campaigns as the 14th, should have many trophies of such service. But, as may be seen from the following notes, not a few that once existed, in some unaccountable way have disappeared.

Concerning these trophies it is difficult to get direct corroborative evidence, but it may be that some of those who read these pages may recollect most, if not all, of those here described.

The Bhurtpore Gongs.

These were captured by the 14th at the taking of Bhurtpore. The large one—which according to an old 14th man now living, who remembers it well—was supposed to be worth a very large sum of money, on account of the metal of which it was made containing an alloy of precious metals. Another 14th veteran writes: "I remember "the story of the Bhurtpore gong. It was lost, with *many other* "*valuable relics*, at the wreck of the head-quarters and two flank "companies, on Christmas Day, 1837, off Guadaloupe. This I heard "from men who were present and wrecked. A brave soldier of the "light company (Private Lane) swam ashore and gave information of "the wreck to the French authorities, although the water was full of "sharks. They were rescued just in time; and I heard that the good "old Colonel Matthias Everard stood with a loaded pistol to shoot "the drunken skipper if any lives were lost through him."

Both the Bhurtpore gongs have now gone, and the writer can gather no further information respecting them than is given above.

The Russian Drum,

according to an officer who joined the regiment in the early forties, was in use by the drummers of the 1st battalion at Malta, in 1858, and was presented by Sir J. Alexander, in the Crimea. It was understood to have been picked up by General Hammersley, at the battle of Tchernaya. This has also gone, no one knows where. It disappeared in Aldershot.

The Sevastopol Bell,

which was taken at the fall of Sevastopol, disappeared many years ago, and was presumably left behind by the regiment when on the move, as being to heavy too carry. There is at present no trace of it.

The Tralee Handkerchief

was presented to the officers of the regiment on leaving that station. It was an embroidered lace handkerchief, with a regimental device and tear-drops, and was given at the railway station in June, 1867. The tears were supposed to be worked—a tear from each—by the sorrowing fair ones of Tralee left behind. This handkerchief, for some years, used on guest nights to be pinned to the regimental colour, but it has also disappeared.

The Galway Kidney.

This was merely a reminiscence of a disturbance which took place in Galway, in September, 1885. There was a Nationalist demonstration in the town, and some soldiers returning to barracks were assaulted by the civilians, and had to fight their way back. They were followed by a large crowd, and stones were thrown at the barracks, which broke most of the windows. Many came through the mess room, where the officers were at dinner, and the next morning the one referred to—a kidney-shaped pebble—was found on the mantelpiece of the ante-room, and was kept for several years in memory of this event. During the time the mess was repapered, and everything turned out, in 1890, at Fermoy, this stone was lost.

The Afghan Standard

was a large triangular flag, captured at Mazeena by Colonel Noyes, and given by him to the company of which he was then in command ("D"). *See p.* 231.

This flag was lost a year ago, on the march down country, and doubtless was stolen from among the baggage by an Afghan, and is perhaps waving among the hills and over the heads of those from whence it was taken.

Regimental Relics at Claydon.

Sir H. Verney (Calvert), at Claydon House, Buckinghamshire, has the following mementos of his father's old corps:—The ancient colours of the 1st, 2nd, and 3rd battalions 14th Regiment; a drum-major's staff, dated 1786, being a heavy silver-mounted malacca cane with 14 on the head; a drum-major's sash, with CORUNNA, 14 on it; and a suit of mail armour taken by a 14th man in India in some engagement about 1830. A monument, in the church, records the services of Sir H. Calvert, and thus winds up: "As a further reward for his "faithful services, he was made colonel of the 14th Buckinghamshire "Regiment."

X.

PARTICULARS OF MESS PLATE, PICTURES, Etc.

OFFICERS' MESS. — FIRST BATTALION.

IN spite of serious losses in the course of service from time to time, the 1st battalion can still boast of having good plate.

The oldest piece of silver (in point of manufacture) is a Snuff-box (Fig. vii.), presented to the old 2nd battalion, after his promotion out of the 14th Regiment, by Lieutenant-Colonel William Wood, c.b., 85th (Duke of York's) Light Infantry, afterwards General Sir William Wood. (See "*Biographies of Colonels*," App. A.) It was presented as a token of personal attachment and regard, and was accompanied by the following letter, which is entered in the Regimental Records :

"London, 11th November, 1816.

"Gentlemen,—I have requested Captain Temple to take charge of a "snuff-box, of which I beg your acceptance as a token of the sincere "regard that I have for the 2nd battalion 14th Regiment, in which so "many of my happiest days have been passed. Allow me to return my "most sincere thanks for the kindness shown me by the corps when my "promotion was notified. Be assured that it will ever be recollected by "me with gratitude, and I look forward with the greatest pleasure to the "day when we shall meet again. With every wish for the future welfare "of the regiment,

"I am, gentlemen,
"Your very obedient and faithful servant,
"William Wood,
"Lieutenant-Colonel, 85th Regiment.

"To Colonel Burrows and the officers,
2nd-14th Regiment."

The 2nd battalion, on their disbandment, 24th December, 1817, presented this snuff-box to the 1st battalion.

A SNUFF-BOX (Fig. VI.) of curious design was given to the regiment in 1796 by General Watson. The following names are inscribed on this box:

Major Burnett
Captain Meard
„ Clapham
„ Powell
„ Watson
„ Stewart
„ Graves
„ Smith
„ Palmer
„ Hotham

Lieutenant Hutchinson
„ Hamill
„ Colman
„ Morris
„ Owen
„ Tegart
„ Beavan
Ensign Garvey
Surgeon Shea

A handsome and unique pair of SILVER TONTINE CUPS (Fig. III.) or Trinidad Urns, which were presented to the mess whilst the regiment was in Trinidad by the undermentioned officers, whose names are inscribed upon them, upon the condition that they were to become the property of the last survivor of those named serving with the regiment. They fell to Sir James Watson, whose services with the regiment covered a period of seventy-five years, and who left them to the mess. It was remarked that a regiment with "Hope" for a colonel, "Quick" for an adjutant, "Ready" for paymaster, and "Goode" for quarter-master, must surely be of exceptional merit!

Honble. Colonel A. Hope
Lieut.-Colonel Burnett
Major - Dowell
„ Clapham
Captain - Watson
„ Stewart
„ Graves
„ Burrows
„ Hotham
„ Burslem
„ Mill
„ Riddell
„ Costello
„ Fraser
Lieutenant Morris

Lieutenant Shea
„ Carpenter
„ Wood
„ Wren
„ Henry
„ Gosselin
„ Watson
„ Stewart
Ensign - Morris
„ Moore
Surgeon Jackson
Assistant-Surgeon Draper
Paymaster McReady
Quarter-Master Goode

A fine CUP (Fig. II.), given in 1813 by Mr. Hope, Civil Commissioner of Eastern Java, "as a mark of his esteem and regard," on the

Mess Plate, 1st Battalion, 1892.

departure of the regiment from Java in 1813. [Hugh Hope, Esq., B.C.S., of Craighill, afterwards Collector at Patna and Mirzabad.]

Two handsome SILVER CLARET JUGS (Fig. IV.) were given by Lieutenant-Colonel Maurice Barlow in 1849.

Another CLARET JUG (Fig. VIII.), capable of holding a "Jeroboam," was presented in 1852 by General Sir James Watson, colonel of the 14th Foot (see "*Biographies of Colonels*," App. A.), "as a token of the esteem and affection he feels for a corps in which he has served so many years."

A 100-GUINEA CUP (Fig. I.), won by the 14th Regiment in 1877-78 as the best shooting regiment in Bengal. Thirty regiments competed. It was presented by General Sir Frederick Haines, G.C.B., commander-in-chief in India.

The crests of the regiment, viz.—the White Horse, the Royal Tiger, and the Prince of Wales's Plume, are represented by the following pieces:

A SILVER HORSE, standing nineteen inches in height and twenty inches long, presented by General Sir William Wood, K.C.B., in 1866.

A SILVER TIGER (like dimensions), presented in 1873 by the following officers:

Lieut.-Colonel	H. W. Hawley	*Lieutenant*	R. Morrison
Major -	D. S. Warren	"	C. A. Morris
"	E. W. Saunders	"	J. M. Morrison
Captain -	J. Laing	"	J. Reed
"	H. McL. Hutchison	"	H. H. Young
"	B. W. C. Firman	*Surgeon-Major*	J. Moffatt

Both the above are triumphs of Messrs. Hunt and Roskell.

Eight MENU-HOLDERS, each formed of a plume of three feathers, presented by the following officers:

Lieut.-Colonel	C. E. Grogan	in 1877
Captain - -	Le Breton Butler	"
"	H. McL. Hutchison	"
Lieutenants	H. Fenning	"
"	J. Hosack	"
"	T. M. Robinson	"
Captain - -	C. A. Morris	in 1878

The Prince of Wales's Plume is, however, best represented by the magnificent CENTRE-PIECE, which stands 3 ft. 10 in. high, and weighs over sixty-four pounds. The pedestal has representations in relief of Waterloo, Bhurtpore, and New Zealand. In reply to an enquiry on behalf of the mess, the following letter was received from the makers, Messrs. Elkington:

"22 Regent Street, London, S.W.
"April 2, 1892.

"Sir,—In reply to your favour of 29th ult., we have referred to our "order for the Centre-piece, and find that the design was our own.

"The group represents the finding of the body of the King of Bohemia "by the Black Prince (the mounted figure). The prince's attendant is in "the act of opening the dead king's visor and the prince recognizing him.

"We are, sir,
"Your obedient servants,
"ELKINGTON AND CO. (LIMITED.)"

The Centrepiece was presented in 1879 by the undermentioned officers:

Lt.-Colonel T. P. Cosby
" A. F. de B. Dixon
Major - B. W. C. Firman
" G. Van Heythuysen
Captain - G. Grant Dalton
" F. D. Walker
" E. G. S. Schuyler
Lieutenant C. J. M. Heigham
" H. Walker
" G. W. Swaine
" F. R. P. MacAdam
" H. B. C. Critchley-Salmonson
Colonel - G. Grogan
Major - A. A. Le Mesurier
Captain - R. G. Pigott
" W. S. Purkis
" P. Crosbie
" A. Ruttledge
" R. H. Atkinson
" T. M. Robinson

Lieutenant W. S. Hewitt
" W. R. Ravenshaw
" C. R. Molesworth
" H. R. Merrit
" W. R. Rae
" C. A. Morris
" F. Lowry
" G. Wemyss
" G. Mitchell
" W. Fry
" C. S. Orr
" F. W. Thurston
" A. C. Adye
" H. Fenning
" T. H. Ansley
" E. C. Mills
" A. W. St. George
" W. R. C. Baird
" C. R. C. Robinson
" C. H. Cox
" H. R. B. Carter
" J. C. Romilly

A WELCOME HOME CUP (Fig. v.), presented by Major J. D. Bradley, on the return of the battalion from foreign service in 1879. In the sides are inserted three crystal-covered medallions, representing Father Christmas and the Crests of the Regiment.

Fourteen SILVER CUPS, presented on promotion by the under-mentioned officers:

1	*Captain* G. Harwood Cope	1866
	Lieutenants Tredenham and F. Carlyon ...	1865
2	W. Carden, Esq., *Paymaster*	1867
3	*Lieutenant* R. Graham Pigott	1867
	" John Hosack	
4	*Captain* E. T. Briscoe	1867
5	*Captain* J. L. Davids	1867
6	*Lieutenant* R. H. Atkinson	1867
7	*Lieutenant-Colonel* J. Dwyer	1867
8	*Captain* T. Lawrence	1868
9	*Lieutenant* C. O'H. Trench	1868
10	*Captain* H. Metcalfe	1868
11	*Lieutenant* W. H. Ridgway	1870
12	*Lieutenant* R. W. Richardson	1870
13	*Captain* R. S. Lemon	1870
14	*Lieutenant* W. S. Purkis...	1871

A CIGAR AND CIGARETTE BOX, with regimental crest in silver on the lid, presented by Surgeon-Major Martin, A.M.D., in 1884.

A SUGAR BASIN and CREAM JUG, presented by Lieutenant G. Freer in 1884.

A CIGAR-LIGHTER GRENADE, presented by Lieutenant W. Mills, Sligo Artillery Militia, in 1882.

A CIGAR-CUTTER, presented by Lieutenant Murray, 4th West-Yorkshire Militia, in 1885.

An INKSTAND, presented by Colonel Whitting in 1886.

A glass and silver CHAMPAGNE JUG, presented by 2nd Lieutenants Barff and Champain in 1890.

PICTURES ETC.

Photographs of H.R.H. the PRINCE OF WALES, PRINCESS OF WALES, H.R.H. the DUKE OF EDINBURGH, and of LORD ALBEMARLE, presented by themselves.

A portrait in oils of GENERAL SIR JAMES WATSON, who served seventy-five years in the regiment.

Oil painting of WINDSOR CASTLE, painted and presented by Lieutenant O'Toole, of the 14th regiment, in 1854.

A picture of THE 28TH AT QUATRE BRAS, presented by Captain Pilkington, 28th regiment, in 1888.

Two military pictures, presented by Captain Barchard in 1887.

Two military pictures from the officers 49th field battery, R.A., in 1891.

An engraving, "FLOREAT ETONA," by the officers of a detachment of the Northamptonshire Regiment who were attached to the battalion in 1892.

A picture, by De Beaurepaire, "French Infantry on the March," given by Captain Alexander, 1892.

A picture, with representations of the various uniforms which have been in use in the 14th Regiment; from Lieutenant O'Donnell, 1888.

Picture of H.R.H. THE DUKE OF CAMBRIDGE, from Lieutenant-Colonel Harington, 1888.

Picture of HER MOST GRACIOUS MAJESTY THE QUEEN and one of H.R.H. THE PRINCE OF WALES, presented by Captain C. H. Cox in 1889.

Captain Ward, in 1888, gave two well-carved NOTICE BOARDS.

CLOCKS were given by Lieutenant Charles Monk Wilson in 1853; by Lieutenant H. Critchley-Salmonson in 1881; and by Colonel T. P. Cosby in 1886.

Silver-mounted ivory PAPER KNIFE, from Lieutenant Bomford in 1891.

Silver-mounted BLOTTING BOOK, from Captain Trafford, 3rd West Yorkshire Regiment, in 1885.

Silver-mounted oak INKSTAND, from Lieutenant Tottie, 4th West Yorkshire Regiment, in 1885.

Silver-mounted BOOKSTAND, from Captain O'Donnell in 1891.

LETTER-BOX, from Captain C. M. Lester in 1887.

HEAD OF BARAH SING, from Kashmir, shot and presented by Lieutenant H. Walker in 1880.

A CABINET, with lockers, from Major G. Grant-Dalton in 1891.

Brass CANDLESTICKS, from Captain Heigham in 1886.

BAROMETER, from Captain C. H. Cox in 1880.

Three PHOTOGRAPHIC ALBUMS, given by Captain Pilkington, 28th regiment, in 1888; by Lieutenant D. Alexander in 1891; and one, giver unknown, probable date—186 ?. This is adorned with a fine silver crest of the regiment, and contains many old portraits.

Various SKINS adorn the walls, the most noticeable being that of a tiger, presented by Mr. Carmichael, Commissioner of Benares, in 1876.

The TELESCOPES.—The Officers' Mess at one time was in possession of two telescopes, the gifts of various friends of the regiment. One, a very large one, was presented by Captain Wood, R.N., as a token of regard and esteem, and its outer case (but without lenses) was in the mess a few years ago. Another, ornamented with flags of all nations in miniature, has long since disappeared.

OFFICERS' MESS.—SECOND BATTALION.

Two ELECTRO-PLATED CANDELABRAS (Fig. 1), bought by the Mess fund in 1858, before the battalion proceeded to New Zealand.

Two medium-sized ELECTRO-PLATED SALVERS (Fig. 2), bought by the Mess fund in 1858, before the battalion proceeded to New Zealand.

Two small ELECTRO-PLATED SALVERS (Fig. 3), bought by the Mess fund in 1858, before the battalion proceeded to New Zealand.

Two ELECTRO-PLATED WINE COOLERS (Fig. 4), bought by the Mess fund in 1858, before the regiment proceeded to New Zealand.

A Chillingham wild bull's horn, silver mounted, SNUFF-BOX (Fig. 5), presented by Colonel Sir James E. Alexander, C.B., F.R.S.E., in 1859.

A PHOTOGRAPHIC ALBUM, unmounted (Fig. 6), bought by the Mess fund in 1859.

Two embossed SILVER CLARET JUGS (Fig. 7), bought by the Mess fund in 1860.

A CIGAR BOX (Fig. 8), made of different kinds of Australian wood, presented by Colonel W. Cosmo Trevor, C.B., in 1868. The box was partly renewed with Australian wood and restored by Captain G. W. Swaine in 1889.

A SILVER SNUFF-BOX (Fig. 9), with the Turkish, Medjidie, Legion of Honour, C.B., New Zealand, Crimean, Peninsular, and Indian Medals—worn by the officers of the regiment at the time—presented by the officers in 1870.

Two ELECTRO-PLATED round stand LAMPS (Fig. 10), bought by the Mess fund in 1870.

The WHITE HORSE (Fig. 11), bought by the Mess fund in 1871. (The original model was presented to the 1st battalion 14th Regiment by Colonel Sir William Wood, K.C.B.)

5

13 23 9 34 21 12 13

7 20 33 37 32 20 7

25 39 2 39 17

1 19 10 10 19 1

11 24 18

KEY TO ILLUSTRATION OF MESS PLATE, 2ND BATTALION.

Mess Plate, 2nd Battalion, 1893

A SILVER CUP (Fig. 12), presented by the Mayor of Liverpool to commemorate the opening of Sefton Park by His Royal Highness the Prince of Wales in 1872.

Two small SILVER MUGS (Fig. 13), presented by Lieutenant Moss, 2nd Cheshire Militia, on being attached to the regiment at Chester in 1872.

Two WINE GLASSES (Fig. 14.), used by the officers of the 3rd battalion 14th Regiment on the field of Waterloo; presented to the 2nd battalion by Lady Vansittart when the battalion was quartered in Chester, in 1872.

Five WATER-COLOUR PAINTINGS (Fig. 15), presented by Mrs. Burton when the battalion was quartered in Chester, in 1872.

A WATER-COLOUR PAINTING (Fig. 16), of the battalion attacking a Maori Pah in the New Zealand war; presented by Lieutenant Aubrey L. Patton in 1872.

A two-handled SILVER CUP (Fig. 17), presented by Captain Harrington, Lieutenants Lapenotiere, Crosbie, Butler, Ruttledge, Lye, Gordon, and other officers in 1875 on promotion.

The "ROYAL BENGAL TIGER" (Fig. 18), presented by Major T. P. Cosby and R. H. Vivian, Captains R. Hall, A. L. Patton, Lieutenants T. W. Penno, T. R. Mills, G. Grant-Dalton, L. C. de Trafford, and other officers in 1876 on promotion.

Two ELECTRO-PLATED square pillar-shaped LAMPS (Fig. 19), bought by the Mess fund in 1877.

Two SILVER CUPS (Fig. 20), presented by Captain J. C. D. Bradley on exchanging to the 49th (the Princess Charlotte of Wales' Regiment) in 1878.

Prince of Wales' Plume PROGRAMME STAND (Fig. 21), presented by J. C. D. Bradley as a New Year's Gift in 1878.

A SILVER CIGAR-LIGHTER (Fig. 22), presented by the Officers' Depôt Battery, third brigade Royal Artillery, on the battalion leaving the Curragh Camp for India in 1878.

A SILVER CUP (Fig. 23), presented by Major W. Young on his joining the battalion from the 49th (the Princess Charlotte of Wales' Regiment) in 1878.

THE CENTRE PIECE (Fig. 24). A group representing the scene of Edward the Black Prince, at the battle of Crecy, finding the body of the blind King of Bohemia, whose plume and motto, "*Ich Dien*," he adopted thenceforward as his own. A knight is raising the helmet from the head of the dead king, a secretary is making a list of the slain in his tablets, and the prince on horse-back, with the banner of England behind him, forms the central figure of the group.

On the triangular stand are embossed silver plates representing "Waterloo," "Bhurtpore," and "New Zealand."

Presented by the officers, past and present, to the Mess, 1878-79.

An INDIAN CUP (Fig. 25), presented by Major H. A. Burton, on his leaving the regiment, in January, 1882.

A SILVER CIGAR-LIGHT HOLDER (Fig. 26), presented by Lieutenant and Adjutant F. W. Kitchener in 1882.

An ELECTRO-PLATED BAND PROGRAMME STAND (Fig. 27), presented by Lieutenant and Adjutant F. W. Kitchener in 1882.

A PHOTOGRAPHIC ALBUM (Fig. 28), presented by Colonel D. S. Warren, C.B., on leaving the regiment, in 1883.

AN ENGRAVING OF THE REGIMENT, taken in the year 1826, (Fig. 29) assaulting the fort of Bhurtpore on the 18th January, 1829; presented to the regiment by General Chapman, C.B., Quarter-Master General in India, in 1887.

Four PRINTS OF THE FRANCO-PRUSSIAN WAR of 1870 (Fig. 30), bought by the Mess fund in 1887.

A CHIMING CLOCK (Fig. 31), presented by Major A. W. Noyes to commemorate the Jubilee of the Queen Empress, in July, 1887.

A SILVER CUP (Fig. 32); won and presented to the officers by Lieutenant F. W. Towsey, Mian Mir Divisional Sports, 1888—one hundred yards open race.

A SILVER CUP (Fig. 33); won and presented to the officers by Lieutenant C. H. G. Moore, Mian Mir Divisional Sports, 1889—one hundred yards open race.

A SILVER TROPHY (Fig. 34), for annual competition (shooting) by the officers of the battalion; presented by Lieutenant-Colonel H. McLeod Hutchison in June, 1890.

A large ELECTRO-PLATED SALVER (Fig. 35), presented by Captain G. W. Swaine and Lieutenant A. J. Stephens, at Umballa, in 1891.

Two ELECTRO-PLATED CREAM JUGS and SUGAR BASINS, with sugar tongs (Fig. 36), presented by Captain G. W. Swaine and Lieutenant A. J. Stephens, at Umballa, in 1891.

A SILVER CIGARETTE BOX (Fig. 37), presented by Captain G. W. Swaine on 1st January, 1891.

A SILVER CIGAR CUTTER (Fig. 38), presented by Captain G. W. Swaine, Christmas Day, 1891.

Two SILVER FLOWER STANDS (Fig. 39), presented by Lieutenant-Colonel A. W. Noyes to the officers of the 2nd Battalion West Yorkshire Regiment, in 1893.

SERGEANTS' MESS.—FIRST BATTALION.

(BY A VERY OLD MEMBER.)

I remember but little in connection with the 1st battalion 14th Sergeants' Mess—they have, or had when I left, very little property with a history.

About the oldest thing in the mess, I think, is a steel-plate engraving of the Commander-in-Chief, when he commanded the Dublin District, which was obtained when the regiment was in Dublin in 1851. I remember seeing a similar print in the mess of the 2nd Scots Guards when I was up in London, at Lord Albemarle's funeral. The engraving of the Duke of Wellington would probably be obtained about the same time.

The ram's-horn snuff-box tells its own tale. The wine-coolers and a painting of the Battle of Waterloo were officers' mess property, and were given by the officers to the sergeants' mess before I joined, in 1867. "Waterloo" was touched up after the battalion came home from India.

The Scotch "Piper" was given to the sergeants' mess in Malta, in 1857, by the 71st Highland Light Infantry, under the following circumstances. The highlanders were ordered at short notice from Malta to India during the Mutiny, and proceeded overland, being, I think, the first regiment that proceeded that way. The Khedive gave permission for them to go through Egypt, but *not* as an armed body. They had, therefore, to pack up their arms, and marched in smock-frocks, silk handkerchiefs, and straw hats. Their baggage being limited, they handed over the "Piper" to our sergeants' mess, with the proviso that it should be handed back if ever the two regiments were stationed together again. A detachment of theirs relieved our head-quarters at Tralee in 1867.

There is a photographic group of the sergeants, taken in Calcutta, round a big gun which stands in front of the officers' mess, between the north and south barracks. The cannon is said to have been captured at Bhurtpore by the grenadier company of the regiment, nearly all of whom were blown up with the gun.

XI.

ATHLETICS IN THE REGIMENT.

CRICKET and football have flourished in the regiment for many years. The 1st battalion has for several years won a certain reputation for its cricket team, and the team of 1889 and 1890 was, perhaps, as good a one as the regiment has ever had. Owing to the zeal of the captain (Captain Schuyler), and the influx of new blood into the team of exceptional merit, the record of these years—1889 and 1890—stands as under :

1889	... 12 matches played	... 8 won,	2 drawn,	2 lost.
1890	... 16 " "	... 15 "	0 "	1 "

The individual averages in batting ran from 30 downwards ; Mr. Trevor, 21 for 11 innings in 1889, and 22 for 21 innings in 1890, being perhaps the most persistent scorer. Mr. Champain, who had captained Cheltenham and Sandhurst Colleges, was, perhaps, the best all-round.

The football (Rugby) team has also won laurels in the South of Ireland ; and, latterly, in the North of England, where, although it has been beaten when opposed to crack teams, it has always made a good fight of it, and has been highly commended. In Fermoy, in 1889 and 1890, it won most of its matches ; and the presentation of a challenge cup for company competition excited much interest in these matches, which still continues, and many successes have been recorded since.

Shooting.—A great deal of interest has been shown in shooting in the regiment in the last few years, and many prizes were won by members of the 1st battalion at the All-Ireland army meeting—held annually at the Curragh—in the years 1888 to 1890. There is a challenge shield, for which a special competition by companies takes place annually. Since its institution, three years ago, it has been twice won by "D" company and twice by "F" company.

With Polo the 2nd battalion has warmly identified itself during its present tour of service in India. On four occasions it has competed for the Challenge Cup presented by the Earl of Airlie for annual competition among all infantry regiments stationed in India.

In three out of the four attempts the battalion has been unsuccessful, but was more than compensated for its ill-luck by proving victorious in the spring of 1891, and holding the proud position of the best infantry regiment at Polo—for the year—in all India.

The following short accounts of the games leading to this fortunate result are from the *Asian* :—

"On Wednesday, 25th February, 1891, the first tie in the second round of the Infantry Polo Tournament was played here (Umballa), when the 14th Prince of Wales' Own West Yorkshire Regiment, who had drawn a bye in the first round, had to try conclusions with the 1st battalion of the Rifle Brigade. The sides were :—

1st Battalion Rifle Brigade.	*14th West Yorkshire Regiment.*
Captain Alexander.	Lieutenant G. C. Lang.
Captain Sherston.	Captain J. C. Yale.
Major T. G. Talbot.	Lieutenant A. G. Stephen.
Lieutenant H. E. Vernon.	Captain H. E. Vialls.

"The first period was very fast, but the Yorkshires, who were well mounted, and for whom Stephen played splendidly, had the best of it, as they secured two goals. In the second quarter the play was very even; the Rifles scoring, one and the Fourteenth two goals. The third period produced a splendid contest, each side scoring a goal. Vernon, for the Rifles, played exceedingly well. In the fourth period the Rifles penned their adversaries and obtained a goal, while the fifth, which was the game of the meeting, was very fast. Neither side scored, Vernon and Stephen playing magnificently for their respective sides, the Fourteenth winning by five goals to three.

"The Tournament was concluded on the following Friday, when the final was played between the 3rd battalion of the Rifle Brigade and the 14th West Yorkshire Regiment. There was a large and enthusiastic gathering, including a large number of the men of the garrison, who were most enthusiastic, and applauded every good stroke to the echo; the game being very fast and even throughout. The sides were :—

Sergeant Clarke   Colour-Sergeant Gorton   Lance-Sergeant Halliday   Corporal Peters   Corporal Rawlins   Lieutenant and Quarter-Master W. Lowing

Lieutenant R. M. Barff   Lieutenant H. Champain   Captain H. O'Donnell   Captain E. Schuyler   Lieutenant W. S. Carey   Lieutenant H. Trevor

1st Battalion Cricket Team of 1890.

3rd Battalion Rifle Brigade.	*14th West Yorkshire Regiment.*
Mr. B. W. D. Alexander.	Lieutenant G. C. Lang.
The Honble. C. E. Walsh.	Captain J. C. Yule.
Mr. S. G. C. Cosby-	Lieutenant A. G. Stephen.
The Honble. W. D. Cairns (back).	Captain H. E. Vialls (back).

"In the first three periods the West Yorks held their own, and had slightly the best of it, the score being West Yorks, three goals; Rifle Brigade, two.

"In the fourth round both teams failed to score, but play was very fast and close. In the fifth period the Rifle Brigade scored their third goal, thereby equalising matters, and although both sides were striving their utmost, the play was so good that neither scored further. On the goals being widened, Mr. Cairns made a brilliant run three-quarters of the way down the ground, and a magnificent stroke on the near side of his pony resulted in a goal. All in vain, however, as the West Yorkshire team claimed a cross, which the umpires allowed; and on the Rifles hitting the ball out from behind their goal line, the Fourteenth at once made their fourth goal. The Rifle Brigade did their utmost to retrieve their fortunes, but were unable to do so, and the game resulted in the Fourteenth being the holders of the Challenge Cup by the narrow margin of four goals to three."

Besides winning the Infantry Challenge Cup in 1891, the battalion, while at Sitapur in 1892, entered for the Counell Cup Tournament, open to all Polo Teams stationed in the North-West Provinces. As will be seen from the annexed extracts from the *Pioneer* of 30th March, 1892, the result was equally satisfactory although the competition was limited:—

"The one game of the above Tournament was played at Lucknow on Saturday, the 26th instant, on a sweltering hot day, before a fair number of spectators. There is no doubt that the number of recent accidents at polo had some effect in keeping a great many of the residents from swelling the ranks of the onlookers. There were but two entries, the 14th West Yorkshire Regiment and a team from the Batteries at Lucknow. The Fourteenth had lost two of their regular team—Captain Gale, invalided home, and Lieutenant Lang, gone home on leave; their places were supplied respectively by Colonel Noyes and Captain Watts. The Gunners of Lucknow very sportingly got together a team, so that there might be at least one fight for the Cup, and though they did not perform as well as was generally expected of them, still they made a very plucky fight throughout, though at the finish the superiority of the Fourteenth ponies and the more than brilliant play of Stephen could not be denied. Below is a detailed account of the game :—

"The first period consisted in the first half of three successive runs by Stephen into the Gunners' ground, but all the shots at goal failed. Nairne, by a long run, now relieved the Gunners' goal and took the ball into the Fourteenth territory, where the Gunners claimed and obtained a 'foul.' After the free hit, the Fourteenth were considerably pressed, and the Gunners had several shots at goal. Vialls and Stephen now relieved by a good run up the ground, which was responded to by Nairne for the Gunners. Some fast up and down play now followed, but neither side had succeeded in scoring any advantage when time was called.

"The second period was the fastest of the lot, the ball being kept well in play throughout. Stephen commenced by saving an attack on the Fourteenth goal, and next, Vialls, getting possession, drew first blood for the Fourteenth. Shortly after the hit off, Noyes, nipping in, took away the ball from a too slow Gunner and scored a second goal for the Fourteenth. Some good play between Nairne and De Rougemont resulted in the latter hitting a goal for the Gunners. Vialls next made a good run for the Fourteenth, and scored a third goal. A good run by De Rougemont brought the ball back into the Fourteenth territory, where Vialls, again getting hold of it, was well away, when his career was checked by the breaking of the ball. On the throw out the Gunners played well together, and Nairne, De Rougemont, and Graham made a determined attack on the Fourteenth goal, but did not succeed in scoring. When time was called the score stood—Fourteenth, three goals ; Gunners, one.

"In the third period Vialls commenced by adding a fourth goal to the Fourteenth score. Some fast play next ensued, in which Nairne and De Rougemont showed up well for the Gunners. Then the Fourteenth made an attack on their opponents' goal, which was averted by good play on Bannister's part. Vialls now added yet another goal to the Fourteenth score. The Gunners now pulled themselves together and took the ball all down the ground, where Bannister sent it between the flags. At the call of time the score was—Fourteenth, five goals ; Gunners, two.

"The fourth period was chiefly notable for some magnificent runs by Stephen, who scored no less than three goals for his side during it. Nairne started with a good run for the Gunners, to which Vialls responded by a goal. Stephen then made a long run up the ground, but De

Captain Yale | Captain Vialls | Lieutenant A. J. Stephen | Lieutenant G. G. Lang

The Winners of the Polo Cup, 2nd Battalion.

Rougemont, getting hold of the ball, averted the danger for the time. However, Stephen was not to be denied, and making a long run up the ground added another goal to the score. The Gunners' ponies were now getting done, and Stephen had matters pretty well his own way, and added two more goals to the Fourteenth score before time was called, leaving the Fourteenth victorious by nine goals against two. For the Fourteenth, Stephen played brilliantly throughout, and was well supported by Vialls. For the Gunners, Nairne and De Rougemont played best. The teams were :—

Royal Artillery.	*14th West Yorkshire Regiment.*
Major Bannister (back)	Colonel Noyes (back).
Lieutenant Graham (1).	Captain Watts (1).
Lieutenant De Rougemont (2).	Lieutenant Stephen (2)
Lieutenant Nairne (3).	Captain Vialls (3).

"The umpires were Captain Apthorp, Royal Irish Regiment, and Captain Haynes, East Lancashire Regiment."

It may be added, that in 1893 the battalion team proceeded to Allahabad to defend the last-mentioned Cup against all comers. The competition on this occasion was more extensive, five teams being entered. In the first ties the representatives of the battalion defeated an artillery team somewhat easily, and met the 8th Bengal Cavalry in the final, beating them by nine goals to one, a very satisfactory result; and, taking into consideration the impending move of the battalion to Poonah, a happy termination temporarily to its connection with the "Royal Game" in India.

XII.

Lodge of "Integrity," No. 771,

14TH REGIMENT OF FOOT.

HALIFAX, NOVA SCOTIA,
October 20th, 1846.

At an Especial Provincial Grand Lodge, holden on Tuesday, the 20th October, 1846, at the Freemasons' Hall, Barrington Stiect.

PRESENT—

The R.W. THE P. GRAND MASTER, in the chair,

I.W.	James Forman - - - - -	*D.P. Grand Master*
„	Lieut.-Colonel Calder, R.E. -	„ *S.G. Warden*
„	Charles M. Cleary - - - -	„ *J.G. Warden*
V.W.	The Rev. Dr. Twining - - -	*P. G. Chaplain*
„	Adam Gordon Muir - - -	„ *Grand Secretary*
„	John Richardson - - - - -	„ *Grand Treasurer*
W.	Henry C. D. Twining - - -	„ *S.G. Deacon*
„	William Rogers - - - - - -	„ *J.G. Deacon*
„	John Willis - - - - - - -	„ *G. Dir. of Ceremonies*
„	H. Hestin - - - - - - - -	„ *G. Architect*
„	Lawrence McLearn - - - -	„ *G. Pursuivant (?)*
„	George Anderson - - - -	„ *Grand Tyler*

The officers and members of the several lodges working in the city of Halifax and other transient brethren.

Grand Lodge opened in due form and with solemn prayer.

Read and confirmed minutes of last Grand Lodge.

The R.W. the Provincial Grand Master informed the Grand Lodge that this Especial Grand Lodge had been convened for the purpose of constituting the lodge of "Integrity" (No. 771) in Her Majesty's

14th Regiment of Foot, on the Registry of the United Grand Lodge of England, and installing its officers.

The application from the Master Elect and members of "Integrity" Lodge to be formed into a lodge of Antient, Free, and Accepted Masons was read by the Grand Secretary.

A warrant from the M. Worshipful Grand Master of England, in which the lodge is named "The Lodge of 'Integrity,' No. 771," was likewise read by the Grand Secretary.

The Grand Secretary having reported to the R. W. the P. Grand Master that the members of "Integrity" Lodge were in attendance, they were introduced.

The R.W. the Provincial Grand Master then proceeded, with the assistance of the P. Grand officers, to constitute the lodge of "Integrity" with all the solemn and imposing ceremonies customary and necessary on such occasions, and the W. Brother James Hogan having been duly installed as Master, the lodge was proclaimed in the usual manner, and the Grand Secretary instructed to register it accordingly.

The following brethren were also installed as its officers:

Bro. Henry Wing - - - *S. Warden*
" Samuel Robins - - - *J. Warden (pro tem.)*
" I. E. Carte - - - - *S. Deacon*
" R. W. Romer - - - *J. Deacon*
" Elisha Loasby - - - *Treasurer*
" F. Hammersley - - *Secretary*
" William Miller - - - *I. Guard*
" William Horsfall - - *Tyler.*

(Signed) A. G. MUIR,
P. Grand Secretary.

In 1853 I was made a freemason in this lodge, which used to meet, while quartered in Limerick, in the rooms of Lodge 73 (Irish Constitution), and the latter lodge still holds a silver snuff-box presented to them by 771 upon the departure of the 14th Foot from Limerick in 1854.*

* The little lodge of "Harmony," No. 555, in Fermoy, in which several brethren of the 14th Regiment were initiated in the years 1889-91, or held office, also holds mementoes of the regiment in the shape of a poignard and sword, engraved with the date of presentation and by whom presented.

When the regiment received orders at Malta to proceed to the Crimea, I said, if the lodge would allow me I would take the minute-book, the collars, and the warrant in my own baggage, and a small bag was made for them. I took the warrant off the rollers, and the rollers and the above extract went home to Ireland in my baggage.

During the heavy work of the siege of Sebastopol there was no time to give to the work of the lodge, nor was there any place to meet in; but after the surrender of the town, and after everybody began to be settled comfortably for the winter, we resolved to resume work.

I well remember, it got wind that we were going to meet in a certain hut (it must have been in October, 1855), and we had quite a gathering of interlopers, trying to get a peep, and we had to post four *bonâ fide* Outer Guards round the hut to keep off intruders from gazing through the chinks. At this first meeting we made arrangements for a regular meeting, and from that day until the departure of the regiment, in 1856, work was done nearly every week. The meetings were held by arrangement at the camps of regiments in every division of the army and in the village of Balaclava. One noble lord and two "honourables" were initiated in one night, and, as far as I can recollect, we admitted very nearly one hundred members to the Order. I handed over about £40, as well as I can recollect, to my successor in Malta.

After work we had most amusing meetings, brethren from all corps bringing their own seats with them, as well as their eating and drinking implements, and many a tumble and souse in mud-holes had to be endured in finding the way home in the dark, particularly when brethren were quite confident that they had a talent for topography and the science of short cuts.

At these meetings all "lights" used to be placed in empty champagne bottles, and all the working tools were of the most primitive kind, but such rude implements did not seem to depreciate the quality of the work in the least.

Unfortunately, some years after I left the regiment, one of the colonels asked to see the warrant, and, when brought to him, put it in the fire, saying he would not have such meetings in his regiment.

This officer had been a member of 771, but never attended in my time, and left the Crimea shortly after the arrival of the regiment there.

I was applied to some years ago for information about the lodge, and I ascertained these facts.

It is most curious that, although His Royal Highness is Grand Master of England, and Her Most Gracious Majesty patroness of the Masonic Order, some arrangement or understanding has been come to between the Army Authority and the Grand Lodge by which the Grand Lodge of England now declines to issue regimental warrants. I think this is a great mistake, particularly upon the part of the army authorities, as officers when travelling can, as freemason strangers, readily obtain information and entrance into society in any country they may happen to be, and the information so obtained through them might turn out to be of the most vital importance to the Intelligence department of an army.

It is pleaded in excuse that the characters of short-service non-commissioned officers are not well enough known to permit of all ranks uniting in a regimental freemasons' lodge; but, considering the better education of those who join the army as privates in these days, many being of as high or higher social standing than commissioned officers, there would not be the least danger of undue familiarity between the various grades of the members.

I was personally acquainted with all but one of the officers named in the installation of the lodge officers at Halifax in 1846.

C.M.W., 90°, 35°-95°, 33°, VI.°

March, 1892.

There are several masons among the officers of the 1st battalion at the present time, and a few among the warrant and non-commissioned officers, but it has been intimated that the old regimental lodge, which existed on the books of Grand Lodge up to 1890 as 528 in the English Constitution, is now extinct, having ceased work for many years, and that as H.R.H. the Commander-in-Chief objects to military lodges, there is no hope of a new warrant being issued.

XIII.

REMINISCENCES OF DUBLIN DUTY.

FOR military men, the various guard rooms in Dublin have an interest of their own. To those who have served in that garrison, they are full of memories. I venture to hope that these few recollections of mine may recall to the minds of some of my readers, incidents, perhaps forgotten, of pleasant days gone by, and friends lost to sight, but not to memory.

The duties in Dublin, when the 1st battalion of the regiment was stationed there in 1886-87-88, were somewhat severe. In addition to the various regimental picquets and guards, and the numerous small garrison guards, there were four others, in which the subalterns on the Dublin roster were especially interested. These four guards were: the Castle, the Bank, the Mountjoy Prison, and the Magazine. On one or other of the four the subaltern officer found himself, in the summer at least once in ten days, and in the winter regularly every fifth day.

THE CASTLE GUARD.

The general favourite was, I think, the Castle guard, owing chiefly to the comfort and cleanliness there prevailing. The guard consisted of a sergeant, corporal, drummer, and nine men, under an officer; but during the time the Lord Lieutenant was at the Castle, it was increased by a captain, and three privates, and the Queen's colour was carried by the subaltern officer. On these occasions the guard was mounted with great ceremony, and the band of the regiment furnishing the guard played for an hour in the upper castle yard.

THE OFFICERS' GUARD ROOM

was upstairs, immediately above that of the men, and from its four windows afforded a view of the entrance to the Castle, Cork Hill, and the top of Parliament Street. The room itself was a fine large room, and its walls were adorned with some really fine specimens of the artistic skill of that much maligned "idler," the British officer.

Of the many paintings which decorated the walls, one, "A wreck at sea," was reputed to be worth £250, and that such a sum had been offered for it by a past lord lieutenant was the common report. There were several portraits, somewhat of the caricature type, admirably painted by Swinton, of the 71st, and unmistakable likenesses. These included Lord Spencer, General Frazer, Colonel Caulfeild (now Lord Charlemont), and Liddell, the celebrated pianist and conductor. There were several landscapes, and some horrors; a copy of the "Premature burial," taken from a Brussel's gallery, being perhaps the best of the latter. "A coaching party," and "A bivouac in New Zealand," the work of Major Robley, 91st Highlanders, were large pictures, and though old and cracked, bore evidences of good colouring. Barton, of the Scots guards, and Baden-Powell contributed some good pictures in portraits of Colonel Trotter, as master of foxhounds, the face of a housemaid in the castle, peeping through a broken panel in the door, and Cinderella. Of these, the face in the doorway was capital, and apt to startle anyone who might casually look up at the closed door.

Of the smaller pictures, portraits of the Ladies Hamilton as highlander and guardsman, a pen and ink interior of an ancient courtyard, and various groups of soldiers, mostly guardsmen and highlanders, were most noticeable.

There was not much writing on the walls of this room, since the room is painted periodically, and at these times only the best pictures are left, and all writing is painted out.

On the right of the mantelpiece, however, beneath a picture of a woman in scanty clothing, were the following lines:

Surely such lavish loveliness would steal
The thrice locked heart of churlish anchorite or saint,
Yet, gazing on her charms, but makes me feel
One ounce of real flesh were worth a pound of paint.

Again, under another, a sketch of a cat, looking monstrous happy, was this:

Men have often declared, when it comes to the push,
That a bird in the hand is worth two in the bush,
But Puss with advantage may say "I engage"
That a bird in the paunch is worth four in the cage.

There was another verse written underneath, the "Premature Burial," a very clever one, suggestive of madness overtaking the unfortunate occupier of the half-opened coffin.

All these were written by an officer of the 14th, and were, I think, the best in the room.

The excellent catering and attendance on this guard contributed not a little to its popularity.

There was a picquet house in the lower yard, where there were also some good pictures. This room had in past days been occupied by an officer in charge of that guard, in times when the Castle guard was one of considerable strength. During our stay in Dublin it was occupied by a cavalry picquet, and the Ship Street guard consisted of a corporal and three men.

THE BANK GUARD.

Perhaps the bank guard has even more individuality than the Castle. The arched roof, the close-barred windows, the many names scratched on the window panes, the musty smell, the clatter of the guard below, the gloomy corridor, and last, but not least, the clang of the awful bell, rung by the sentry whenever the guard turned out, will, I am sure, never be forgotten by those who have been confined to that room as protectors of the ancient Parliament House of Ireland.

There were evidences of the regiment having furnished officers on the Bank guard many years ago, for when we arrived, in 1886, there were two old 14th names scratched on the windows, *viz.*, Barlow and Bristowe, who probably memorialized themselves in 1851, when the 1st battalion was stationed in Dublin. There were many anecdotes of the old Bank guard room, perhaps more than any other guard room in the garrison. It was told how a subaltern of this guard (now a distinguished commanding officer, who related the story himself), being on guard for the first time, lost his sword, and could not find it when the field officer came to visit at night, and how in despair he shouted to the guard from out of the upstairs window to present arms, and made his report from the same place to the astonished field officer; how a lady who had dined with her husband on guard, finding that he had partaken not wisely, but too well, of the fruit of the vine, donned his clothes and received the field officer in a proper and regular military fashion, escaping detection, whereby she saved her husband from lasting disgrace; of the attacks made on the Bank on many occasions by Fenians and others, and of the days when the Bank was not the Bank of Ireland, but the Irish Parliament House, with all its historical associations.

Who does not remember the old Bank attendant in ancient livery, who used to appear regularly every evening with the half sovereign, which the Bank generonsly presented to the officer on guard, to furnish him with the necessaries of life?

The pictures in this guard-room were, perhaps, not so artistic as those in the Castle guard, but many were worthy of more than passing remark. Some cleverly copied figures from a child's book of funny figures, men with heads out of all proportion to the bodies, adorned the wall over the mantelpiece. On the mantelpiece itself was a very realistic bottle of "Bass." On the opposite wall, a picture of a sentry of the Queen's, about 1760, was good; and a chariot race, a lion in a tangled forest, John Baptist's head in a charger, and several ludicrous pictures of Highlanders, Guardsmen, and French soldiers, were, perhaps, most noticeable. There was the commencement of a picture of a 17th Lancer at Ulundi and his wounded charger, which showed considerable promise. On the back wall the best of the few pictures was undoubtedly the figure of a guardsman, levelling his rifle at the observer's eye. There were also other large and life-sized pictures above the sofa, on which the weary subaltern was allowed to rest, of Peace, the murderer, and Gretchen with a candle.

There were many verses written on the walls, and one on the window scratched with a diamond, which were especially deserving of note.

Some few had been inscribed by members of the regiment.

To a love born poet who thus begins his lay:

> "O critics be kind, for I really well know, it's
> No use my attempting to rival the poets."

The sharp retort is given:

> "Then why the deuce do you, I ask,
> Essay in vain the thankless task?"

And yet again, to another, who called himself "Ye Ancient Sub," and who had been employed in running down the sentimental strain of one whose signature ran "A.S S.," the witty query is put:

> O hoary headed subaltern, say why are you so cynical,
> To all the gentler joys of life remorselessly inimical!
> Pray, don't you think you'd best come down from Pharasaic pinnacle,
> Or doff the red, and don the brown of order called Dominical,
> Since we've no patience in the line, for prudes with morals finical,
> You'd better drop your pride, old cock, and steer by Nature's binnacle.

Poor old Jumbo! He's gone now, and his last writing on guard on the Bank was this:

> They say that life's but a great big joke,
> That the Providence finger does but poke
> Its jests in the faces of young and old,
> Such as love, ambition, and lust for gold;
> You'll confess that a world quite out of joint
> Is quaint, but you'll question then, "Where's the point?"
>
> The answer to that they say you will find
> When you've crossed death's stream and left life behind,
> When you've found the cause of original sin
> And why the death's heads always grin,
> And other things that are equally strange,
> And are rather beyond our mortal range;
> Such as, why the deuce, I'm condemned to stay
> In this cursed room for a livelong day,
> And why, oh why, I should try and scrawl
> Such hopeless bosh on this filthy wall.

There are other reminiscences of the regiment in the Bank guard-room, but perhaps of little interest to any but the persons actually concerned.

The following clever sketch was written for *Our Journal*, by a young officer of the regiment, just a month before his lamented death, in September, 1888.

"We, that is the 14th, had been quartered in dear dirty Dublin "just three days, when I found myself on the Bank Guard. Are you, "kind reader, acquainted with the interior of the officer's room on "that guard? In case you are not I will describe it briefly for your "edification. It is approached by a flight of stone steps, and a "dingy narrow dark passage. The room itself is large, so it is but "scantily lighted by two small closely barred windows. The ceiling "is very low and vaulted; in fact the whole place, with its musty "odour and dim religious light is not unlike an old family vault. But "what most attracted my attention in the place were the paintings, "which half-covered the surface of the blotched, discoloured walls. "Truly, I thought to myself, my artistic predecessors do not seem to "have been cheerfully affected by their surroundings. Indeed I "have never seen a more gruesome selection of subjects than those "which they had chosen to depict.

"In one corner of the room was an uninviting leather-covered sofa, "over one end of which towered a life-like presentment of that mis-"named ruffian, Charles Peace. On the wall above the side rose the "figure, life-size, of a woman, a cruel, sensuous, leering woman, "holding in her hand a lighted candle. On the same wall, and close "beside her, were ranged three fantastic and hideous abortions— "monsters that were worthy to have emanated from the fertile brain "of Edgar Poe himself. The opposite wall of the room was much "less decorated, but in the centre, just above a bare table, was a "painting of a head, a head but newly severed from the body, the life "scarce yet extinct in the ghastly features, whilst the blood still "seemed to be gushing forth from the torn arteries.

"I passed the long lonely day as best I might; I wrote letters, "made futile attempts at reading a particularly futile society novel, "drank occasional pegs, smoked numerous pipes, lunched and dined, "went round my sentries, and so forth. At last the clock struck ten, "and I thought I might as well succumb to the drowsiness against "which I had been struggling all day. So I lay down on the "adamantine sofa, and muttering to myself—as I took a last look at "the burglar's portrait—R.I.P., dozed gently off, notwithstanding the "stony nature of my couch. How long that doze lasted I cannot "say. I was awakened again by feeling fingers, long, long, clammy "fingers, pressing against the back of my neck as though to thrust "me from my resting place. I half rose and looked round, but my "gaze met only the glittering wolfish eyes of Charles Peace; all was "still, all was as before. Stay—am I mad? No! the arm of that "repulsive Jezebel, which holds the lighted candle, is projecting "straight out from the wall, while a gleam of triumphant malice on "her face is made visible by its light. I spring up and stand erect "upon my feet; as I do so, I see that that candle sheds its rays also "straight on the lurid death's head opposite. At the same moment "I hear, I can hear it now, a discordant chuckle from behind me. I "turn round, but only to stand face to face with those three gibbering "shapeless imps. I turn again. Urged by some uncontrollable "impulse, I draw nearer, guided by the sight of that she-devil's candle, "nearer to the HEAD!

"Nearer, still nearer, there is only the table between me and it "now! Oh God, it is my own face that I am looking upon—the "features are my features—the blood must be my blood—the life that "is fading slowly from the flesh, it must be my life!

"There is something sharp and glistening lying on the table.

"Who hissed, take it? 'I will not!' yet my fingers close upon it; "they are lifting it slowly—despite my will—up towards my throat.

"Again the mocking laugh!

"Suddenly arises a wild jangling of bells, a clanging of bells—roof, "ceiling, floor, seem resonant with their mad violence. I turn, I "pause, I stagger back, all becomes dark and void, and I fall prone "upon the sofa.

"'Good evins, sir, there's the field-officer awaiting outside, an "axing where the hofficer of the guard is, and I've been ringing that "there alarum bell these last five minutes to try and waken you.'

"It was the voice of my sergeant that greeted me thus. I stagger-"ed to my feet. As I did so, *something* fell from my hand. What "was it? A glittering poinard? No, but a brand new nickel-plated "CORKSCREW!

CONROFT."

MOUNTJOY PRISON.

Well would it be if we could draw a veil over this, the filthiest and most obnoxious of all the guard rooms. A wretched room, in the old hospital, approached through the old "dead house;" fifthy passages, eleven rooms more or less covered with dust, windows encrusted with dirt, an awful smell, and smoke everywhere—such was the Mountjoy Prison guard. The guard and picquet, for both were combined, used to consist of an officer, sergeant, three corporals, drummer, and thirty privates.

The guard during the day was located in two portions of the prison, and the picquet in a third. The wretched officer was in a room over the men's guard room, in the mortuary before alluded to, with no outlook but the bare prison wall within a short distance of his window. It was possible, by putting a chair on the table and standing on the chair, to see over the wall, and even as far as the road two hundred yards away. The round of sentries was nearly a mile in length, as they were placed all round both prisons, and this was really a relief to enable one to breathe freely the purer atmosphere of the outer prison.

As the walls are whitewashed annually (they need it) and no regard to any artistic talent is permitted in a prison, the guard room showed but few proofs of painters' skill.

One or two good representations of uniforms of the old 5th, "The Gosling Greeners," remained when we first came, as also one or two figures of mythology. These, however, soon vanished. I often regret not having copied some of the many verses which from time to time appeared on the walls. However, I did not, and some were exceedingly good ones. The only regimental one I did copy is a description of the guard room, by a brother officer. Here it is:

THE lights are lit, the day is done,
By gloomy walls I'm here surrounded,
So slowly have the hours run,
Retreat has but this moment sounded.
I hear the sentry's even tread—
The challenge that he's always crying—
I almost wish that sentry dead,
His tramp, tramp, tramp, is awful trying.

I see, beneath my half-closed lids,
  The mice from out their holes are peeping ;
They know that this d——d place forbids
  The smallest chance to catch me sleeping.
I try to sleep ; my fancy flies
  To where my friends are quartered cosy,
I find that after all my tries,
  I am not yet a trifle dosey.

Weird forms the single gas jet throws
  Upon the dirty walls and ceiling ;
A horror of the lone place grows,
  I feel it o'er my senses stealing.
The bottle on the table makes
  A lengthy shadow, most fantastic,
My brain but seldom fancy takes,
  This place has made it quite elastic.

I heard folks marvel at the state
  To which their worldly comfort's risen ;
They say, discomfort's out of date,
  I'll swear it's not at Mountjoy prison.
The chairs are hard, the floors are bare,
  The pipes around the room are rusting,
The dirt is thickening everywhere,
  And quite defies your servant's dusting.

But there's tattoo, it's come at last,
  I hear the last relief dispersing,
The subaltern may blast and blast,
  But little comes of all his cursing.
It seems some subs are not aware
  Their lot in life's a thing of beauty ;
It may be so, but this I'll swear,
  It's not upon a Mountjoy duty.

Before quitting the subject of the prison, I might venture to relate a little incident of one of the many guards I mounted in this fearful place. We always had a countersign given to us at night, and the men had strict orders to allow no one to pass without giving the countersign, not even warders, as only special ones were allowed. One morning, about 2 o'clock, the sergeant of the guard rushed up to me to say, "Please, sir, there's one of our sentries has got the governor in a corner, and won't let him pass, as he ain't got no countersign." I immediately doubled off with a file of the guard to see what was up, and found the sergeant's statement was correct. The sentry had his bayonet right up to the governor, who was in a corner, and would not pass him, nor would he allow anyone else near. Of course, when I got up with the guard, it was all right. The sentry had only done his duty. The governor had forgotten the countersign, and it might have been a ruse. The governor was fearfully angry, and wanted to have the sentry confined, but after a time he accepted my explanation, and acknowledged eventually that the man was quite right.

I fancy we were not very sorry that the governor had been put to this discomfort, as he never took any notice of the officer on guard, and, on the other hand, prevented visitors coming in to see him on many occasions; he also did not allow him to have any meals cooked in the warders' kitchen, or to obtain any supplies in the prison. This latter interference was felt to be quite out of place, as at one time the officer on guard was catered for, at a fixed charge, by a married warder, and a very comfortable arrangement it was, too; but owing to this prohibition, all supplies had to be brought from barracks, at great inconvenience and considerable expense.

THE MAGAZINE.

This guard consisted of an officer, one sergeant, one corporal, one drummer, and twelve privates. The Magazine stands in the Phœnix Park, on a high eminence, overlooking the river, and during the summer this was a very favourite guard.

From the cavaliers of the fort itself a good view of the park could be obtained, and the bands of the various regiments could be heard playing at different cricket grounds in the vicinity. Smoking in the Magazine, of course, was prohibited, and at night great care was exercised in the proper custody of so important a place. A patrol was sent round the outer ditch during the course of the night, and sentries were always warned to be particularly alert as to suspicious noises etc., since the Magazine contained vast stores of ammunition of all descriptions. The officers' guard room, as well as the men's and other inhabited buildings, was outside the Magazine itself, and consisted of one very small room, just within the gate of the outer courtyard. The pictures on its walls were few, and of no particular merit; one or two landscape scenes of foreign aspect were about the most striking; and a humorous sketch of an ancient battle between the English and French, was, perhaps, the most entertaining of the scanty stock. The furniture of this room, for which a charge was made, was of a wretched quality, and of considerable antiquity. Food could be obtained in the fort, without the necessity of sending into Dublin for it.

In the summer time, owing to the fresh air and lively aspect of the park, this guard was undoubtedly the first choice, and recalls many memories by no means unwelcome, I venture to assert, to many who have taken their turn of duty within its walls.

ÇA IRA.

www.ingramcontent.com/pod-product-compliance
Ingram Content Group UK Ltd.
Pitfield, Milton Keynes, MK11 3LW, UK
UKHW021837270726
14058UKWH00002B/205
9 781843 42096

ISBN: 9789079889617; 978-90-79889-61-7

Original title: «Дух, душа и тело». The original Russian text can be found on the website http://lib.pravmir.ru/library/readbook/911

Editor: Convent of the Mother of God Portaïtissa, Trazegnies, Belgium, portaitissa@skynet.be
Translators: Rimma Andronova and Guram Kochi MSc
Design: Guram Kochi MSc and Marijcke Tooneman
Front cover: photo of St. Luke
The Scriptural quotes are taken from the King James Bible, with thanks to the BibleGateway.com

Tel.: 00 31 (0) 70 352 15 65
E-mail: gozalovbooks@planet.nl
Website: www.hetsmallepad.nl

Table of Contents

Introduction

The author of this book, Saint Luke the Confessor (secular name Valentine Voino-Jasenetsky) of Simferopol and Crimea was born on 27th April 1877 in Kertch, Crimea and he passed away to rest in Christ on 11th June 1961 in Simferopol, Crimea.

He was the descendant of a White-Russian-Polish impoverished princely line. By the end of his life he was an archbishop of the Russian Orthodox Church and at the same time a prominent physician, surgeon, scientist-researcher, inventor, author of scientific and theological works and painter.

From a young age his ideal was serving the weak and the sick. Therefore he ended his promising studies at the Academy of Fine Arts in Kiev unfinished and he started medical studies at Kiev University where he graduated. He chose to work in the province to help relieve the suffering of ordinary people. At the same time he did scientific research work on the cases of his practice and he learned several European languages in order to follow the development of medical science in the West. His innovative ideas and advanced surgery techniques have received broad acknowledgement in the Russian as well as in the European medical world.

He inherited a deep faith from his father and from his student years on he was known for his spontaneous public speeches on Christian ways and values.

At the time of the civil war after the Bolshevik revolution in Russia, when their consistent morbid and deadly persecution of the clergy was at full swing, Valentine Voino-Jasenetsky consented to be ordained as priest, and later he took monastic vows with the name Luke (after the Apostle and Evangelist Luke) and he was ordained as bishop. Being an intelligent and powerful spiritual pastor he was persecuted by the Soviet authorities. They convicted him on the base of false accusations and sentenced him to banishment in Siberia, above the Polar Circle, for nearly eleven years.

He ended his days as archbishop of Simferopol and Crimea. During all his ordeals he remained faithful to his Russian-Orthodox faith and his principles and he set forth his work as surgeon, physician and scientist, notwithstanding even blindness which struck him in the last decennium of his life.

We offer here a translation of his beautiful treaty on God, man and the universe, based on the Holy Scriptures, his religious experience and the works of scientists, theologians and philosophers.

The reader will even find clues with regard to the secrets of life in the whole universe, on the distant stars and planets that God's Spirit has revealed to Saint Luke.

The publisher
August 2014

Saint Luke the Confessor
of Simferopol and Crimea

"For the word of God is quick, and powerful, and sharper than any two-edged sword, piercing even to the dividing asunder of soul and spirit, and of the joints and marrow, and is a discerner of the thoughts and intents of the heart." (Hebr. 4:12)
"And the very God of peace sanctify you wholly; and I pray God your whole spirit and soul and body be preserved blameless unto the coming of our Lord Jesus Christ." (1 Thess. 5:23)

Chapter 1. What Conclusions can we draw with regard to the existing System of Science

We are going to start our discourse on the interrelation of the human body, soul and spirit from a long time ago. Up to the end of the XIXth century, exact sciences impressed by the explicitness and precision of their treaties. Not long ago the main scientific postulates were taken as unconditionally true. Only a select few could see cracks in the colossal construction of classical science. The great scientific discoveries at the end of the previous and the beginning of this century however suddenly shook the foundations of this construction and forced us to review the main concepts of physics and mechanics. The principles, that seemed to have the most reliable mathematical framework, are being debated by scientists. Every page of such books as the profound work "The Science and the Hypotheses" by Henri Poincaré contains evidence of it. This famous mathematician has shown that even mathematics is based on plenty of hypotheses and conventions. One of his most outstanding colleagues-mathematicians, Emile Picker, in one of his works shows how incoherent are the principles of classical mechanics, the basic science which claims to have formulated the general laws of the universe. Ernst Mach in his "The History of Mechanics" expresses a

similar viewpoint: “The foundations of mechanics, seemingly the most simple, are extremely complicated in fact; based on unrealized or even unrealizable experiments, they cannot be taken for the axioms of the mathematics.”
The physicist Lucien Poincaré writes: “There are no longer great theories, which would [nowadays] enjoy a unanimous acceptance by the researchers; there is an evident anarchy in the field of science, none of its laws are accepted as truly necessary. We witness the breaking of old concepts, and not the completion of scientific works.
The ideas that seemed best well-founded to our predecessors are being revised. The possibility of explaining all phenomena on the grounds of mechanics is abandoned now. The very foundations of mechanics are disputed; new facts are shaking the absolute meaning of the laws considered to be the main ones.”
However, though speaking of physics and mechanics 30-40 years ago it was possible to qualify them as being in a state of anarchy, this is no longer true. Breaking the main principles and concepts of physics has lead to the creation of new concepts, much more correct and profound. Not only those concepts over-rule old classical mechanics but they include it as an approximately valid theory with well-defined limits of applicability.
For example, it is found that classical mechanics cease to be valid for the smallest known objects, molecules, atoms, electrons etc. and must be replaced by a more precise and at the same time more complicated and abstract theory – quantum mechanics. And quantum mechanics is not something completely contrary to classical mechanics but it includes it as a kind of approximation which can be applied to massive objects.
Moreover, the processes characterized by high velocity movement, approaching the speed of light, can’t be truly described

with classical mechanics but must be replaced by the more rigorous theory of relativistic mechanics based on Einstein's relativity theory.

The law of the invariability of elements is no longer valid, because the transformation of one element into another is indisputably proved.

The established fact is the existence of elements with the same atomic weights but not identical chemical properties. A similar statement could have caused derision among chemists a few years ago (T. Svedberg).

There is hope for the complex structure of an atom to be proved, so there is no reasonable doubt in the possibility of heavy atoms being composed of lighter ones. In the last analysis all elements might be composed of hydrogen atoms. According to these hypotheses a helium atom comprises four hydrogen atoms closely spaced. In its turn, a hydrogen atom consists of two particles: an electron and a proton.

An atom is no longer the primary unit of matter, as it was set out that an atom is complex. Now the smallest known particles of matter are an electron and a positron. Both have the same mass but differ in their electric charge: an electron is negatively charged while the charge of a positron is positive. In addition there are heavier particles, protons and neutrons, comprised in a nucleus. The mass of such a particle, 1840 times the mass of an electron, is almost equal; while a proton is positively charged, a neutron is electrically neutral.

It was recently found that cosmic rays coming inside the earth atmosphere from the interstellar space include a series of unknown particles, and their mass varies in a very large range (from 100 to 30 000 times more than an electron mass). These particles have got different names: mesons, or mesotrons, varytrons and others. It was also found that all these particles are not absolutely invariable. Protons may be changed into neutrons and vice versa, electrons in a composition with positrons may cease to exist as particles converting into radiation. On the other hand, under certain conditions an electromagnetic field may "give birth" to the pair, an electron and a positron. The particles found in cosmic rays may significantly alter their mass in the process of interaction with the atoms in the atmosphere.

Modern books on physics call "annihilation" (the destruction of matter) the process of conversion of an electron – positron pair into radiation, and the process in the opposite direction is called "materialization".

Consistent materialists call such terminology just conditionally acceptable, leading to idealistic distortions in the description of reality. They say there is no conversion of mass into energy and vice versa, because both mass and energy are properties of the particular reality called matter, and appearing particles are endowed with energy while energy is endowed with mass.

This last statement for us, brought up on the previous concepts of physics, is absolutely new, though we are very far from celebrating the victory over materialism.

We have neither the right nor the intention to deprecate the very important achievements of modern physics. The fact that particles may change their mass, like the new particles found in cosmic rays, or annihilate themselves converting into an electromagnetic field, like electrons and positrons, may not lead to the conclusion that matter disappears, as the electromagnetic field may be considered as the other form of matter.

Both these forms may be converted into one another the same way as liquids may change their form to solid or gas. These transformations though, are possible on condition that energy is constant according to the law. Energy cannot disappear or be created out of nothing. Only the form of the matter containing the energy may be changed, but the total amount of energy must be constant.

At present physicists reject the hypothesis of an existing weightless and "absolutely elastic" substance, the ether, replacing it with the concept of the electromagnetic field. The electromagnetic field is not a substance in the usual mechanical sense of the word, as it has no weight, it is not hard or elastic in a normal sense, it does not consist of particles, etc. But it is endowed with energy, so in this sense it should be considered as a form of existence of matter. It is generated by movements and interactions of elementary particles, electrons and others. On the other hand, the electromagnetic field itself acts on the particles and may even give rise to them under certain conditions.

Instead of weight, hardness, elasticity etc. the electromagnetic field has other characteristics that determine its properties. These characteristics are the intensity and direction of electric and magnetic forces in different points of the space.

The laws describing the electromagnetic field and its interaction with an electric charge are the subject of a particular field of physics – electrodynamics, and the laws of motion and interaction of material particles comprise a part of mechanics.

All products of matter dissociation are finally "absorbed" into electromagnetic fields. Regardless of the kind of dissociating matter and of the dissociation methods, the products of the dissociation are always the same. In case of nucleus disintegration of radioactive elements, of emission out of a metal surface under light, of interaction products in chemical or burning processes, the result is always the same with variety in quality, amount or speed. The matter disintegrates into elementary particles: neutrons, protons, mesons, electrons, positrons etc. The motion and interaction of these particles generate the electromagnetic field, magnetic and electric oscillations of different frequency, radio-waves, infra-red rays, visible radiation, ultra-violet and gamma-rays.

Electrical phenomena underlie all chemical reactions, and there is an attempt to explain all the other forces on their basis. It is found that light is also a form of electromagnetic energy, and electricity has the corpuscular, or as some wrongly name it, atomic structure. It is certainly not right to call atoms, the corpuscles forming electricity, electrons. Electricity is carefully and completely acceptably determined by Milliken. He says: "I have never tried to answer the question what electricity is; I am satisfied with the established fact that whatever it is by its nature, its amount will always be an exact multiple of a definite electrical unit.

...Electricity is something more fundamental than atoms of matter, because it is a substantial composition of hundreds of different particles. Like matter, it is something consisting of separate units, but it is different from those of matter, because all these units, as far as it could be defined, are absolutely equal."

The corpuscular theory of electricity is the greatest achievement of theoretical physics. However it can't be said that according to its corpuscular nature electricity ceases to be energy, and becomes something like matter. Physicists don't say so either, they just claim that energy has a certain mass, and "mass" is an immanent feature of some reality: matter. This is not an attempt to consider matter and energy as being the same; regardless of how close electricity is by its nature to matter, it remains energy for us, and the most important, the main part of atomic energy.

And yet this basis of the physical life of the world has become known to us only 300 years ago, from the time of Volta. For thousands of years electricity was unknown to people.

Only 50 years ago science was enriched by the knowledge of new and extremely important forms of energy - radio-waves, infra-red rays, cathode rays, radioactivity, and nuclear energy. The latter (we can't even imagine how grand and powerful it is) underlying the whole dynamics of the world, giving birth to unfading inexhaustible thermal energy from the sun, was discovered 300 years later than electricity.

Do we have the right to believe and even say that in the universe there are still other, unknown forms of energy, even more important to the world than nuclear energy?

34 % of the solar spectrum is invisible. Only a very small segment of those 34%, ultraviolet and infra-red rays, are studied, and the forms that underlie them are described now. And are there any reasonable arguments against the idea, even the certainty, that the numerous Fraunhofer lines hide many secrets, forms of energy unknown to us, maybe even more delicate than nuclear energy?

From a materialistic point of view these still unknown forms of energy should be some special forms of existence of matter.

Let it be so, we can say nothing in argument, believing in the power of science. However electricity cannot be called matter, and most certainly should be considered as energy; this energy can generate particles of matter having a definite mass and physical properties. These particles can annihilate that is turn into energy again. Following the logic of these facts we can assume that in future new forms of existence of matter (or rather, energy), will be discovered which according to their properties, may be called "semi-material" even more than is done for electricity.

The very concept of "semi-material" includes the recognition of the existence of "non-material".

Is there any reason to deny the legitimacy of our faith and confidence in the existence of a purely spiritual energy, or power, which we believe to be the primary source of all physical forms of energy, and through it the source of matter itself? What is our idea of this spiritual energy?

For us it is God's almighty love. Love may not exist for itself only, for its immanent feature is the need to pour out over somebody or something, and because of this need God has created the Universe.

"By the word of the Lord were the heavens made; and all the host of them by the breath of his mouth." (Psalm 32: 6).

This energy of love, which outpoured according to the will of the most good God: God's Word, gave rise to all the other forms of energy, which in turn generated particles of matter, and then, with their help - to the entire material world.

Pouring in other directions, God's manifested love created the entire spiritual world too: the world of angelic creatures gifted with reason, of human understanding and the entire world of pneumatic and psychic phenomena (Ps. 104:4; 32:6).

The fact that we know nothing about many doubtlessly existing forms of energy, is due to the obvious lack of sensitivity of our poor five senses for discovering the world, and to the lack of new scientific methods and reagents for the detection of those forms which are not available to our senses.

But is it true that we have only five senses and no other organs and means of direct perception? Is there any possibility of a better perception of selected forms of energy adequate to them? The eagle's visual acuity, and the dog's sense of smell excel to a great extent that of humans. Pigeons are endowed with a sense of direction, which is beyond our comprehension, infallibly guiding them in their flight. The growing acuity of hearing and sense of touch of the blind is well known too.

I believe that the certainties of mental order, which will be described further on, not only oblige us to admit the possibility of a heightened sensitivity of our five senses, but also to add our heart to them as a special organ of sense: the seat of emotions and the organ of our cognition.

Chapter 2. The Heart as Organ of Cognition of the Highest

Already in the time of the ancient Greeks the word "heart" meant not only a carnal organ, but also the human's soul, mood, view, thought, even prudence, intelligence, conviction , etc. "Folk wisdom" gave a right value to the heart in human life already long before: "My heart stops beating – my life is over". That's why people call the heart "the engine of life". We know now quite well that the physical and spiritual well-being depends to a great extent on a proper function of the heart.

In our everyday life we may see that somebody's heart is suffering, aches, etc. It is possible to find expressions in literature and lyrics like: "My heart longs for... is happy...is feeling something" etc. Thus, the heart is like a sensory organ and, moreover, an extremely precise and versatile one. We should dwell upon this because all these phenomena have a profound physiological basis, described by I. P. Pavlov [a famous Russian physiologist]. In distant times, when our ancestors were in the zoological stage of development, they responded to all the stimuli they were receiving almost predominantly by muscular activity, which prevailed over all other instinctive acts. The entire muscular activity is in its turn closely connected to the cardiovascular activity.

Though the muscular reflexes of the modern civilized human being are reduced almost to a minimum level, the changes of the heart's activity, induced by them, have remained fully present.

The modern civilized man is accustomed to hiding his muscular reflexes, by working on himself. Only variations in his heart's activity may make his feelings evident. Thus, the heart still is an organ of feeling, precisely indicating our subjective state and always exposing it.

Being a doctor I should remark that as well-regulated as the cardio – activity is, caused by muscular activity, not excessive, of course, as badly-regulated is the cardio – activity caused by various emotions, which are not necessarily expressed in muscular activity. That is why the hearts of people in the learned professions, who perform just an easy manual labour, but who are excessively subjected to experiencing life's agitation, are affected much more.

Thus is a judgement on the functioning of the heart, of a pathologist-anatomist ("On the death of a man") and a great physiologist, Academician Ivan Pavlov, ("Course of Physiology," edited by Prof. Savich, 1924).

Moreover, the innervation of the heart is stunningly rich and complex; the heart is covered with a network of fibres of the sympathetic nervous system and through it, is very closely connected with the brain and spinal cord. The heart is provided with a complex system of cerebral fibres, which transmits a great number of highly complex impulses of the central nervous system from the vagus. It is quite possible that the centripetal sensory impulses of the heart are sent back to the brain through that very nervous system.

The functions of the sympathetic and vegetative nervous system are poorly studied and are almost unknown, but it is quite clear now that these functions are very important and versatile. And the most substantial thing for us is that these nerve fibres and nodes certainly play a very significant role in the physiology of sensitivity. Thus, our knowledge of the anatomy and physiology of the heart not only do not hinder but rather prompt us to consider the heart as being the most important sensory organ, not just the central pump of the blood circulation.

The Scriptures, however, tell us much more about the heart. Almost every page of the Bible refers to the heart, and those reading it for the first time can't help noticing that the heart is considered not only as the central organ of sense, but also as the most important organ of cognition, reflection and receptacle of the spiritual impulses. Moreover, according to the Scriptures the heart is man's organ of communication with God, and therefore, the organ of cognition of the Highest.

Truly all-embracing is the role of the heart in the realm of feelings according to the Scriptures. The heart feels joy (Jer.15:16); "rejoicing" (Prov. 27:9), "pained" (Jer.4:19), "grieves" to such an extent that the Psalmist is crying (Ps. 73:21), is "cut" from malice (Acts 7,54), is "burning " with a trembling anticipation in Cleopas (Luke, 24,32), "fretteth against the Lord" (Prov. 19:3), is "full of evil" (Eccles. 9:3), "adultery" (Matt. 5:28), envy (Jac. 3:14), "pride" (Prov. 16:5), "courage" (Ps. 27:3), "faintness" (Lev. 26:36); "uncleanness through the lusts" (Rom. 1:,24), "reproach" breaks it (Ps. 69, 20).

However it also perceives consolation, sense and comfort (Philem. 7), capable of a great feeling of trust in God (Ps. 27:7) and be broken to be near the Lord (Ps. 34:18); it may be a repository of meekness and lowliness (Matt. 11:29).

Besides the wide range of emotions and feelings, the heart has the higher ability of feeling God, about which Paul the Apostle says in the Areopagus of Athens: "That they should seek the Lord, if haply they might feel after him, and find him,.." (Acts 17: 27)

Many pious ascetics and reverend fathers tell about feeling the Lord or about gracious influences of the Spirit of God on the heart. All of them could feel to a greater or lesser extent the same as the Jewish prophet Jeremiah: "... his word was in mine heart as a burning fire shut up in my bones..." (Jer. 20:9)

Where does this fire come from? St. Ephraim the Syrian, the great repository of the Divine Grace, gives us the answer: " Incomprehensible to the mind, It comes into the heart and dwells in it, thus the Innermost of the Radiance is to be found in the heart. The Earth exalts His feet, and the pure heart bears Him inside"; and, we may add, the heart contemplates Him without the eyes according to the Lord's word:
"Blessed are the pure in heart: for they shall see God."(Matt. 5:8). We may find a similar discourse by John Climacus: "the fire of the Holy Spirit coming into the heart revives the prayer; after the prayer has been revived and ascended to the Heavens, the descent of the heavenly fire into the chamber of the heart can take place. "
And these are the words of Macarius the Great: "The heart governs all the body's organs and, when the [God's] Grace would fill all parts of the heart, it dominates all man's thoughts and all the parts of his body for the mind and the soul's designs are there...For it is there we should look whether the grace of the law of the Spirit is written there."
What do we mean by "there"? We mean the main organ, where the throne of grace is, and the mind and the soul and all the soul's designs dwell, i.e. the heart.
Let us not cite too much from the works of those who lead the deepest spiritual life. Many can be found in "The Philokalia". All the holy Fathers say that, according to their own experience, when in a good and graceful state of soul they feel in their heart a quiet joy, deep inner peace and warmth in the heart, which always grow stronger after a steadfast and fervent prayer or good deeds. On the contrary, the influence of the spirit of Satan and his servants on the heart raises there a troubled uneasiness, a kind of burning pain, coldness and an unaccountable anxiety. These are the feelings in the heart according to which the ascetics recommend to assess one's

spiritual state and to distinguish between the Spirit of Light and the spirit of darkness.
However the ability of the heart to communicate with God is not limited to this kind of more or less vague feelings. No matter how doubtful it may seem to unbelievers, we state that the heart is able to perceive certain suggestions directly as God's words. Moreover, this ability doesn't belong to the domain of saints solely. Even I, like many others, experienced it more than once with a tremendous power and deep emotion. While reading or listening to the Scriptures I sometimes suddenly was stunned realizing that God's words were addressed directly to me. They sounded to me like a thunderbolt, and pierced my brain and heart like lightning. Certain phrases, quite unexpectedly for me, just stood out from the context of the Scriptures, in a bright dazzling light and were indelibly imprinted in my mind. And these lightning phrases of God's words were always suggestions, instructions or even prophecies, most important and necessary for me at that moment and consistently proving true in the future.
Their strength was sometimes colossal, stunning, and incomparable with the power of any kind of common mental influences.
In part due to circumstances beyond my control I left my Episcopal ministry for a few years, but once during the all-night vigil service, waiting for the Gospel reading, I suddenly felt a shiver, a vague foreboding that something terrible was about to happen. These were words which I myself had often read calmly, "Simon, son of Jonas, lovest thou me more than these? ... Feed my lambs." I was shaken so suddenly and vigorously by God's reproach, His call for the resumption of the deserted service, that till the end of the vigil I was shivering all over, couldn't sleep the next night and for approximately a month and a half was shaken with sobs and tears each time I recalled this extraordinary event.

Let sceptics not think that this feeling was an impact of my own dreary memories of the sacred ministry I had deserted and reproaches of my conscience. On the contrary, I was focused at that time on my illness and the impending surgery, and I was in the most ordinary state of mind, very far from any exaltation.

The holy prophets could listen directly to God's words and take them into their hearts. "Moreover he said unto me, Son of man, all my words that I shall speak unto thee receive in thine heart, and hear with thine ears." (Ezek. 3,10)

"When thou saidst, Seek ye my face; my heart said unto thee, Thy face, Lord, will I seek." (Ps. 27:8)

The prophet Jeremiah qualifies the call from above as God's direct speaking to him.

The prophet Ezekiel , having described his marvellous vision of God's glory, continues: "And when I saw it, I fell upon my face and heard a voice of one that spake. And he said unto me, Son of man, stand upon thy feet, and I will speak unto thee. And the spirit entered into me when he spake unto me, and set me upon my feet, that I heard him that spake unto me." (Ezek. 1,28 – 2,2)

All the prophets speak in the name of God: "And The Lord God said to me", "This is what the Lord Almighty says", "The word of God came unto me". They received revelations from God in visions and dreams (Ezekiel: chapters 40 - 48; Daniel's dream: chapter 7; his visions: chapters 8 – 10; Amos's visions: chapters 8 - 9; Zechariah's visions: chapters 1 6.)

Here is what the Fathers say about these different ways of receiving revelations from God.

"If someone suggests that the visions, images and revelations were the work of fantasy and took place in a natural way, he should know that he is far from the truth and the right facts. For the prophets and also holy hermits living now, who are initiated into the sacred mystery, could see God's revelations not according to the laws of nature, but these were displayed and presented to them supernaturally by the indescribable power and grace of the Holy Spirit, as Basil the Great says, "The prophets could receive revelations in their mind which was pure and not distracted, by an unspeakable power, and they heard the word of God as if proclaimed inside them." Moreover, the prophets could see visions by the action of the Holy Spirit, impressing the images inside their mighty minds. And Gregory the Theologian says, "The Holy Spirit acted first in angels and heavenly creatures, then in fathers and prophets, among whom some could see and know their God, others could predict the future as with the images sent into their mighty minds by the Spirit they could see as if it were in the present." (monks Callistus and Ignatius)

In this excerpt from their writings Callistus and Ignatius do not refer to the prophets' perception of God's revelations through their heart, but through their mind. However we will make it clear later that the Scriptures ascribe to the human heart those very functions which are considered as being of the mind in psychology. It is exactly the heart that is called in the Scripture the organ of the cognition of the Highest.

The Scriptures speak not only of the heart's ability of perceiving the impact of the Spirit of God, but consider the heart to be the core of our spiritual life and Theology, which God Himself improves and corrects.

Here are some fragments which clearly show it.

"And I will give them one heart, and I will put a new spirit within you; and I will take the stony heart out of their flesh, and will give them an heart of flesh..." (Ezek. 11:19)

"For Thou gave our heart fear for Thee so that we call upon Thy Name and we praise Thee in our exile, as we repulsed the untruth of our fathers who have sinned before Thee." (Baruch 3:7)

"...the work of the law written in their hearts..." (Rom. 2:15)

"Cast away from you all your transgressions, whereby ye have transgressed; and make you a new heart and a new spirit..." (Ezek. 18:31)
"That the God of Our Lord Jesus Christ, the Father of glory, may give unto you the spirit of wisdom and revelation in the knowledge of him; the eyes of your understanding being enlightened; that you may know what is the hope of his calling..." (Eph. 1:17-18)
"Make the heart of this people fat, and make their ears heavy, and shut their eyes; lest they see with their eyes, and hear with their ears, and understand with their hearts, and convert, and be healed." (Is. 6:10)
"God hath sent forth the Spirit of his Son into your hearts..." (Gal. 4:6)
"That Christ may dwell in your hearts by faith..." (Eph. 3:17)
"And the peace of God, which passeth all understanding, shall keep your hearts and minds... " (Phil.4:7)
"I will put my fear in their hearts, that they shall not depart from me..." (Jer. 32:40)
"Saith the Lord, I will put my laws into their hearts, and in their minds will I write them..." (Heb. 10:16)
"...The love of God is shed abroad in our hearts..." (Rom. 5:5)
"For God...hath shined in our hearts..." (2 Cor. 4:6)
In the Parable of the Sower the Lord himself says that the seed of the Word of God is sown in the man's heart and it is kept by the heart, if the heart is pure, or the seed is snatched by the devil, if the heart can't or is not worthy to keep it.
The highest functions of the human spirit, such as the human faith in God and love for Him, are carried out by the heart.
"For with the heart man believeth unto righteousness; and with the mouth confession is made unto salvation." (Rom. 10:10)
"thou ...shalt believe in thine heart..." (Rom. 10:9)

"Thou shalt love the Lord thy God with all thy heart,.." (Matt. 22:37)
"...For the Lord your God proveth you, to know whether ye love the Lord your God with all your heart and with all your soul." (Deut. 13:3)
"And thou shalt love the Lord thy God with all thine heart, and with all thy soul, and with all thy might." (Deut. 6:5)
"But sanctify the Lord God in your hearts..." (1Pet. 3:15)
We pray with our heart, and one of the greatest kinds of prayer is a soundless cry to God. So Anne, the mother of Samuel the Prophet, prayed to be granted her this great son.
At Mount Sinai, God told Moses: "Wherefore criest thou unto me?" (Ex. 14:15) And he prayed without words, without moving his lips.
"Their heart cried unto the Lord...", Jeremiah the Prophet says (Lam. 2:18)
It is well expressed by Landry in his book "The Prayer": "One day an angel said to one of the ardently praying souls: "What are you doing actually? You are shaking the palace of heaven, and nothing can be heard there except your screams." However, the soul did not say a word, only her heart fluttered, and this invisible movement was enough to shake the heights of heaven. "
The Lord Jesus Christ said to us that the heart is a repository of both good and evil: "O generation of vipers, how can ye, being evil, speak good things? For out of the abundance of the heart the mouth speaketh. A good man out of the good treasure of the heart bringeth forth good things: and an evil man out of the evil treasure bringeth forth evil things." (Matt. 12:34-35)
"But those things which proceed out of the mouth come forth from the heart; and they defile the man.

For out of the heart proceed evil thoughts, murders, adulteries, fornications, thefts, false witness, blasphemies." (Matt. 15:18-19)
Our heart is the seat of our conscience, which is like our guardian angel: "For if our heart condemn us, God is greater than our heart, and knoweth all things." (1 John 3:20)
Speaking about an awesome ability of the heart, Elisha the Prophet said to his servant Gehazi: "Went not mine heart with thee, when the man turned again from his chariot to meet thee?" (II Kings 5: 26). So the hearts of deeply loving mothers always follow their children in all their ways though with less prophetic clairvoyance than Elisha's heart had been following Gehazi.
The heart is meant not only for feeling and for communion with God. The Scriptures indicate that it is the body of desire, the source of the will, of good and bad intentions.
"...the Lord come, who both bring to light the hidden things of darkness, and will make manifest the counsels of the hearts..." (1Cor. 4:5)
"My heart is inditing a good matter..." (Psalm 45:1)
"...My heart's desire and prayer to God for Israel is..." (Rom. 10:1)
"He did it with all his heart,.." (2Chron. 31:21)
"And he went forwardly in the way of his heart..." (Is. 57:17)
"This evil people...walk in the imagination of their heart..." (Jer. 13:10)
"...till he have performed the thoughts of his heart..." (Jer. 23:20)
"...and he shall give thee the desire of thine heart..." (Psalm 37:4)
"...[heart] that deviseth wicked imaginations..." (Prov. 6:18)
"People hath a revolting and rebellious heart..." (Jer. 5:23)
"Thou hast given him his hearts' desire..." (Psalm 21:2)
"It is a people that do err in their heart..." (Psalm 95:10)

"Speak peace to their neighbours, but mischief is in their hearts..." (Psalm 28:3)
"There are many devices in a man's heart..." (Prov.19:21)
These texts make it evident with perfect clarity that human behaviour, the choice of a way of life is totally determined by desires and aspirations of the heart. We will see later on that the way of thinking is determined by feelings and desires too. But the heart not only determines our thinking; although it may seem strange to all who consider indisputable the teachings of psychology of the mind as an organ of thought and cognition. It is exactly the heart that perceives, thinks and cognizes according to the Scriptures. Let readers not close this book hastily after having read the statement above, which is indeed unacceptable for many. The philosopher Bergson, who in fairness should be considered as one of the greatest thinkers, proclaims the heart to have a prominent role in the process of cognition. Let us however refer to the Scriptures once more:
"Yet the Lord hath not given you an heart to perceive..." (Deut. 29:4)
"every imagination of the thoughts of his heart was only evil continually..." (Gen. 6:5)
"He hath scattered the proud in the imagination of their hearts..." (Luke 1:51)
"...Perceiving the thought of their heart..." (Luk.9:47)
"...that thou mightest know the thoughts of thy heart." (Dan. 2:30)
" and the meditation of my heart shall be of understanding..." (Psalm 49:3)
"The preparations of the heart in man..." (Pr. 16:1)
"For the word of God is...a discerner of the thoughts and intents of the heart..." (Heb. 4:12)
"Commune with your own heart..." (Psalm 4:4)
"Wherefore think ye evil in your hearts?" (Matt. 9:4)

"What reason ye in your hearts?" (Luke 5:22)
"There were certain of the scribes...reasoning in their hearts..." (Mark 2:6)
"All people mused in their hearts..." (Luke 3:15)
"...and in the hearts of all that are wise hearted I have put wisdom..." (Ex. 31:6)
"...wisdom entereth into thine heart..." (Prov. 2:10)
Yet he knew not...,yet he laid it not to heart..." (Is. 42:25)
"...And ye know in all your hearts..." (Josh. 23:14)
Wisdom resteth in the heart of him that hath understanding..." (Prov. 14:33)
"...and madness is in their heart while they live..." (Eccles. 9:3)
"The heart also of the rash shall understand knowledge..." (Is. 32:4)
"...The thoughts of many hearts may be revealed..." (Luke 2:35)
"Why do thoughts arise in your hearts?" (Luke 24:38)
Let us ponder the last sentence. How do these thoughts enter your hearts? Where do they come from? If the thought "enters", it means that it is not born in the heart.
Certainly, the Scriptures do not contradict indisputable physiological facts and do not deny the role of the brain in thinking; not only in thinking but in all psychical processes. In the above mentioned words of the monks Callistus and Ignatius, St. Basil the Great and Gregory the Theologian, the prophecies and the visions are seen as being caused by the graceful impact of the Holy Spirit on the mind of the prophets, and the mental processes take place in the brain. However the process of thinking is not limited by the activity of the cerebral cortex only and takes place also elsewhere. We know which parts of the brain are responsible for the motoric, sensory, vasomotoric, respiratory, thermal and other functions, but there is no area in the brain which is responsible for the feel-

ings. Nobody could find in the brain the centres of joy and sorrow, anger and fear, the aesthetic and religious feelings.
Although all the sensors and all organs of the body in general are connected to the brain cells by means of the nerve fibres, these are the visual and auditory, olfactory and gustatory, tactile and thermal, locomotory, and many others kinds of sensations only. They are however, just sensations. Not making a distinction between sensations and feelings results in the greatest psychological mistake.
If we could, by way of speaking, stop the most high-speed and complex dynamics of mental processes and analyse the individual elements in their static state, then sensations would appear to us only as initial impulses for the emergence of thoughts, feelings, desires and volitional movements. And a single thought isolated from the brain would be just an unfinished raw material destined for the deep final tempering inside the heart, which is the crucible of emotions and will.
We do not know how exactly the thoughts arising in the brain are transmitted to the heart, but the thought being a purely psychological action as opposed to sensation, which is a physiological action, does not need the anatomic paths of transmission. The feelings arising in the heart, depending on one or other thought, and forming these thoughts to a major extent do not need these paths either.
The heart receives these processed thoughts and sensory perceptions not only from the brain but itself has a wonderful and most important ability to receive exogenous feelings of the highest order from the spiritual world, not at all appropriate to the organs of the senses.
And these feelings from the heart are transmitted to the mind, the brain and to a large extent determine, direct, and modify all mental processes, which take place in the human mind and spirit.
Let us review some of the previous quotations:

"...The thoughts of many hearts may be revealed." (Luke 2:35)
"Wisdom resteth in the heart of him that hath understanding..." (Prov. 14:33)
"Madness is in their heart while they live..." (Eccles. 9:3)
If it is possible to speak of "the thoughts of the heart" and of the heart as the accumulator and dwelling of wisdom, it means that not only the thoughts born in the mind become sensually and volitionally completed in the heart; and not only the exogenous spiritual impacts are perceived by the heart and transmitted to the brain. In addition to it those exogenous impacts give rise to thoughts and reflections in the heart, just as the impulses of the organs of the senses are the stimulants and at the same time the material for the mental activity of the brain. Therefore the heart is the second organ of perception, cognition and thought, wisdom abides in it. However if the heart is deprived of the grace of God and does not perceive the promptings of the Spirit of truth and goodness from the transcendent world, and is inclined to perception of the spirit of evil, lies and pride, then insanity arises in the heart and lives in it.
Intellectuals take for an indisputable truth the statement that we cognize reality through the mind, of which the anatomical and physiological body is, naturally, they say, the brain.
But already in the XVIIth century, at the height of Cartesian dogmatism, when intellectualism was all-powerful, a brilliant mathematician and philosopher Blaise Pascal was able to discover the limit and the impotence of the mind, and proposed to replace it with cognitive ability, which would distinguish itself by directness and the ability of studying the truth.
An ability to which Bergson later gave a definite name "intuition", Pascal called a sense of subtleties, flair of judgment, a feeling, an inspiration, the heart, an instinct. In his "Thoughts" all these terms signify identically the direct cognition of reality, awareness of a living reality, which is opposite to the

knowledge through reason and rational calculations. Pascal in his very first works set the new difference between a "geometrical mind" and flair for subtleties.
A "geometrical mind" is exactly what we call the rationalistic or logical way of thinking, while intuitive thinking is a sense of subtleties.
"Reason", Pascal writes, "acts slowly, taking into account so many of the principles that should always be [actively] present that it is constantly tired and distracted, unable to hold them together. Sense works differently: it operates instantaneously and it is always ready to act. So we should lay our hope in feelings; otherwise our hope will always be unsteady."
Then his famous statement follows: "The heart has its reasons of which reason knows nothing". And then Pascal says in addition, "It is the heart which perceives God and not the reason."
The idea of cognition and of the great variety of our spiritual life, which is given to us in the Scriptures, is not compatible at all with intellectualism, that is the philosophical doctrine asserting that every reality can be cognized, and this can be accomplished through the cognitive ability of the mind only.
Intellectualism considers free, speculative cognition as a human activity, which belongs to the domain of perfection, and even as the only activity, worthy of man. But what is even more the core of intellectualism is that it recognizes the reality of things only in so far as it can be accepted by reason.
What on earth can be compared to the pretentiousness of this proud doctrine which denies the reality of everything that does not fit in our poor and very limited reason? The intellectualists ignore all that is so clearly and undoubtedly perceived by our heart from the transcendental world, everything that is cognized through Pascal's "sense of subtleties".

It was however the ancient philosopher Epicurus who said that all the objects of perception are true and real, which is the same as saying that something is true and that it exists. Why would the perceptions of the Highest through the heart be not true then?
It is only the brain which is considered as the organ of reason and will, while the spinal cord is said to be a system of pathways and an organ for instinctive and trophic activity.
However, a beheaded frog, when its skin is irritated, will act in a way to avoid irritation. If the irritation continues, it tries to run away and to hide in the same way as a frog with a head.
In the wars of ants which have no brain, a premeditation is clearly revealed, and therefore a reason, which is no different to the human one.
It is perfectly obvious that not only the brain but also the ganglion of insects as well as the spinal cord and the sympathetic nervous system of vertebrates are the body of the will.
It is impossible to expound even the basic idea of the amazing and deep philosophy of life of Henri Bergson in a small theological treatise.
I can only say that he has shown a completely new way of understanding life. At the same time he exposed the complete inability of the philosophy of rationalism (intellectualism?) with regard to this.
Not only Pascal was a great forerunner of Bergson on this revolutionary path of philosophy. Similar to Bergson's method of cognition is Maine de Biran's method of introspection for the investigation of reality in the human consciousness. He thinks that it is impossible to "grasp" reality in any other way than in "one's living self". Neither keen observations [of the outside world-Translator], nor rational thinking can give us this.

Schopenhauer was the first to prove that a concept, invented by reason, working idly as in a vacuum, can't be anything else but meaningless chimeras, fit only to meet the demands of the "professors of philosophy"; that the mind has forms only, and is "a meaningless ability." He considers intuition to be a counterbalance to reason.

Bergson expressed amazing and quite new statements about the idol of the intellectualists, the brain. He believes that the difference between the spinal cord, which reacts reflexively to the impulses, and the brain, is not in the nature of their functions, but in the degree of their complexity only.

The brain just registers the perception coming from outside and selects the appropriate method of response. Bergson says, "The brain is nothing but a sort of telephone exchange: its role is merely to issue a message or to clarify it." It adds nothing to what is receives.

All the organs of the senses are connected to the brain by nerve fibres, it shelters the motorial system, and it is a centre, where the external stimulation comes in contact with one or another mechanism of motorial reflection.

The brain shows by its very structure that its function is the responding in an appropriate way to a stimulus from outside. Afferent nerve fibres, which conduct the impulses from the organs of sense, end in cells of the sensory zone in the cerebral cortex. These cells are, in their turn, connected through the other fibres to the cells of the motorial zones, thus transmitting the impulses to them. As the number of these interconnections is countless, the brain has an endless ability to modify its reaction to an external stimulation, and acts as a kind of switchboard.

The nervous system and particularly the brain are not devices of ideal thinking and cognition, but only the tools that are meant for use in action.

"The brain is not an organ of thought, feeling, consciousness, but it is what chains the mind, feelings and thoughts to real life, forcing them to listen to the real needs and make them capable of expedient action."
The brain, in fact, is the organ of attention to life, of adaptation to a reality. (The Body and the Soul. You and the Life. 1921 December 20)
This is amazing enough, but these overwhelming thoughts of the great metaphysician almost completely coincide with the new doctrine of higher nervous activity, created by our brilliant physiologist Ivan Petrovich Pavlov.
More so, we can say that Henri Bergson just before Pavlov, had by pure philosophical thinking, realized the essence of Pavlov's physiological teachings based on experimental study of conditioned reflexes of the brain.
To substantiate this position, I shall give a few excerpts from Pavlov's book "Twenty years of experience of objective study of higher nervous activity in animals", but first I need to explain what are what Pavlov calls "conditioned reflexes" , and "analyzers". Each animal has a permanent set of congenital reflexes that Pavlov calls "unconditional". For example, an animal immediately rushes to the food it sees, it pulls back its leg if the leg is irritated; a snail retracts into its shell when touched; a newborn baby makes sucking movements when touching the mother's breast with his lips.
But besides these unconditioned reflexes, higher animals, namely dogs, with which Pavlov experimented , can also gain new, artificially developed reflexes, which Pavlov calls "conditional" , and which can be called also temporary or acquired. For example, if in a series of experiments a dog is irritated with something a short time before receiving food, such as a sharp sound , a light signal, scratching its skin, then soon this arbitrary, "conditioned" stimulus will effect the animal in the same way as the sight and the smell of meat (an "uncondi-

tioned" stimulus): the dog immediately starts salivating and has a motorial reaction, appropriate to a dog when seeing food. The "conditioned" signal led to the formation of a new, temporary, "conditional" reflex.
How exactly are these [conditional] reflexes formed?
When the cells of the eye, sensitive to light, are touched by light, the sensitive cells of the organ of hearing by sound, the sensitive ends of the nerves of the skin by touch, the corresponding sensations are transmitted by the nerve fibres into the areas of the cerebral cortex, the nerve cells of which are meant specifically for the perception of these stimuli (e.g. the optic nerve is located in the occipitalis parts of the hemispheres, the sound nerve in the temporal part, etc.)
The nerve cells of the cortex, upon receiving a stimulus, analyse it, and according to the results, transfer impulses to the lower centres of the brain and spinal cord for a corresponding executive action (effect): the motorial or vasomotorial one, secretion etc.
These downstream executive centres are called effectors.
Pavlov gave the name "analyst" to the entire system consisting of specific, perceiving cells of a sensory organ, the fibres of sensory nerves connected to them, and their extensions, the fibres of the white matter, which end in the nerve cells of the sensory parts in the cortex. There is a countless multitude of such analysts in the cortex.
Among them, besides these analysts, which have their beginning in the cells of each of our five senses, there are plenty of other analysts, which are connected to all the organs of our body and thus transmit to the brain cortex in the hemispheres information about what goes on inside the body.
Thus, the brain is responsible for the huge task of analysing all these stimuli and responding to them by the reactions of the effectors.
The following excerpts from Pavlov's book are now clear:

"In terms of conditioned reflexes, the cerebral hemispheres are a set of analysts which have the task of decomposing the complexity of the inner and outer world into separate elements and moments and then relate this multitude to the diverse activities of the organism."

"The large hemisphere is an organ of the animal organism, which specializes in continuous implementation of an ever more perfect balance of the organism with the environment; it is the organ of an adequate and immediate response to different combinations and variations of the phenomena of the external world; it is to some extent a special organ for the continuous development of an animal organism".

"The motorial part of the cerebral hemispheres is a receptor area, or the main area, and motorial effects during stimulation of the cortex are reflective by nature. The cerebral cortex, thus, is only a receptor apparatus, which analyses and synthetizes the incoming stimuli in a variety of manners. The stimuli reach the effectors through the downward leading connective fibres."

There are no mechanisms in the front lobes that are supreme with regard to the totality of the hemispheres. There can be no question of any general mechanisms which are settled in the front lobes. There are no particularly important devices there which would be responsible for the highest perfection of nervous activity". Pavlov, like Bergson, suggests that the difference between the brain and the spinal cord is in their complexity only, not in the nature of their functions. He and his school believe it is possible to attribute the deciphered patterns of the higher nervous activity of dogs, which was obtained during the experiments with conditioned reflexes, to the physiology of the human brain.

A dog can only get secondary conditioned reflexes based on the primary ones, whereas a monkey has a number of secondary reflexes more; for humans, without doubt, a very large amount of acquired reflexes, overlapping the previous ones, is possible. This ongoing formation of new, ever more complex interconnections inside the brain in the course of a human life, gives the opportunity of improving the mental performance and expanding consciousness. Nevertheless this complicated brain activity remains in essence just reflexes of the brain. This new view on the physiology of the brain, we think, should replace the current doctrine of associations in psychology.

But it is as Bergson said: "The brain is nothing but a sort of central telephone exchange: its role is just to deliver a message or to clarify it. It adds nothing to what is received," isn't it?

The research by Pavlov and his staff on the physiological significance of the frontal lobes of the brain hemispheres has an overwhelming importance. These were hitherto regarded by all as the most important part of the brain, the centres of higher mental activity, the organ of thought "par excellence", even "the seat of the soul." But Pavlov did not find there "any particularly important devices which could provide the highest perfection of the nervous activity," and the cortex in the front parts of the cerebral hemispheres, as well as the rest of the cortex, is just a sensor area. The entire cortex, the most perfect part of the brain, consists of an innumerable multitude of analysts, analysts and more analysts. And if there is no room for any kind of centre of feelings in the cortex, it is even more improbable that it can be situated in the gray nodes of the brain stem, which, as we partly know, have purely physiological functions. The cerebral cortex analyses not feelings, but sensations.

And the fact that the brain can't be considered an organ of feelings, confirms to a large extent the teaching of the Scriptures on the heart as the organ of feelings in general, and especially the higher feelings.

These conclusions of Pavlov's studies match the observations of surgeons on many wounded with abscesses on the frontal lobes. They usually are not accompanied by any noticeable mental disorder or changes in the higher cognitive functions. I will just give two clear observations from my own experience.

I removed about 50 cubic centimetres of purulence from a huge abscess, which undoubtedly destroyed the entire left frontal lobe of a young wounded man, and I noticed absolutely no psychiatric defects after this operation.

I can say the same about the other patient operated on for large cysts of the meninges. When opening wide his skull I was surprised to see that almost the entire right half of it was empty, and that all the right hemisphere of the brain was squeezed and almost impossible to discern.

So, if the brain can't be regarded as an organ of feelings and the exclusive authority of higher cognition, it confirms to a large extent the teaching of the Scriptures about the heart as an organ of feelings in general, and especially of the highest feelings.

Chapter 3. The Brain and the Spirit. The Spirit in Nature

"We are still convinced that the mind is an attribute of the brain, and subordination of the mind to the brain seems to us so obvious that we can't think of one without the other. And hence, by authority of a mistaken synthesis, we conclude that body and spirit are mutually subordinate to one another. This is because we are used to confuse the mind and the spirit. The mind, of course, is not the spirit, but only a manifestation of the spirit. The mind refers to the spirit as a part to the whole. The spirit is much more extensive than the mind, but due to the invariable intellectualistic concept we see the whole spirit in the mind"(Frank Granhman).

"The spirit projects outside the brain from all sides. Activity of the brain is limited to the transformation of a small portion of what is happening in the mind, into motion" (Bergson, "Creative Evolution").

What do we know about the spirit? We know much from the Scriptures, and not a little about the phenomena of the spirit in nature and man.

1. St. Sergius of Radonezh was having a meal with the brethren of his monastery. Suddenly he stood up, bowed to the west and said, "And you too rejoice, pastor of Christ's flock, and the blessing of the Lord be with you." The brethren asked in amazement: "Who are you speaking to, Holy Father?" St. Sergius answered, "About eight miles away from our monastery Bishop Stephan of Perm, stopped on his way to Moscow. He bowed to the Holy Trinity and said," "Peace to you, brother in spirit." So I answered him." Some of the monks rushed to the place and caught up with St. Stephen. He confirmed what was said by St. Sergius.

2. An engineer and staunch materialist, K.I. Pearl, parted from his friend K., who left for Moscow, while he lived in Tashkent with another engineer. One day at three o"clock in the night he was awakened by a loud shout: "Karl Ivanovich!" He lit a candle and woke his lover. They both searched the whole apartment, but found nobody. In extreme astonishment he noted the date of this strange event. A week later he received a message about the suicide of his ex-friend K. that had taken place that very night and hour he had heard the mysterious call. The death call of the suicide, full of anguish and love instantly covered a distance of 3300 km and was perceived by the brain of the sleeping K.I. Pearl.
3. Mrs Green, sitting in the verandah of her house after dinner in a town in England, suddenly had a vision: a cabriolet with two young women in it was approaching the steep bank of a lake; one of the wheels of the cabriolet suddenly broke, and the cabriolet with horses and passengers fell into the lake; they were drowning, and at the same moment Mrs. Green heard her name, cried out in an appeal full of despair. A month later she received a letter from her brother, describing the death of his daughter and her friend just as she had seen it in her vision.
I have borrowed the last example from a very interesting book of academician Charles Richet, a famous physiologist and physicist, one of the most prominent figures in the field of metapsychology, a new science that emerged at the end of the last century. It is now being developed by renowned scientists in different countries. The aim of metapsychology is the investigation by all the scientific methods available of the unexplained and mysterious psychic phenomena labelled by the official psychology as superstitions and fairy tales.
But it can only be rejected by those who are biased or not familiar with it or haven't studied it with the deep scientific

objectivity of C. Richet, Oliver Lodge and other eminent scientists, who have created a wide literature on metapsychology. There are very many similar facts to be found in Richet's book, each more surprising than the other. He analyses the facts with great scientific rigour, and he comes to the following conclusion: "There are vibrations (forces) in the universe, which excite our sensitivity and thus give us a trustworthy knowledge of reality, which our normal senses can't give. These forces, new and strange, will create a revolution in psychology and rebuild it from the ground."

Dr. Kotik, who has written a significant book on his experience with thought transmission ("The Emanation of Psychic-Physic Energy", Wiesbaden, 1908), explains them as follows: "A thought is an energy radiated outwards. It has physical and psychic properties, and it can be called psychic-physical energy. This energy born in the brain spreads itself through the entire body and the limbs. It can pass through metal conductors but it is difficult to transmit it through the air. Apparently, it is not specifically attributed to the thought; it may be, that all things emit some kind of vibrant energy, for sensitive subjects not only perceive what the experimenter thinks, but they also recognize material objects , which do not think."

What gives Dr. Kotik reason to believe that the invisible radiation of energy in a thinking brain is not only mental, but also psychic-physical? It is just the fact that when you connect the experimentist to the sensitive subject with a metal wire, the sensor reads thoughts a little better.

But a huge number of other facts described by Richet show that even without any contact the transmission of an unknown energy takes place. It is only due to our ineradicable habit to explain the incomprehensible and the unknown with the understandable (intellectually) and the known that this energy is thought to be material vibrations, which are caused

by the molecular fluctuations of the brain substance, which fluctuations are also still unknown.
Those who accept the primitive explanation are quite satisfied when they are told that their brain emanates some electric vibrations.
Why, however, should we not recognize, along with Richet, that inexplicable, even bizarre, as he says, forces, quite unknown to us in the present state of science, are the basis of the phenomena of telepathy (the transfer of thoughts and feelings at a distance) and clairvoyance? Anyone who reads his big book gets the shock of his life.
I believe that for the present we have the right to come to just one but very important conclusion: in addition to normal stimuli, which are adequate to our sensor organs, our brain and heart can receive much more important stimuli coming from the brain and heart of other people, animals, and all of nature around us and, what is the most important, from the transcendental world unknown to us.
By what, if not by forces of transcendental order, can such facts as reported by Richet, be explained?
1. M. Hialon confirmed the trustworthiness of a strange story which had happened to M.M. Greedy, the director of the Daily Telegraph. When one Sunday he was standing in St. John's Church, he suddenly felt a very powerful suggestion. It was something like a voice saying to him: "Return to your editorial office." The order was so overbearing that M.M. Greedy ran through the whole church like a madman, through the streets, rushed into the editorial office to the amazement of his employees, and flung open its door. The kerosene lamp on his desk was shooting large flames and the whole room was filled with smoke.
2. Mrs Tomeli in San Martin, one night trying to fall asleep, suddenly saw her son being run over by a carriage and dying.

She ran during the storm 5 miles down the road in Costa de Borge and found her son in the ditch next next to the road.

3. The famous William James quotes the following case: Bertha, a young girl, disappeared on the 31st of October 1898 in Enfield. More than 100 people were sent to search the woods and the lake shore. It was known that she went to the bridge, and since then nobody had seen her. The diver who had searched for her in the lake could not find her. But in the night of 3rd of January, a woman who lived 8 km away from Enfield saw in her dream Bertha's body in a certain place. In the morning she went to the bridge and showed the diver the exact spot where he would find Bertha's body. According to her instructions the diver found her lying upside down and well hidden. The body was at a depth of 6 meters in the snags, and at first he could see only a rubber shoe on her foot. "I was shocked", the diver said, "I'm not afraid of dead bodies in the water, but I was afraid of the woman standing on the bridge. How is it possible that she came here from somewhere 8 km away to show where the body was? It was lying in a deep hole, upside down. The water was so dark that I could see nothing. "

The last two examples belong to the area of clairvoyance, to which Richet gave a new name - kriptosteziya. This term describes well the following extraordinary facts.

1. In Blueau prison a prisoner hanged himself with his tie. Dr. Dufali cut a piece of this tie, wrapped it in several layers of paper and gave it to a clairvoyant (not a professional one) called Mary. She said there was something wrapped in a paper that killed a man, a rope... not a rope but a tie; a prisoner who had killed a man hanged himself with it. She said the man was killed with an axe, and pointed out the place where the axe was thrown. Indeed, the axe was found in this place.

2. Charles Richet met his friend Stella, a young girl on the 2nd of December and said to her, "I'm going to give a lecture on

snake venom." Stella, showing an astonishing clairvoyance several times before, replied immediately: "Last night I saw serpents, or rather, eels." Without of course mentioning the reason I asked her to tell me the dream. And these were exactly her words: "It was more likely two eels than serpents, because I saw they had a bright white belly and a sticky skin. And I said to myself, I really do not like these animals, but I hate them being tortured."

This dream coincided amazingly with what I had done the day before, on the 1st of December. On that day, I was experimenting with eels. To take their blood, I put two eels on the table. I remembered their glossy white belly and sticky skin. They were attached to the table to take out their heart. I had not seen Stella for a long time and certainly had not talked to her about it, and she had had no contact with people who visit my laboratory.

Discussing many facts similar to the above mentioned Richet concludes that all people, even the apparently least sensitive, have other capacities of cognition in addition to the common ones. But these capacities are extremely weak, almost imperceptible, to non-sensitive people. The thoughts of a person may in a mysterious way appear in the minds of others. We are not isolated but we are, in an unfathomable way all connected together. And certainly there is some truth in what is called "mass-hysteria". A powerful stream of sympathy or anger, indignation or enthusiasm leads to an almost complete unity of people gathered in a theatre, a market or a parliament. It is a kind of stream which sweeps everything away. Why couldn't we consider the unifying work of emotion in a crowd and the transmission of ideas in Richet's experiments as the same phenomenon? The selfless act of one brave man can inspire a whole army. This powerful flow of the spirit of bravery and courage, emanatingfrom one ardent heart, kindles the hearts of hundreds of others who receive it as an antenna receives radio waves. What, if not a powerful spiritual energy, should we call this all-prevailing power, which generated global psychic epidemics in the Middle Ages and recklessly and irrepressibly gathered hundreds of thousands of people in the crusades?!

Isn't it clear that a flow of an evil power, an evil spirit is poured into our hearts and brains when we see the face of our enemy, distorted by hatred, his eyes casting sparks and making our heart shrink in fear?

The love of a mother flows quietly and sweetly into her child that is clinging to her chest; the passionate feelings of a loving husband flows powerfully onto his wife. Quiet and joyful light illuminates the soul of the man who does the works of love and mercy, when the grace of God touches it.

What is this but the spiritual energy of love?

"I will pour out my spirit upon all flesh." (Joel 2:28) God is spirit. God is love, and the outpouring of His Spirit is the outpouring of His love onto all the living. Love creates. The all-creating and all-consuming, endless stream of spiritual energy of God's love created the Universe. It was created out of nothing, in that sense that there was no such thing as the primary... (may be "matter", an omission in the original text. ed.). There is no eternal matter, the same as there is no matter at all, and there is only energy in its various forms, and when condensed it is matter.

Matter represents a stable form of inter-atomic energy, and heat, light, electricity are the unstable forms of the same energy. The core of the process of disintegration of atoms, i.e. the break-up of the matter, is a modification in the state of the inter-atomic energy from stable equilibrium to an unstable one, which we perceive as electricity, light, heat, etc. The matter is thus gradually transformed into energy.

In the first chapter we have said that by the atomic break-up the still more subtle forms of energy, close to something immaterial are liberated.
What hampers us taking the final step and recognizing the existence of a fully immaterial, spiritual energy, and considering it as the primary form, the mother and the source of all forms of physical energy? Only an a priori rejection of the Spirit and the spiritual world, a stubborn and incomprehensible denial, in spite of all the facts that imperatively compel us to reckon with them and recognize a limitless, far more important spiritual world along with the material one.
"Thine incorruptible Spirit is in all things..." (Wis. Sol. 12:1)
"Do not I fill heaven and earth? saith the LORD." (Jer. 23:24)
"If He set His heart upon man if He gather unto Himself His spirit and His breath, all flesh shall perish together, and man shall turn again unto dust." (Job 34:14-15)
"...Thou takest away their breath, they die, and return to their dust. Thou sendest forth Thy spirit, they are created: and Thou renewest the face of the earth." (Psalm 104:29-30)
"It is the spirit that quickenth..." (John 6:63)
"In whose hand is the soul of every living thing, and the breath of all mankind." (Job 12:10)
"The stars shined in their watches, and rejoiced: when He calleth them, they say, Here we be; and so with cheerfulness they shewed light unto Him that made them." (Baruch 3:34)
"The pastures are clothed with flocks; the valleys also are covered over with corn; they shout for joy, they also sing." (Psalm 65:13)
And many of the Psalms and the Song of Hananiah, Mishael and Azariah, are full of the spirit of hylozoism.
The idea of the universal realization and quickening by the Spirit of God is shown quite clearly in all these texts from the Scriptures. It is impossible to speak of a "dead nature." There

are no distinct boundaries between inorganic and organic nature. This is also the point of view of modern science.

A vivid evidence of this is given by the philosophy of Fichte and Lotze, and the great profundity of Leibniz. These are Leibniz's words, coinciding with the texts of Holy Scripture: "There would be a gap in the creation, if the nature of matter were the opposite to that of the spirit. Those who deny the existence of a soul in animals, and imagination and life in general in the other bodies do not acknowledge the Divine power, because they invent something incongruous with God and with nature, an absolute lack of forces, a kind of metaphysical void, which is as absurd as the empty space or a physical emptiness. "

Countless stars and planets rush in eternity through the space, never slowing down their movement. It is only through the power of this movement that incredibly heavy bodies are kept in space, like a heavy 40-inch shell can keep rushing along its trajectory through the air.

Myriads of stars, planets, asteroids, meteors and comets rush through space. The face of the earth is changed by blowing winds, water flowing, the friction of sliding glaciers, temperature fluctuations, the waves' surf. The powerful movement of underground volcanic forces creates new ranges of mountains and precipices on earth.

During the centuries of movement innumerable worlds and stars are destroyed and created again; there exists a great process of evolution, the highest order movement in the universe. Atoms and electrons are moving infinitely almost with the speed of light, like X-rays, ions, and all the products of the dissociation of matter, which always takes place. Life of the organisms is sustained by the continuous motion of molecules in the organism's cells. The thought of a man is followed by molecular motion in the nerve cells. There is no motionlessness even in death, which is nothing but a certain change in the instantaneous states of equilibrium, which are of a short duration also.

If it is so obvious that motion is the essence and the fundamental law of material nature, it is hardly possible that this universal law would not reign also in the spiritual life.

The principal law of the entire nature, the whole law of motion wipes away the boundary between the living and the dead. Motion is the essence of matter. And if we want to assume that motion in the living organisms, which is the basis of their psychic phenomena, is generated and determined by the energy of the spirit of life, we must admit then that the motion of inorganic nature is a derivative of the same spirit. Spiritual energy emanated by the Spirit of God, the energy of love drives forward the entire nature and "quickens" all and everything.

It is the source of life, and nothing is dead.

Motion in the inorganic nature is a manifestation of life, though at the lowest level, about which we know little. The existing genetic link between inorganic and organic nature proves it. For the life of plants starts in the soil, and serves as food for the entire animal kingdom. Both inorganic and organic nature are created from the same chemical elements and according to the same physical laws.

One great universal law of development governs the whole of Universe, and the development can not be stopped, and there can not be a sharp boundary between the "dead" nature and the world of living beings. And the most striking feature of living nature, its being animated is not something which appears suddenly at its boundary with inorganic nature. Spiritual energy permeates all inorganic nature, the entire Universe. But this energy reaches the level of a free, self-conscious spirit in the higher forms of development (creation) only.

Sensitivity is inherent in all living things. However, the exact methods of research show that matter is not only extremely mobile (mercury in the thermometer rises at the touch of a hand), but that it has an unconscious sensitivity, superior to the conscious one of living beings.

The bolometer, comprising a platinum plate as its substantial part, is so sensitive that it reacts to the impact of a very weak beam of light, which raises the temperature only one hundred millionth fraction of a degree. Stele showed that touching an iron wire with a finger is enough for an electrical current to appear. It is known that the Hertz waves deeply influence metals at a distance of hundreds of kilometres, causing electrical fluctuations in them. The wireless telegraphy is based on this phenomenon. In his ingenious experiments Stele showed that metals may become "tired" and that after "relaxing" this "tiredness" disappears, that poisons can "irritate" and "supress" them.

Living things differ from the "dead" nature because of the fact that they feed themselves and multiply. Science didn't discover these properties in the minerals., but there are indications that these functions are inherent in them. There is nevertheless reason to expect that in the future science will find evidence of it in nature. An enzyme has already been found in certain minerals, which is similar in its effect to sexual hormones, accelerating the growth and sexual maturation of newborn animals. And we know that in living organisms the enzymes mainly serve the purposes of feeding. What is so strange about the assumption that the mineral enzymes are necessary for the feeding of minerals, of course not as in a living organism, but in some way, not yet known to us?

Doesn't the chemical metabolism which always takes place in the inorganic nature have the role of nutrition? Isn't the soil thirsty, isn't it in need of water like all living beings? A mineral substance is characterized by its crystalline form the same as a living creature by its anatomical structure. Before reaching its definite shape, a crystal passes through consistent evolutionary steps like a living creature and a plant: it begins in a granular state, which becomes fibrous, and finally, homogeneous.

Like animals and plants, disfigured crystal corrects its damage. Every crystal is an organized being. Crystals have two forms of reproduction. In certain circumstances, for example at a certain pressure, concentration of the solution, etc., the liquid can crystallize only when the crystal nucleus is added. The resulting crystals then can be considered descendants of the crystal, just as bacteria that develop in the solution are the descendants of the bacteria that we have introduced into this solution. However, there may be such conditions in an environment which allow crystallization to take place without the initiating "embryos".

It is necessary to think about the depth of Joel's words: "I will pour out my spirit upon all flesh." (Joel 2:28)

Precise and deeply important is the relationship between the spirit and the form. The spirit inherent in matter is clearly reflected in the forms generated by this matter. And more than that, the spirit creates the forms.

This is clearly expressed in the forms of the human body, although they do not always correspond to the spiritual nature of a man. Not only the eyes are the mirror of the soul, but all forms of the body and its movements correspond to the soul, the spirit, the same as the image of an evil man described by Solomon (Prov. 6:12-14). A man's entire appearance clearly reflects his spiritual essence. A rough and brutal spirit as early as the embryogenesis directs the development of somatic cells and creates gross and repulsive forms which correspond to it. A pure and gentle spirit creates a dwelling for itself which is full of beauty and tenderness. Think of the Madonnas by Raphael and Leonardo da Vinci's Mona Lisa.

What, if not the shaping influence of the spirit, can explain the amazing subtle differences between two very similar faces, especially between women's faces? They might have the same outlines of nose, mouth, almost an equal size and proportions of face and head, while one face is vulgar, and the other is refined and beautiful.
A careful analysis shows that this difference in the spiritual image of almost identical physical forms depends on very small and delicate variations: a slightly different outline of the eyebrows, of the fold of the lips, an almost indiscernible difference in the size of the eye socket and shape indicates a very different spiritual image. A striking image of Mona Lisa was created by the subtle features that Leonardo da Vinci was adding to this woman's face in the course of several years.
This is exactly how the spiritual energy inherent in the chromosomes of germ cells acts in the embryonic development and creates vivid images of beauty or ugliness; tenderness, purity and love or rudeness, repulsively dominating animal instincts and malice. These innate external forms become still more evident in the postembryonic life corresponding to the development of the spirit in one direction or another. The value of beauty in nature is versatile, and, of course, it is not limited only to the purposes of sexual selection. Male birds sing beautifully and wear bright and beautiful plumage not only to attract females. The shining beauty and fragrance of flowers is not only to attract the insects which bear the fertilizing pollen. The great beauty of nature doesn't only have utilitarian purposes of course.
Compositions and elements of beauty and ugliness in nature are perceived and transformed by the human spirit into works of art and science: the two great engines of the spiritual development of mankind. From the simplest forms of beauty, from a roundness and smoothness of lines pleasant to the eyes, proportion and symmetry of form, beauty and

strength of light and shade, a harmonious combination of colours and sounds, nature reaches the heights of the grand picture of beauty, full of spiritual power and greatness.
Dark clouds covering the mountain chains and cliffs, huge ocean waves driven by the wind, storming the coastal cliffs, are full of the spirit of an immense power. The spirit of eternity and infinity is poured into our souls from the myriads of stars in the night sky. The gentle colours of the dawn and the fields and the lakes, illuminated by moonlight evoke joy and peace. Nature shows the supreme value of moral beauty in the meek and pure eyes of good men and the loathsomeness of disgraceful things in the repulsive appearance of villains and dishonest people.
And if it is so obvious that in the forms of moral beauty or disgrace we actually perceive the emanations of the spirit of beauty or the spirit of evil that stir our hearts, then why can't we state that the perception of beauty and ugliness in inorganic nature , is a similar effect of spiritual energy inherent in all nature, on our soul?
It does not matter that shapeless amorphous matter does not produce such an impression of spiritual order on us; it is important that the Spirit is associated with form. We say that the Spirit governs the development of human bodies in the forms respective to their kind. All forms of the Universe, including inorganic nature, are built under the creative influence of the spirit. Therefore the purpose and the meaning of beauty should be seen in the deep spiritual, even moral influence on human souls, produced by the spiritual energy of nature's beauty. If the universe were what it appears to be to the materialists, there would not be the beauty of forms created by the spirit in it.
I was brought a bouquet of flowers. O how much subtle, charming beauty is there in these marvellous little creatures of God! They are also charming in their small, meek simplic-

ity. The finest lace of delicate white blossoms, little pink, purple and blue creatures look at us with the pure eyes of their petals and corolla, and pour their wonderful fragrance over us.

Isn't it obvious that this is a silent sermon on the purity of the soul? One must have a very rough heart in order not to hear this voice of God, so clearly sounding in the beauty of the material forms of nature. It is natural that women are particularly fond of flowers, and this fact speaks in favour of their hearts.

Chapter 4. The Spirit of Plants and Animals

The fact that animals possess a spirit is clearly testified by the Scriptures. Here are the texts to prove it:
"Who knoweth the spirit of man that goeth upward, and the spirit of the beast that goeth downward to the earth?" (Eccles. 3:21).
"O God, the God of the spirits of all flesh." (Num. 16:22)
"And out of the ground the Lord God formed every beast of the field and every fowl of the air, and brought them unto Adam to see what he would call them, and whatsoever Adam called every living creature, that was the name thereof." (Gen. 2:19).
"For the life of the flesh (animal's soul) is in the blood: ... therefore I said unto the children of Israel, Ye shall eat the blood of no manner of flesh." (Lev. 17:11-14)
Blood of an animal sacrificed is holy and sanctifying, because the animal's soul, breath of the Holy Spirit, dwells in it. Therefore it is forbidden to use it in food.
"It is the Spirit that quickeneth..." (John 6:63)
"The spirit of life from God entered into them..." (Rev. 11:11)
In our prayer to the Holy Spirit we call Him the giver of life. And if even in inorganic nature the presence of the Spirit is so evident, then, of course, all plants and animals must be considered spiritualized. Of all the gifts of the Holy Spirit the most common in nature is the spirit of life, and it is, of course, inherent not only in animals but in plants, too. Indians and other peoples of Asia differ from the Europeans in their views on plants. They recognize deeply the spirituality of the plants. Plants with their entire being take in avidly light, air, moisture, which their whole life depends upon. They clearly enjoy light, wind, dew and rain. Why not admit that they brightly perceive and feel these sources of their life and joy, but the way they do it may be quite different from a man or an ani-

mal, whose need of light is not so vital. The plant feels probably much deeper than an animal all the finest qualities of the soil, in which rich branches of its roots exist, on which, together with the light and air, depends its whole life. We know how delicately different plants select from the soil nutrients necessary specifically for them and not for the other plants. Isn't it irrefutable that this plant should have a very special kind of sensitivity, which neither animals nor humans have? (Fechner)

"Nerves can't be considered a necessary substratum of spiritual life. The strings are the nerves of a violin or a piano. But even without strings, the wind instruments emit a wonderful melody. Plants have no autonomic nervous system, without which processes of nutrition, respiration and metabolism of humans and higher animals are impossible, and yet all these processes occur in plants. "(Fechner)

If you turn the underside of a vine leaf to the light, it will bend and turn persistently to have its upper side in the light. Amazing are the instinctive movements of climbing plants. The plant first grows in height, and then it bends its stem horizontally and makes a circular movement, trying to find a support. The longer the stem grows, the bigger is the circle, that is, the plant is continuously searching. Finally, the stem can't carry its own weight, falls to the ground and crawls over it, looking for another support. Also in this case it is governed by a certain guideline. For example, a convolvulus never winds around inorganic or dead organic supports, but only around living plants, to which it adheres avidly because its own roots in the soil quickly die, and it drains the nutrients with the help of the special nipples out of the entwined plants.

Plants are known to sleep when the leaves bend or fold up, and the flowers hang their heads and close. Surprisingly expedient are the pistil's movement in certain plants to fertilize the stigma with pollen.

In the evening a lot of flowers in a meadow turn their heads to the sun, as if saying their evening prayer to it, and after sunset they sleep quietly till the next morning, turning to the east to meet the sun with their joyful morning prayer. The fragrance of flowers is their burning incense to God, and the flowers are the censers. Water-lilies open wide enjoying the light and air under the blue sky; they fold their petals and sink into the water when it gets dark.

There is no definite border between the worlds of plants and animals , because in protozoa there are many almost completely similar forms, of which some are the beginning of flora, the others of fauna, and it is almost impossible to find any difference between them. The simplest forms of animals, such as the river hydra, the volvulus, are quite similar to plants, and their vital functions are almost indistinguishable from one another. The class of protozoa is the starting point of the two magnificent worlds of creatures - plants and animals. The plant world in its gradual development has reached such magnificent, impressive forms as wonderfully fragrant flowers, slender palm trees and cypress trees, majestic cedars of Lebanon, mighty oaks and giant sequoias, which live for three thousand years. In comparison such primitive forms of animal life as polyps, sea cucumbers, starfish and worms are quite insignificant, and it would be strange to recognize the spirituality of these lower forms of animals, yet refuse to recognize the spirituality of highly perfect, and even grandiose, plant forms.

There is no doubt at all that the entire plant and animal world is endowed with at least the most basic of the gifts of the Holy Spirit, the spirit of life.

For the vast majority of naturalists, the "vitalists" and "neo-vitalists" doctrine of vital force is odious and absurd. But just think about the following facts.

According to the observations of Spalantsani, in swamps and in the sand of gutters live rotifers, gelatinous in the normal state. They can dry up while still remaining in the sand to such an extent that, if pressed with the end of the needle, they break like a grain of salt. However if in four years time the sand would be moistened, the dried rotifers would come to life again. They can withstand drying at 54° C, whereas in a living state, they die if the water temperature reaches 25° Celsius.

John Franklin in his first trip to the North American coast of the Arctic Ocean saw that fish, frozen immediately after they were dragged from the water, turned into a mass of ice so that they could be hacked to pieces with an axe, and that their frozen intestines were like solid ice chunks. However, some of such undamaged frozen fish came to life after they were defrosted near the fire. These facts indicated that, although every trace of life in the body had disappeared, yet the ability to come back to life might remain under favourable conditions, if only there were no changes in the anatomical or physiological spheres, which would have made impossible the retrieving of vital functions. It is known that wheat, barley and mustard found in Egyptian mummies, stored for 3000 years, if they had not been subjected to harmful effects damaging their enzymatic processes, give excellent germination when put in the favourable conditions of moisture and heat.

The next experiment was made by J. Becquerel in Paris in 1909. Seeds of wheat, alfalfa and white mustard were dried in vacuum for 6 months at 40°C and then sealed in vacuum in glass tubes. These tubes were sent to London and they were held there for three weeks in liquid air at about -190°C, and then another 77 hours in liquid hydrogen at -250°C. In Paris, the tube was opened again, and the seeds were put in a moist bath at 28°C. It was found that germination was completely normal. No difference was observed in comparison

with samples of seeds stored in the usual way. At such a low temperature as -250°C, every trace of life is excluded. Even the most energetic chemical reactions wouldn't occur at such a low temperature.
These experiments show us that temporary death is possible if the action, inhibiting life functions, does not extend to the destruction of the body ("Degeneration of energies" by Svedberg).
If it is clear that the temporary death of seeds and animals does not prevent their coming to life once again, surely we have no right to say that all this is possible without some unknown but evident force, vital energy, not amenable to harmful agents that destroy the life of seeds and plants? And this energy can certainly be only a spiritual energy, the life-giving power of the Holy Spirit.
The above-mentioned amazing facts of life and spirituality of plants allow us to agree with Edward Hartmann when he argues that plants have an unconscious imagination and an unconscious will. Leibniz, in his turn, ascribes to monads a vague imagination and aspiration. A plant is a monad.
Our belief in the spirit in plants, of course, doesn't contradict at all the opinion of St. Anthony the Great on the impossibility of plants having a soul. Here are his words:
"I have written the present paragraph for the information of those who are simple, against men who assert that plants and grasses have a soul. Plants have physical life, but have no soul. Man is called a rational animal, because he is endowed with mind and capable of acquiring knowledge. Other animals – those on the ground and in the air who possess voice – breathe and have a soul. All things that grow and decrease can be called alive because they live and grow, but it cannot be said that all such things have soul. There are four kinds of living beings: some of them are immortal and have souls, such as angels; others have mind, soul and breath, such as

men; yet others have soul and breath, such as animals; and others only have life, such as plants. Life in plants is maintained without soul or breathing, without mind or immortality. But all the rest, too, cannot be without life. Every human soul is very changeable." (Early Fathers from the Philokalia, tr. E.Kadloubovsky and G.Palmer, ed. Faber & Faber, 1973)
There is no contradiction of course. We do not ascribe a soul to plants in the same sense as we understand a human's and an animal's soul, but only like the unconscious imagination and the unconscious will.
Anthony the Great, speaking of the soul of animals, of course, was referring to the higher animals, and not to such as coelenterates, molluscs, sponges, even slipper animalcules, of whose existence or the question whether they belonged to the animal world or not he could have had no idea. Of course, these simplest animal forms, as to their spirituality, are not higher than plants, and might be even lower, and they are endowed with the spirit of life only, like any animal.
So, let us investigate thoroughly our knowledge of the spirit of the higher animals and humans.

According to physiologists' point of view, the activity of consciousness, that is, psychic activity should be seen as an enormously complex system of unconditioned and conditioned reflexes, initially formed and constantly re-formed, as a huge chain of perceptions, brought into the brain by receptors, analyzed by the brain to generate a motoric response.
Our brilliant scholar on higher nervous activity, I. P. Pavlov, defines consciousness as follows: "Consciousness seems to me as the nervous activity of a certain area of the cerebral hemispheres, which has an optimal (probably an average) excitability at a certain moment under certain conditions. At the same moment the rest of the cerebral hemispheres are in a state of more or less reduced excitability.
New reflexes are easily formed and differentiations are successfully worked out in the area of the cerebral hemispheres with optimal excitability. It is at that moment, so to speak, the creative department of the cerebral hemispheres. Other parts of the hemispheres with a reduced excitability are not able to do the same, and their functions in this case are mainly previously developed reflexes, arising stereotypically under appropriate stimuli.
The activity of the other parts is what we subjectively call the unconscious, automatic activity.
The part with the optimal activity is not, of course, a fixed one, on the contrary, it constantly moves around the cerebral hemispheres, depending on connections existing between the centres, and the area with reduced excitability varies under the influence of stimuli.
If we could see through the skull of a consciously thinking person, and if the part of the cerebral hemispheres with increased excitability could emit light, we would see an oddly

shaped bright spot moving about the cerebral hemispheres, constantly changing in shape and size, surrounded by a more or less considerable shadow on all the rest of the hemispheres.

We fully accept this deep scientific understanding of the activities of consciousness, but we do not consider it complete. We are even prepared to subscribe to the basic thesis of materialism: "Being determines consciousness", but only under the condition that the term "being" signifies both material and spiritual realms.

In order to present our understanding of consciousness, we should divide it into acts, states and volumes of consciousness. Thought flows like a stream. Thought flashes like a bright light. Thought digs into the depths of existence.

Calm and deliberate volitional actions; sudden outbursts: stabbing the heart of the offender; the constant effort of will in the course of life, directed to achieving important plans and objectives; a quiet love, devoid of passion; profound, calm, aesthetic pleasure; the stormy passions of anger, and fear; deep constant devotion to God, which governs my entire life – these are acts of consciousness.

They are caused by: 1) perceptions of the sensors, 2) the organic sensations of our bodies, 3) the perceptions of our transcendental being, 4) the perceptions of the higher spiritual world, 5) the impacts of our spirit.

Acts of consciousness are not isolated: thought is always accompanied by feeling; feeling and will by thought, and feeling by volitional movements; acts of will are always associated with feelings and thoughts; the complex of these simultaneously occurring acts of consciousness determines the state of consciousness. These states of consciousness are constantly changing, for the acts of consciousness are in constant motion.

The volume of consciousness, also constantly changing, usually increasing, is determined by the richness, diversity and depth of acts and states of consciousness. Our spirit is always involved in the acts and states of consciousness, identifying and directing them. In its turn, the spirit grows and changes owing to the activity of consciousness, through its individual acts and states.

This is our idea of the full range of human psychic activity. But not only people have a soul, the animals have it as well. "The soul of the animal is in its blood." And an animal, like man, comprises the spirit, the soul and the body.

What is the soul? In its simplest kind, in animals, it might be the complex of organic and sensual perceptions, thoughts and feelings, traces of memory, united by self-awareness (the mind in higher types of animals), or only the complex of organic sensations (in the lower type of animals). The primitive spirit of the animals is just the breath of life (in the lower type of animals). When climbing the ladder in the hierarchy of creatures spirituality increases, and the breath of life is enriched with rudiments of the mind, will and emotions.

The human soul is much higher in its essence, for the spirit participating in its activities is incomparable with the spirit of animals. It may have the highest gifts of the Holy Spirit, which St. Isaiah the Prophet (11:1 - 3) calls the spirit of fear of God, the spirit of wisdom and understanding, the spirit of strength and might, the spirit of light, the spirit of reason, the Spirit of the Lord, or the gift of supreme piety and inspiration. The spirit and the soul of a man are inseparably connected in a single entity during his life, but people possess different degrees of spirituality. There are those whom the Apostle Paul calls "spiritual" (1 Cor. 2:14). As we have said, there may be people - beasts, people - plants, but on the other hand: people - angels. The first are not much different from the beasts, because their spirituality is very low, and the last ones are close to the bodiless spirits, who have neither body nor soul. So, the soul can be described as a combination of organic and sensual perceptions, traces of memories, thoughts, feelings, and acts of the will, in which the highest manifestations of the spirit are not obligatory, for they are not present in animals and in some people. The Apostle Jude spoke about them: "These be they who separate themselves, sensual, having not the Spirit." (Jude 1:19)

The life of the spirit during the material life is closely intertwined with those psychic acts which are common to humans and animals, i.e. with organic sensations and the perceptions of the five senses; the latter, in turn, are inextricably connected with the life of the body, of the brain in particular; and they disappear at the death of the body. Therefore, the primitive soul of animals is mortal, as those elements of human consciousness are mortal, that have roots in the physical body (organic and sensory perceptions).

However those elements of self-consciousness that are associated with the life of the spirit are immortal. Immortal is the spirit, which, as we show below, may exist without any connection with the body and the soul. Materialists deny the immortality of the soul, precisely as they do not want to know anything about the spirit. But we acknowledge the mortality of that kind of self-consciousness which rests solely on a physiological basis.

Let us see now if the Scriptures give us reason to understand the spirit and the soul as we have just outlined. We believe that our understanding of the soul and the spirit is in perfect accordance with the Revelation.

The word "soul" is used in the Scriptures in various ways.

As in common parlance, it may simply mean the person: "Not any soul". "No soul of you shall perish," Saint Paul says to his companions on the ship.

"The soul that sinneth it shall die." (Ezek. 18:20)
In other parts the soul is synonymous with life.
"...Their bread for their soul shall not come into the house of the Lord." (Hos. 9:4)
But a number of texts clearly refer to what one might call the "animal soul."
"For he satisfieth the longing soul, and filleth the hungry soul with goodness.
Their soul abhorreth all manner of meat." (Psalm 107)
"... And he shall feed ... and his soul shall be satisfied upon mount Ephraim and Gilead." (Jer. 50: 19)
"The liberal soul shall be made fat." (Prov. 11:25)
"The soul of a sluggard desireth, and hath nothing: but the soul of the diligent shall be made fat." (Prov. 13:4)
"An idle soul shall suffer hunger." (Prov. 19:15)
"The full soul loatheth an honey-comb, but to the hungry soul every bitter thing is sweet." (Prov. 27:7)
"Let your soul delight itself in fatness." (Is. 55:2)
"Their soul shall be as a watered garden." (Jer. 31:12) (This is about earthly goods).
"His soul clave unto Dinah the daughter of Jacob." (Gen. 34:3)
"... His life abhorreth bread, and his soul dainty meat." (Job 33:20)
"And I will satiate the soul of the priests with fatness." (Jer. 31:14)
Let those who are used to the concept of the immortality of the soul not be confused with our words about the immortality of the spirit. It isn't a novelty, for in most fragments of the Scriptures, when speaking of death, it is said that the spirit is leaving the body and not the soul.
"So as the body without the spirit is dead, so faith without works is dead also." (James 2:26)

"Who knoweth the spirit of man that goeth upward, and the spirit of the beast that goeth downward to the earth." (Eccles. 3:21)
"...He went and preached unto the spirits [not unto the souls!] in prison." (1 Pet. 3:19)
"... And to the spirits [not souls] of just men made perfect." (Heb. 12:23)
"And her (the daughter of Jairus) spirit [not soul] came again." (Luke 8:55)
"Into thine hand I commit my spirit." (Ps. 31:5)
"Father, into thy hands I commend my spirit." (Luke 23:46)
"Lord Jesus, receive my spirit." (Acts 7:59)
"His breath [spirit] goeth forth, he returneth to his earth; in that very day his thoughts perish." (Ps. 146:4)
"Then shall the dust return to the earth as it was: and the spirit shall return unto God who gave it." (Eccles. 12:7)
The last two quotes are particularly important to support our view that mortal are those elements of the soul that are associated with the life of the body, i.e. the five senses and thought processes, inextricably connected with the activity of the brain.
"...in that very day his thoughts perish... ", i.e. the activity of consciousness, which requires all the perceptions of the living brain, will stop.
Not a soul, but the spirit, goes forth and returneth to his earth, i.e. into eternity. Ashes will return to the earth as they were, but "the spirit shall return unto God who gave it."
And the spirit of an animal, of course, must be immortal because it also has its source in the Spirit of God, the immortal Spirit.
The idea of the immortality of the spirit of animals is clearly present in the famous words of Apostle Paul on the hope of all creation (Rom. 8:20-21): "...who hath subjected the same in hope, because the creature itself also shall be delivered

from the bondage of corruption into the glorious liberty of the children of God." In a few places of the Scriptures death is defined as the departure of the soul (not the spirit) from the body (Gen. 35:18; Ps. 15:10). This is easily explained by the fact that in the Bible, and particularly in the Psalms, the word "soul" is often used in an conventional sense, i.e. as the sum of all mental and spiritual acts. But we say also that the spirit and the soul of man are combined in a single entity, which can be simply called the soul.

One should understand in this way the texts speaking about the soul of the Lord Jesus Christ.

"When thou shalt make his soul an offering for sin..." (Is. 53:10)

"His soul was not left in hell..." (Acts 2:31)

"My soul is exceeding sorrowful, even unto death..." (Matt. 26:38)

"Now is my soul troubled..." (John 12:27)

"He shall see of the travail of his soul..." (Is. 53:11)

The Lord suffered and died in His human nature, and therefore we can understand these words. But the soul of God Himself is mentioned in the following texts:

"And my soul loathed them, and their soul also abhorred me... (Zech. 11:8)

"And his soul was grieved for the misery of Israel..." (Judg. 10:16)

"...But the wicked and him that loveth violence His soul hateth..." (Ps. 10:5)

But, of course, it is only a metaphor. It is impossible to speak of the soul of the Spirit in absolute terms as of the human soul which is a limited, incarnate spirit. Here we can speak only about the analogy with the human spirit, according to which we ascribe the mind, thought, will and emotions to God. In this way we understand the image of God in every man.

We have already mentioned self-consciousness, self-awareness. How should we understand this? Man forms the consciousness of his own personality through organic sensations received from his body, through perceptions derived from his senses, through the entire set of memories, the understanding of his spirit, character and mood.

How is self-consciousness formed as the sum of those elements, and what is its subject ? It is not the mind, as is generally understood, but the spirit. For the mind is only a part of the spirit, and not the whole spirit. A part can't grasp the whole. This is an important conclusion to which we shall refer later when discussing immortality. And it is not arbitrary, but based on the words of Apostle Paul:

"For what man knows the thoughts of a man save the spirit of man which is in him? Even so the things of God knoweth no man, but the Spirit of God." (1 Cor. 2:11)

We learn the deepest essence of our being not by our mind, but by our spirit. Self-awareness is a function of the spirit, not of the mind. We learn about God's Grace granted to us by Him not by the spirit of this world, but by our spirit, given to us by God, too.

The same thought is in Solomon's words of Wisdom: "The spirit of man is the candle of the Lord, searching all the inward parts of the belly." (Prov. 20:27)

A lot can be found in the Scriptures on the Spirit as the supreme power of our spiritual activity. Here are some examples:

"For he that soweth to his flesh shall of the flesh reap corruption; but he that soweth to the Spirit shall of the Spirit reap life everlasting." (Gal. 6:8)

"Howbeit that was not first which is spiritual, but that which is natural, and afterward that which is spiritual." (1 Cor. 15:46)

This means that spirituality is the highest achievement of the human soul.
"The fruit of the Spirit is love, joy, peace, longsuffering, gentleness, goodness, faith, meekness, temperance." (Gal. 5:22 - 23)
"Fervent in Spirit..." (Rom. 12:11)
"Howbeit in the Spirit he speaketh mysteries..." (1 Cor. 14:2)
"There is a spirit in man and the inspiration of the Almighty giveth them understanding..." (Job 32: 8)
"The spirit indeed is willing, but the flesh is weak..." (Matt. 26:41)
"And my speech and my preaching was not with enticing words of man's wisdom but in demonstration of the Spirit and of power..." (1 Cor. 2:4)
"Cast away from you all your transgressions whereby ye have transgressed; and make you a new heart and a new spirit." (Ezek. 18:31)
Here is the idea of the close connection between the heart and the spirit, which confirms the primary role of the heart in consciousness as we mentioned earlier.
"And my spirit hath rejoiced in God My Saviour..." (Luke 1:47)
The human spirit rejoices in God, worships God, seeks God and comes closer to God. And this, of course, is the highest ability of a man's soul.
Of course, this most perfect manifestation of the spirituality of the human soul can be only a gift of the Holy Spirit. This is quite clearly demonstrated in Revelation:
"I will put my spirit within you..." (Ezek. 36:27)
"And because you are sons, God hath sent forth the Spirit of His Son into your hearts, crying, "Abba, Father"."(Gal. 4:6)
"God hath not given us the spirit of fear, but of power, and of love..." (2 Tim.1:7)
"But God hath revealed them unto us by His Spirit..." (1 Cor.2:10)

"For to one is given by the Spirit the word of wisdom; to another the word of knowledge by the same Spirit, to another faith by the same Spirit..." (1 Cor.12:8-9)
"In whom ye also are builded together for an habitation of God through the Spirit..." (Eph.2:22)
"A new heart also will I give you, and a new spirit will I put within you: and I will take away the stony heart out of your flesh, and I will give you an heart of flesh. And I will put my spirit within you, and cause you to walk in my statutes." (Ezek.36:26-27)
"The Lord... formeth the spirit of man within him." (Zech.12: 1)
"That which is born of the Spirit is spirit..." (John 3:6)
"For God gives not the spirit by measure unto him..." (John 3:34)
"He hath given us of his Spirit..." (1 John 4:13)
"Now if any man have not the Spirit of Christ, he is none of His..." (Rom. 8:9)
"Now we have received, not the spirit of the world, but the Spirit which is of God..." (1 Cor. 2:12)
The descent of the Holy Spirit upon the apostles really confirmed the truth, which is stated by all of these texts: the human spirit has its source in the Spirit of God.
All who are familiar with the Scriptures know how many texts in the Old and the New Testament are about the devil and unclean spirits and the evil effect of them on the human spirit. St. John the Evangelist in his first epistle (1 John 2:22) speaks directly of the spirit of Antichrist. The influence of this spirit, hostile to the spirit of God, on the human heart is huge.
Defining above what are the sources for the acts of consciousness , we talked about the perceptions of the higher spiritual world. In this case we are referring to effects of the Spirit of God and the spirit of Satan on the human spirit.
Whence and how was the spirit of Satan born?

Pride is the antithesis of humility, rage and hatred are the antithesis of love, blasphemy is the antithesis of love for God, vanity and greed are the antithesis of love for people. The negative occurs and is growing steadily at the cost of the positive. Darkness appears by the loss of light, cold appears by the loss of warmth, motionlessness and stagnation appear by the loss of movement. The spirit of Satan was born out of the loss of love for God.

The Spirit of God and the spirit of Satan are everywhere and influence all living beings. Something that lives is capable of perceiving one akin to it. Animals take in oxygen and plants take in carbon dioxide from the air. There are beings that do not require light and avoid it. Some bacteria (called anaerobes) can live only in the absence of oxygen and they die in the air. People who are akin in spirit to Satan, perceive that spirit and are steadily developing in it. Paul the Apostle says about it in images: "To the one we are the savour of death unto death; and to the other the savour of life unto life. And who is sufficient for these things?" (2 Cor. 2:16)

People who are akin in their spirit to God-Love, receive the Holy Spirit and are steadily improving in goodness and love. Jesus Christ Himself and the Prophet Ezekiel testify to the spiritual influences from outside on the human spirit:

"The wind bloweth where it listeth, and thou hearest the sound thereof, but canst not tell whence it cometh, and whither it goeth." (John 3:8)

"It is the spirit that quickeneth, the flesh profiteth nothing..." (John 6:63)

"The words that I speak unto you, they are spirit, and they are life..." (John 6:63)

"And the spirit entered into me when he spake unto me, and set me upon my feet, and I heard him that spake unto me..." (Ezek. 2:2)

The spirit came from outside into the man who has a soul. The same is said by the Apostles:
"Paul was pressed in the spirit, and testified to the Jews..." (Acts 18:5)
"And now, behold, I go bound in the spirit unto Jerusalem..." (Acts 20:22)
Through God's work the spirit of one person can be transferred to another.
"The spirit of Elijah doth rest on Elisha..." (2 Kin. 2:15)
"I will take of the Spirit which is upon thee, and will put it upon them..." (Num. 11:17)
What should we say about the spirit of animals?
They, like people, are carriers of a certain spirit according to their nature. In one breed there may be a bold and a cowardly animal, an evil and a moody one, an affectionate and a funny one. They do not possess the higher forms of spirituality, religiosity, moral sense, philosophical and scientific thinking, fine art and musical sensitivity. But love and the beginnings of altruism, and aesthetic sense are inherent in animals too. Not the highest form of love, not Divine love, but only love for family, in which swans and doves are, perhaps, even superior to men. Suicide is known among swans having lost their female: they fly high up, fold their wings and drop like a stone on the ground.
The lower on the ladder of perfection of zoological forms animals are, the lower degree of spirituality they have. The exception to this rule is love among birds. We can to some extent put this in parallel with the fact that the highest form of love and religion are often found among simple, uneducated people. Higher animals, having spirituality, even a limited one, must have self-awareness in a primitive form. Couldn't a dog express, "I am cold, I am sick, my master treats me badly"? The degree of consciousness in animals is determined by

the development of their mind and the degree of spirituality available to them.

Chapter 6. The Spirit is not indubitably connected with the Body and the Soul

What is the relationship between the spirit, the soul and the body? The materialists who do not recognize the spirit as real, reduce all manifestations of the psyche to processes occurring in the brain, and above all, in the cortical substance of its hemispheres, and all psychic acts are considered to be functions of the brain. To a large extent this is true. Physiologists have determined precisely the dependence of psychic acts and states on normal or pathological functions of the nervous system in general, and especially the brain, and consequently, on functions of the body, on internal secretion, on the entire complex hormonal system that has a powerful influence on the brain and nerves.

Everything that happens in an organism, and even its anatomical structure, causes a deep imprint on the psyche. Various structures of the body correspond to some form of character, and the character is one of the most important manifestations of the soul and the spirit.

But can we say that a purely materialistic concept of the psyche is quite justified in those undoubted physiological data? By no means!

After all, the same physiology, and especially the great discoveries of Pavlov and his school, found that the central nervous system predominates over all the somatic processes, it determines and directs the work of all organs, their growth and trophic state; it influences powerfully the course of physiological processes. But the nervous system is the organ of the psyche; according to concepts of vulgar materialism, even thoughts and feelings were treated as the secretions of the brain. This primitive idea was left behind a long time ago, but

the modern materialists attribute the psychic functions to the brain.
And if so, along with the lower functions there should be included the higher ones too, which dominate all lower functions, and inseparable from them in the same manner as the function of contraction can't be separated from the muscle tissue. Therefore, we may consider all the impacts of the central nervous system on organs and tissues as psychic impacts.
And if there is no doubt that somatic processes largely determine the course of mental processes, it is equally certain and should be recognized that there is a psychological impact on all somatic processes.
The powerful influence of the psyche of a patient on the course of his illness is well known. The patient's state of mind, his trust or distrust of the doctor, the depth of his faith and hope for healing or, conversely, mental depression, caused by careless doctors talking about the seriousness of his illness in his presence, determine to a great extent the outcome. Psychotherapy, which consists in the doctors' verbal, or rather spiritual impact on the patient, is a widely recognized method of treating many diseases, often giving excellent results.
Charles Richet explains the true miracles at Lourdes by the powerful influence of the brain (we would say the spirit). Of the three miracles which he takes as examples, the most wonderful one is the healing of Derruder, a worker who in 1875 had an open fracture of the tibia complicated by infection. There was a heavy suppuration, the bones had not knitted, and the lower part of his shinbone with the foot was dangling loose. Eight years later he went on a pilgrimage to the holy city of Lourdes to pray for his recovery, and there he was suddenly healed. He could stand and walk, using both legs while for eight years he could only walk with crutches.
No less surprising is the other case. In 1897 a person called Gargam had a fractured spine in a train accident, with as

result paralysis of both legs with muscle atrophy and incipient gangrene. In Lourdes, Gargam was cured almost instantly: having just entered the cave, he was able to take a few shaky steps. The next day the festering sores of his foot were healed. He could walk without a stick, in spite of the atrophy of muscles. Three weeks later, he had gained 10 kg in weight and was able to return to work.

We could cite many similar examples of healing at the opening of the relics of St. Seraphim of Sarov, from the hagiography of St. Pitirim of Tambov, and many other saints.

We cannot agree with the opinion of C. Richet that these surprising facts are only physiological impacts of the brain, albeit powerful ones. It is impossible because all of the brain processes and all the impacts of the brain take time, and these miraculous healings have occurred almost beyond time, in a fraction of an instance. Such speed of action is possible only for the spirit. In addition, Gargam had a severe spinal cord injury, as evidenced not only by paralysis of the legs and muscle atrophy, but by incipient gangrene. According to our medical knowledge all these were irreversible changes, and no physiological impact of the brain, even the most powerful one, could possibly have reversed them.

With this chapter we substantiate our statement in the third chapter that the spirit creates forms. The assertion that the spiritual is determined by the material is narrow-minded and unreasonable, because we must admit that there is a reverse effect of spiritual influence on the matter of the body through the nervous system, the organ of psyche. The Spirit not only creates the forms of material bodies, directing and defining the process of growth, but it can take these forms directly, through materialization.

No matter how we feel about spiritualism, some facts certainly can't be denied. Only those who judge the spiritualistic phenomena offhand, can lightly and indiscriminately deny

them. Everyone who has read the extensive chapter on this subject in C. Richet's "The Treatise on Metaphysics," will be convinced of the reality of facts of materialization of the spirit in some unknown, special forms of matter.
I must however, give a few examples of materializations from the great number of them. When witnessing them even the most stubborn sceptics, prominent scientists, regretted their mistrust.
While Sir William Crookes was experimenting with the famous medium Home during daylight, he saw a graceful hand rising up from the table and giving him a flower. "It appeared and disappeared three times, giving me the opportunity to make sure it was as real as my own hand, and all that time I was firmly holding the medium's hands and feet. The hand and fingers did not always seem to me dense and just alive. Sometimes they seemed the coagulation of a cloud of steam; a white cloud, as it seemed, was forming and transforming itself into a perfect hand. It seemed like human flesh, like the hands of those present. Around its wrist and shoulder it became like steam and was lost in a cloud of light. I tried to hold the hand in my hand firmly, determined not to let it disappear, but it released itself without even the slightest effort, and turned into steam."
Richet obtained copies of such fluid hands. The ghost dipped its hands into paraffin molten at 43°C. When the hand came out it left a paraffin mould and disappeared. The mould was filled with gypsum and the paraffin was removed. On the photograph of this plaster cast the smallest details of the skin, the lacework of veins can be seen. To avoid any possibility of the medium exchanging the cast cholesterol was secretly added to the paraffin. Cholesterol becomes a purple colour when sulphuric acid is added, and a piece torn off the cast had indeed become purple when the sulphuric acid was

added. The hands and feet of the medium were tightly held during all the experiments.

For all those who believe that the Holy Scriptures are all truth, there can be no doubt about the ability to materialize the spirit, because they know that the Medium of Endor evoked the spirit of the Prophet Samuel on request of king Saul "And the king said unto her, Be not afraid: for what sawest thou? And the woman said unto Saul, I saw gods ascending out of the earth. And he said unto her, What form is he of? And she said, An old man cometh up, and he is covered with a mantle. And Saul perceived that it was Samuel, and he stopped with his face to the ground and bowed himself. And Samuel said to Saul, Why hast thou disquieted me, to bring me up?" (1 Sam. 28:13-15)

If the facts of evoking the dead, that is, materialization of the spirit, were not generally known in antiquity, then why would Moses forbid contact with those evoking the dead, saying "Regard not them that have familiar spirits, neither seek after wizards, to be defiled by them" (Lev.19: 31; ibid 20:6). "There shall not be found among you... a charmer, or a consulter with familiar spirits, or a wizard, or a necromancer..." (Deut. 18:10-11)

Is it possible that Isaiah, the great prophet would say meaningless words? And he said: "... And they shall seek to idols, and to the charmers, and to them that have familiar spirits, and to the wizards..." (Is.19:3) How explain the appearing of Moses and Elijah to the Lord Jesus Christ at His Transfiguration on Mount Tabor, if not as the materialization of the spirit? And what about the appearing of angels in human form? An angel had appeared to Gideon and his wife, and then disappeared in the smoke of burning meat and broth in the same way as the materialized spirits disappear whenconjured up by spiritualists (see Judg. 6:19–21; 13:20).

During the First World War, a professor of physics, N., a materialist, was staying for a while in a Ukrainian village in summer. In the evening he came out onto the porch when the owner of the hut went to the gate to bring her cow in. Suddenly she seemed stunned, clasped her hands, exclaimed, "Peter!" and fainted. Later, she told the professor that she saw her son, who was at war, smiling and joyful. That day he was killed.

The appearance of ghosts at the moment of a person's death is a well-known and indisputable fact. Richet gives many examples of this kind in his book. I will mention only a few of them.

1. Colonel N. was sleeping in his room in London. At dawn, he suddenly woke up and saw Pool, his comrade in the army, in a khaki uniform and a helmet on his head, with a thick black beard, which he hadn't had when N. knew him.

N. knew that Pool was in the Transvaal, on the front. The vision was so clear that N. almost mistook it for the reality: he saw the face, the lively eyes, the khaki uniform, the helmet. N. sat on the bed, staring at Pool who spoke to him: "I was killed by a stab wound in the chest" and saying this, he slowly raised his hand to his chest. "The General ordered me to go."

N. spoke about this appearance to some of his comrades and the next day he got news that Pool was killed in battle. He wore a khaki uniform and a beard, and he was killed by a stab-wound in the chest.

2. Once at night a man called Panchi, who lived in Pisa saw his father, pale and dying, pronouncing the words: "Kiss me one last time, because I'm leaving forever", and he felt the cold touch of lips to his. Although there was no any other, common reason to suspect any misfortune, he went to Florence and there he heard that his father had died last night at the hour when he had had the vision of the ghost.

A similar case took place in my family. My sister died in a wing of the house, where our elder brother lived. He dozed off while sitting on the couch, and awoke at one o"clock at night, feeling clearly a breeze in front of his face and a kiss on his cheek. At this moment our sister breathed her last.

3. Dr. Marie de Thiele, who lived in Lausanne, heard a knock at the door at 6 o'clock in the morning. Someone came in, wearing a black dress, and wrapped in a white translucent fabric like in a veil. The cat in the room arched her back, her fur bristling, growling terribly and shaking. After some time, Madame de Thiele got to know that one of her best friends, whom however, she hadn't thought of at the time of the appearance had died of acute peritonitis in India.

Here's another example of this kind. Miss C. was stroking her cat which was lying on her lap. Suddenly the cat jumped up in horror, arched her back with her fur bristling, and began hissing frantically. Then Miss C. saw an old woman with a pale face sitting in a chair near her and staring at her. The cat rushed like mad with a tumultuous leap through the door. Miss C. cried out for help in terror. Her mother entered the room, and the ghost disappeared. Miss C. had the vision for about five minutes. She was told later that an old woman had hanged herself in that room.

Richet gives many vivid examples of the collective seeing of ghosts. Here is just one of them.

In 1896 Mrs Teleshov was in her living room in St. Petersburg with her five children and the dog Mustash. Suddenly the dog began barking loudly, and everyone saw a little boy of six wearing a shirt. They recognized Andrew, a son of the milkman, who they knew was sick. The ghost came out of the oven, went over the heads of those present and disappeared through the open window. It lasted for about five seconds. Mustash did not stop barking and chasing the moving phantom. At that moment little Andrew breathed his last.

Here's how Richet qualifies seeing ghosts: it is impossible to think that these images, the noise, these ghosts, sometimes having been seen by a group of people, do not represent an objective reality (mechanically objective). And yet we can't prove it absolutely and unquestionably.

It is just the same with all knowledge based on observations. If it were impossible to explain the phenomena of those materialized objects otherwise than by collective hallucinations, then due to the strangeness of these phenomena we should consider them unreal. However, experimental data of the materializations are quite convincing. Ghosts can be observed. This observation can't be exactly as in the [common scientific. Translator] experimental methods, for no photographic plates, no microphone, no weights, no galvanometer can be used by those watching. The only real evidence of the materialization, which is material and luminous, may be its simultaneous perception by several persons, and, moreover, exactly in the same way. If two normal, reasonable people describe exactly the same vision, exclaim simultaneously, inform each other of their experiences in the presence of the phantom, it would be absurd to think of a completely identical hallucination of both of them.

Somewhat different from the appearance of ghosts, but very close to it is the dead calling the living, seen in the form of ghosts only by those whom they call and invisible to everyone else. In other cases these calls are perceived as a voice, without the appearance of ghosts. Facts of this kind are extremely numerous and undoubtedly trustworthy.

Here are a few striking examples.

1. The case reported by Bozzano refers to Ray, a child 2 years and 7 months old. His brother of eight months had just died. His ghost appeared regularly to little Ray, who often saw his dead little brother sitting on a chair and beckoning to him. He said, "Mum, Ray's little brother is calling him, he wants

Ray there with him." On another occasion he said, "Do not cry, mother, little brother is smiling. Ray will go to him". Ray, wise for his age, died two months and seven days after his brother. This case is even more surprising, for children at such an early age do not understand what death is.

2. Louise C. 45 years old died after laparotomy. During her illness she had always asked that her three-year-old niece Lily, whom she loved, should be brought to her village after her recovery. Little Lily was very smart, very healthy; a month after the death of her aunt she would often suddenly interrupt her game, go to a window and stare somewhere. Her mother asked what she was looking at. "Aunt Louise is holding out her hands to me and calling me." Frightened, the mother tried to entertain her, but the child, not paying attention to her, took a chair to the window and for a few minutes did not take her eyes off the ghost of her aunt, visible to her only and calling her. She asked her sister, "How is it you can't see Tata? ("Tata" was a name their aunt was called in childhood and by her relatives). But her sister saw nothing. The visions stopped a few months later. On the 20th May little Lily became ill. Lying in bed and staring at the ceiling, she said that she saw her aunt, surrounded by little angels. "How beautiful it is, Mum," she said. Day by day she grew worse and worse, but she kept repeating, "This is my aunt. She has come to take me and she is holding out her hands to me." And she said to her weeping mother, "Do not cry, Mum, that's fine, the angels are around me." She died on the 3rd July 1896, four months after Louise.

3. The dead father of Philaret, the Metropolitan of Moscow, appeared to him in his sleep three months before his death and he said, "Remember the nineteenth." He died on the 19th November.

4. Mrs. Morrison, lying in bed in India, suddenly heard a voice: "When darkness comes, death will come." She sat up in bed frightened. The same voice slowly repeated the same words.

Two days later her daughter became seriously ill. During that week there had not been a cloud in the sky, but on the eighth day there was suddenly a terrible storm. A few minutes before eleven o"clock the house was completely dark. Her little daughter died at 1 p.m..
The so-called externalization of living people is close to the phenomenon of ghosts of dead people. Examples may be found in the lives of many saints. In the Roman-Catholic hagiology the case of Alphonsus Liguori is well-known. On the 17th September 1774 he became motionless and silent in his cell. He didn't eat and spoke to no one. Then on the 22nd September in the morning he woke up and told his brethren that he had been with the Pope, who had just died. That same night 21st to 22nd September Pope Clement XIV died, and Alphonsus Liguori was beside him.
We can mention the case of externalization of Ambrose the Elder of Optina, our contemporary (died 1891). Avdotya, a peasant, who suffered from bad legs, went on foot to the Optina Monastery, expecting to be healed by Elder Ambrose. Seven miles from the monastery, she lost her way and in tears she fell on the ground. Soon she was approached by an old man in a cassock and cap who asked her why she was crying. The old man showed her the way to the monastery with his crutch. When she reached the hermitage where Ambrose lived she wormed her way in the crowd of women waiting to see the Elder. In a few minutes Fr. Ambrose's cell-monk came onto the porch and said loudly: "Where is Avdotya from Voronezh?" Avdotya was so astonished that she only reacted when the cell monk called a second time. Fifteen minutes later she came out in tears from the Elder in whom she had immediately recognized that very old man who had pointed out the path to the hermitage in the woods. Ambrose had very poor health, would leave his cell only in summer, and he would often fall asleep, always lying on the couch. Neither in

the hermitage nor in the monastery there was anybody who looked like him. He had a striking appearance and Avdotya could not be mistaken.
Exteriorization of the human spirit often takes place in a hypnotic sleep.
Dr. Pierre Janet, while in Le Havre, put Leonie B. in a hypnotic sleep and suggested that she should go (in this hypnotic sleep) to Paris, to his apartment. After this suggestion Leonie suddenly stirred and cried: "Fire, fire"! Dr. Janet tried to calm her down. She woke up and then fell asleep again and woke up with the words: "Janet, I assure you that there is a fire in your apartment". Indeed, that day fire destroyed the Paris laboratory of Dr. Janet.
Let's remember how St. Basil the Blessed at the feast of Ivan the Terrible three times poured the cup of wine poured for him onto the floor, and replied to the angry shouting tsar, "I am putting out the fire in Novgorod." Indeed, at that same time a terrible fire raged in Novgorod!
What is so incredible then about the fact that the enlightened spirit of the saints always possess fully the transcendental powers which ordinary people manifest only in a state of somnambulism ? We can draw extremely important conclusions from these mysterious and utterly inexplicable facts. They are inexplicable not only in the present state of science, but are unlikely ever to be explained by psychic-physiological methods. These phenomena are, of course, of a very special order, radically different from those that are available to science. These are not psychic-physiological effects, but the effects of the spirit, temporarily or permanently separated from the body.
Exteriorization of the spirit of the living in the normal state (Ambrose the Elder) or under hypnosis (Leonei B.), of course, differs from the appearance of the dead in the form of materialized ghosts or mysterious voices predicting death or

misfortune. But for all their differences, these unexplained phenomena suggest that the relationship between the spirit and the body is not unconditional, and that the spirit can exist separate from the body.

The appearances of the dead are, however, very important, even conclusive, proof of the existence of the spirit. The fact that the spirit can exist separately from the body may also be proved by the transfer of inherited spiritual qualities of parents to children. I'm talking about the inheritance of spiritual traits, because only the basic traits, their moral direction, their tendency to good or evil, the higher abilities of the mind, emotions and will are inherited, but the memories of parents' lives, their sensory or organic perceptions, their private thoughts and feelings connected rather with their heart than with their spirit, could never be inherited. This indicates a separation of the spirit from the body and the soul. The facts of heredity of the spirit are known and undeniable. In the twenties of the last century in America there lived a very corrupt young woman. She had already been sentenced to the gallows in her early youth but she escaped punishment through marrying and she had many children. After sixty years the number of her direct descendants had reached eighty. Of these, twenty were condemned for crimes, and the remaining sixty were drunks, lunatics, idiots, and paupers.

In a French family, called Lemoine, known in the history of the end of the XVIIth century, the hereditary transmission of the noblest qualities was noticeable. This is one of those families where the members are born, it seems, only for justice and mercy, in which virtue is transmitted by blood, supported by advice and is stimulated by great examples. (Flechet)

The history of the ancient Roman imperial families, the Spanish and French royal families shows many clear well-known examples of moral and mental degeneration.

The Lord Jesus Christ came into the house of Jairus, to resurrect his dead daughter. He was pressed by the crowd. A woman having an incurable bleeding secretly touched him in the firm hope that she would be healed. Her blood flow immediately stopped. The Lord Jesus stopped too and asked, "Who touched me?" His disciples were surprised with this question: "Master, the multitude throng thee and press thee." Jesus' response was amazing, "Somebody hath touched me: for I perceive that virtue is gone out of me." (Luke 8:46) Let's compare another fact to this one.
In the surgical clinic of Leipzig dr. Hansen made the following experiment in the presence of many professors. He asked dr. Hermann to turn his back to him, facing the wall, so that he could not see what dr. Hansen was going to do. After dr. Hermann did so, dr. Hansen put his right hand onto dr. Hermann's head while with his left hand he took a steel pen, dipped it into ink and touched his own tongue with it. At the same moment dr. Hermann felt the taste of ink in his mouth, which remained for an hour and was impossible to get rid of with any food.
The holy apostles passed on the grace of the priesthood by the laying-on of hands on the ordained heads. They healed the sick by the laying-on of hands. And now the sacrament of the priesthood is given by the episcopal laying-on of hands.
What has happened and is happening in all these cases? We are very far from being able to explain these striking facts; it is a matter for the future, if it is possible at all. But the facts are indisputable. Observations of mediums, who were in a trance (this so-called state like the most deep hypnotic sleep), but sometimes even in their normal condition, definitely established that a part of the motive power of the medium comes

free, i.e. exteriorizes, is separated from him. This phenomenon of exteriorization explains the amazing spiritualist phenomena of movement, even different objects flying about, knocking, self-propelled automatic writing by pen or pencil. Suggestion in hypnosis or in the awakened state can be explained only by the exteriorization of the hypnotist's thought and its reception by the person who is being hypnotized.

After a séance where heavy objects have been moved about, the medium feels great fatigue, because muscle power has gone out of him.

The Lord Jesus Christ felt that virtue was going out of Him, healing the bleeding woman. These are facts of the same order, which surpass the scope of physiological phenomena: these are facts of a transcendental order.

There are many striking facts where we have to admit that, apparently, acting agents are the simultaneous exteriorization of thought and motive power and their telepathic transfer. I will limit myself to citing two examples:

1. At 7 A.M. Mrs Severi jumped out of bed, suddenly awakened by a strong blow to her face. She felt that her upper lip was split and put her handkerchief to stop the bleeding. But to her surprise, there was no blood on the cloth. As it became clear later, at the same very moment her husband, who had taken the boat early in the morning around the lake, was surprised by a sudden strong gust of wind. The rudder shot out of his hands and split his upper lip. He lost a lot of blood.

2. In 1848 in India, Mrs. Richardson saw in a dream that her husband, a general, who was fighting in a battle about 150 miles away, fell badly wounded, and she heard his voice: "Take this ring off my finger and send it to my wife." At about this hour (11 P.M.), the general, very badly wounded, transferred the command to major Lloyd and said, "Take this ring off my finger and send it to my wife." The general survived his injury and recovered.

We know from the lives of many saints, that they knew the names of people whom they were meeting for the first time. Basil the Great recognized immediately through his enlightened spirit that St. Ephraim of Syria was entering the church, and he sent a deacon to call him by name. He had never seen Ephraim before. Through his spirit he knew that in the secret room of the house of the holy presbyter there was a man with a repulsive disease, of whom the presbyter has been selflessly taking care. He called the patient by his name.
We might consider it a legend. But what about a case that struck C. Richet?
Working at the Hotel-Dieu, he hypnotized a recovering girl. Once his friend, an American student who had never been to the hospital before, went to the Hotel-Dieu with him. Richet asked the sleeping girl, "Do you know the name of my friend?" She began to laugh. "Would you tell us the first letter of his name?" She replied, "N, then E; I don't know the third one, and the fourth letter is K". The student's name was Neak. Wasn't this the transcendental power of the human spirit? Who can explain it by physiological reasons?
Hypnotizing Leonie V., Dr. P. Janet once asked her "What happened to my friend N.?" Quickly, not even polite, she replied: "He burned his paw. Why was he so careless to overturn something?" "When did he overturn what?" "A red liquid in a small vial. His skin swelled immediately." Janet writes, "As I heard later, it was quite accurate as two hours before, the head of Janet's laboratory called N. Langua, preparing an alkaline solution of bromine, overturned the vessel with this solution of a red colour. He burned his hand, and blisters soon formed on the skin."
A deacon's wife came to Sarov from Penza to see St. Seraphim. She was standing at the back of the crowd waiting for her turn. Suddenly the Elder, leaving the others, called her "Evdokia, come here quickly." Astonished by the fact that St.

Seraphim, never having seen her, called her by her name, Evdokia approached. "Go back home quickly, or you will not find your son." Evdokia hurried back to Penza and she found her son, who had graduated cum laude from the seminary and obtained a post at the Kiev Academy, on the point of leaving, as he was afraid of arriving late.

Is there any difference between this case and Leonie B.'s clairvoyance and the misfortune of J. N. Langua?

What is so incredible about the fact that the enlightened spirit of the saints always have in full the transcendental abilities that some ordinary people manifest only in a state of somnambulism?

We possess more than the common five senses. We have an ability of perception of a higher order, unknown to physiologists. Charles Du Pröll in his book "The Philosophy of Mysticism" mentions such a remarkable fact. In 1845 Berzelius, a famous chemist, Reichenbach, a physicist, and Hochberger, a physician at a mineral water resort, conducted a highly interesting experiment. They spread out a set of chemicals wrapped in paper on a table before one of the so-called sensitive girls, called von Sakkendorf. She was asked to touch them with the palm of her right hand; after doing so, she reacted differently to the chemicals: some of them did not have any effect on her, but some caused a peculiar trembling in her hand. Then she was asked to sort them out and put in one group the substances which didn't have any influence on her and all the others in another.

Reichenbach says, "The founder of the electrochemical system Berzelius expressed considerable amazement when one group of the substances were without exception electropositive, and the other had a solely electronegative character: not a single electropositive chemical was among the electronegative ones and vice versa... Thus, the electrochemical classification of the chemicals, the building of which took centuries of labour and ingenuity was produced in 10 minutes by a simple girl, endowed with the talent of sensitivity just by touching them with her hand without special devices."

It is very likely that physics and physiology in the future will find an explanation for this special sensitivity. But it is important for us that this ability to distinguish the electrical properties of substances does not belong to all people, but only to a very few, who are called "sensitive". This term, of course, explains nothing, it just marks the extraordinary sensitivity in a certain field. We shall speak of it later.

Completely incomprehensible and even more extraordinary is the ability, found only among the most powerful mediums, to learn a lot about people by touching objects belonging to them. An example of such clairvoyance is given in the third chapter. Touching a tie wrapped in paper a somnambulist found out that a murderer in jail had hanged himself with it. Mrs Pieper, one of the best mediums, showed this ability more than once. Mrs X. gave her three strands of hair, marked with the letters "A", "B" and "C". She herself knew only the origin of strand C. Mrs Pieper said about strand A, "This is from Frode Smoggins. Who is this Smoggins?" Indeed, a person called M. had cut this strand off Mrs Smoggins and handed it to X. Mrs Pieper said about strand B, "This is a very sick person." The woman, whom this strand belonged to, died in the same year. She said about strand "C", "She takes great care of her hair." Mrs X. had secretly cut her mother's hair. "This is your mother. She has four children, two boys and two girls." All this was true.

Widely known are facts of creativity in dreams. Condorcet, Franklin, Michelet, Kandilyak, Arago give evidence of it. Voltaire imagined a big part of his "La Henriade" in his sleep, while La Fontaine created during his sleep the fable "The two doves." Maignan discovered two important theorems during his sleep.

Burdach says, "Often such important scientific ideas were born in my dream that I would suddenly wake up. In many cases they were dealing with subjects, with which I was busy at that time, and they were quite new in their content."

Coleridge fell asleep reading and woke up feeling that he had created two or three hundred verses, ready to be noted down. He wrote down fifty-four verses freely and as quickly as the pen could keep pace. But someone came to see him about some business or other and stayed for about an hour. And Coleridge, to his great chagrin, felt that he had only had a vague recollection of his vision, and only eight or ten verses remained in his memory, and all the rest had gone forever.

De Rossi used to lay a paper and pencil beside his bed. Waking suddenly he would note down the important thoughts that had come to him in a dream.

A very important subconscious activity may take place in reality and in a state on the border between sleep and wakefulness. What is called inspiration, very often comes in a state of more or less total eclipse of the consciousness of reality.

Théophile Gauthier said of Balzac, "He was like a frenzied somnambulist sleeping with open eyes. Immersed in deep meditation, he did not hear what was said." Hegel was calmly finishing his "Phenomenology of Spirit" in Jena, on the 4th October 1806, not realizing that the battle was raging around him. Beethoven, in the throes of inspiration, once appeared half naked in the street of Neustadt. He was taken to jail as a vagrant, and in spite of his indignation , no one believed that he was Beethoven. Schopenhauer said about himself, "My philosophical tenets appeared [in my mind] by themselves without my interference at moments when my will was as if asleep, and my mind was not focused on a predetermined direction... My personality was as if astranged to my work..."

Sometimes the subconscious movement is so clear that it seems a suggestion from outside. This is expressed in Musset's verse:

I do not work, I am listening, waiting...

As if someone unknown speaks into my ear...

Similar examples of Socrates (his daemon), Pascal, and Mozart have become classic.

Prophetic dreams may be called the supernatural abilities of the spirit, inexplicable by modern science. Here are two examples.

1. In 1885 in St. Petersburg a certain Lukaevsky, one of the top officials at the Ministry of Naval Affairs (which, however, does not mean that he often sailed at sea), dreamed that he was on board of a large ship, which collided with another ship; and he fell into the water along with other passengers and was drowning. After this dream, he was convinced he would die in a shipwreck, and in anticipation of imminent death he put his affairs in order. A few months later, when the memory of this dream had already faded, he received orders to go to one of the ports of the Black Sea. He remembered his dream, and on departure he said to his wife, "You will not see me anymore. When I die, put on mourning, but not that black veil, which I hate." Two weeks later the ship "Vladimir" by which Lukaevsky sailed collided with another ship, and Lukaevsky was drowned. Hanicke, one of the "Vladimir"'s passengers who was rescued, said that he was holding onto the same lifebelt with Lukaevsky for a few moments.

2. An intelligent Uzbek K., a former prominent member of the Tashkent city council, told me about an unusual occasion in his life.

A year after his father's death he dreamed that he was riding a horse across a deserted and hilly place. On one of the hills he saw his sister, who had died long ago. She angrily asked him why he never prayed for their father. Saying this she took him to a deep black pit, pushed him into it and said that he would stay in it for forty days. Shortly after this dream K. was arrested and taken to jail. During the interrogation, the police officer showed him two letters with his signature, which were addressed to the Emir of Bukhara. One of the letters contained K.'s appeal to the Emir to rise against the authority of the Russians and the other one contained the detailed plan of the rising. K. admitted that the signatures were indeed similar to his, but that he never wrote these letters or signed them. However he could do nothing to prove his statement, and it was clear that he would be hanged. In despair, he prayed to God for salvation, and, remembering his dream, he began praying for the repose of his father's soul. So about a month passed. One day while praying he fell asleep and he heard a voice in his dream saying, "Write your signature on three separate sheets of paper, put them together, one on top of the other, and look through them under the lamp." When K. woke up, he did as he was told and he found that his signatures were not identical. He repeated this experiment many times, and he could establish undoubtedly that in no case were his signatures identical. He demanded a new expertise, and they discovered that the signatures on both incriminating letters were perfectly identical and on this basis, the experts acknowledged that the signatures were forged, and subsequently it was found that the letters were written by enemies of K. to ruin him. He was acquitted and released on the fortieth day after his arrest. This was exactly the period that his sister had mentioned in his first dream when she pushed him into a black pit.

Some facts are known about the human mind and memory which cannot be explained by science. Under certain conditions things long forgotten may be remembered. It has since long been known that in the mind of the dying man just before death his whole life may unfold with a remarkable clar-

ity and an incredible speed. Fechner tells us about a woman who had fallen into the water and nearly drowned. Two minutes passed after all movements of her body had stopped till she was taken out of the water, during which time she, in her words said she once again lived through all her past life unfolding before her innner eyes in the smallest details.
Admiral Boffre gives another example of such a stream of images in which the memories of many years pass in a short time through the human mind. He fell into the water and lost consciousness. He says, "In this state, one thought was followed by another so fast that it was neither possible to describe nor conceivable for anyone having not experienced this." First, he imagined the direct consequences of his death to his family, but then his spiritual eyes turned to the past: once again he lived through his voyage and shipwreck, his school days, his years of study and the time he had wasted; even all his childhood, his journeys and mischief. He says: "While going further into the past, in my memory I saw facts in my life in a reverse order of their natural sequence, and not in vague outlines, but in a very clear picture in the smallest details. In short, my whole life passed before my soul as a panorama, and each step was accompanied by the awareness of its right or wrong, the precise understanding of its causes and effects. Many minor adventures of my life, in fact, already forgotten, appeared before my spiritual eyes with the same clarity as if they had happened recently."
In this case, at the most two minutes passed from Boffre's falling into water till getting him out.
If in these cases T. Ribot tries to explain the uncanny speed of the flow of consciousness by the state of asphyxia (suffocation), this explanation however is totally not applicable to the following case reported by him.
A man who had an amazingly bright mind, was stepping over train rails at the moment that a train was suddenly approach-

ing at full speed. He had no choice but to lie down between the rails. And while vans were passing over this man, who was almost paralysed from fear, the sense of danger made him remember all the events of his life as if the pages of the Doomsday book were unfolding before his eyes. In the same way, when people die, the "film" of their life from beginning to end, unfolds itself within an instance in their consciousness.
Sekkendorf saw in his dream those events of his past life, which he barely remembered, and with such clarity and vividness, as if they just happened. With extraordinary clarity he saw himself as a three-year-old child and all the smallest details of his upbringing were revived in his memory. Each mark at school, every unpleasant incident were perceived in his consciousness as if just happening. Contemplating his life in the right sequence of events, he saw finally, his stay in Italy, where he had left a lady, whom he would have married if fate had not forced him to leave the country quickly. That sharp feeling of separation from his beloved lively experienced by him in the dream was the cause of his awakening.
If such a sudden remembrance of life is possible for the dying, and even in a dream, it becomes clear how "The Doomsday Book" will be revealed to our consciousness at the Last Judgment.
In the book "Studies of Animal Magnetism" we find a report of John Everdtveger, who after a long illness fell into a death-like state which lasted for several hours. When he opened his eyes he told his confessor that he had seen all his life, all his sins, even those that had long been erased from his memory. The vision was so vivid, as if he experienced it for the first time. The ability of such a phantasmagorical re-experiencing of their entire life by the dying, accompanied by the compression of years of their life into just a few seconds, and the contemplation of individual phases of their life as degrees of development of their spiritual being, was already known

in ancient times and was considered a distinctive ability of the human soul. The philosopher Plotinus says in "Enneads": "But over time, by the end of life memories of the earlier periods of existence appear... because, released from the body, it (our soul) recollects that what it didn't remember here." In these amazing facts of the two - three minutes lasting reproduction of the events of a lifetime, which lasted decades, we are struck first by the preternatural speed of recollection, and secondly by the amazing completeness and clarity of them.
Let us discuss the first point. Images of memories rush in our consciousness with a transcendental rapidity when smoking opium or hashish. T. Ribot in his book "Diseases of the Memory" writes the confession of Kepsey, a passionate opium smoker. He says that while intoxicated he had dreams, lasting ten, twenty, thirty, sixty years; even those that surpass apparently, all thinkable limitations of human life. Insignificant events of his youth, forgotten scenes of the early years of his life often rose before him. He could not say that he remembered them, because if he had been told about them when awake, he would not have recognized his past life in them. But when they passed before him like a dream, in a long-forgotten atmosphere and feelings , then he immediately recognized them.
How can we explain this uncanny quickness of the recollections? In physiology it is known that all processes in the nervous system require a certain time, although very small, measuring fractions of a second. The longer the nerve the greater is the time required for the passage of stimuli from the receptor through the sensitive nerve. Some time is required for the formation of a response in nerve cells, which received this stimulation; some time is required for the transfer of the reaction along the motor nerve. All mental and sensory processes that take place in the brain need time. And if it were possible to summarise and calculate the time of all the men-

tal processes that occur throughout our lives, it would have been a very considerable amount of time. Consequently, it is impossible that this recollection of an entire life which takes just a moment can occur inside the material brain.

Therefore we may conclude that it does not occur inside the brain.

Where else then? As we have said, the life of the spirit is inseparably and intimately connected with all the neuro-psychic activity. It is there, in the spirit that all our thoughts, feelings, volitions, i.e. everything that occurs in our phenomenal consciousness, is imprinted. And this is something different to those tracks and impresses in the nerve cells, which physiologists and psychologists consider to be the mechanism of memory.

We, of course, are far from denying the existence and necessity of such traces and impresses in nerve cells and their rightful explanation of many, perhaps even all the usual functions of the memory. The gradual decline and even disappearance of the memory caused by senile dementia, of course, depends on the atrophy and dying of nerve cells of the cerebral cortex which may be reduced to half or even to one-third of their number. We also know that memory can fade through damage to the brain medulla or cortex after a trauma or an infection. Therefore, the explanation of memory in its most complicated forms by the theory of molecular traces in the brain cells and associative fibres does not satisfy us at all. Although the nerve cells do not proliferate and are not replaced by new ones, like all other organs and tissues, but only die, yet there are continuous exchanges in them, and very likely changes in their molecules. How could we imagine any adequate mechanism for fixing and maintaining forever the traces of all mental acts? And how possibly do we have the right to talk about saving these tracks forever if we know how fragile the memory is and how much of it disappears irretrievably?

The other aspect is even more important. It is impossible to consider the anatomical substrate of memory to be the impresses in just one cell, for in the memory there should be imprinted the impresses of mental acts, which are always complex and involve the participation of the set of cells and association fibres. In the brain there must be kept the impresses not of individual changes in single cells, but the chains of "dynamic associations", according to T. Ribot. During a lifetime quite an incalculable number of such dynamic associations, which constantly change each other, occur in the brain. Their number is as immense as the number of metres from the earth to Sirius. A number of brain cells however, though very large (6 billion according to Meinert), are still quite insignificant compared to the number of mental processes which are supposed to be impressed in them. That is why only some impressions , the most vivid ones are stored in the memory, and it is impossible that the brain could preserve forever the impresses of all the smallest events of our lives with all their details, their sensuous tinge and moral evaluation.

Therefore, we must recognize that, beside the brain there must be another, much more significant and powerful substrate of memory. And we believe such a substrate to be the human spirit, where all our psycho-physical acts are impressed forever. For the manifestation of the spirit there is no time limit, it does not need any sequence or relation of cause and effect for the recollection of the experiences in the memory as are necessary for the brain function.

The spirit embraces all at once and instantly reproduces it in its integrity.

In addition, we believe it is appropriate to remember Richet's valuable words: "The Spirit can operate without consciousness knowing about it: the very complex mental activities pass through our consciousness unnoticeably. A whole world of ideas unknown to us, trembles within us." Probably no recollection of the past is blotted out. Consciousness forgets much, memory forgets nothing. The whole set of images of the past remains almost unchanged, although they have disappeared from our consciousness. For the unconscious stays awake. If we assume that all the richness of memory is kept in full power in the spirit, the amazing phenomena of hypermnesia reported by many authors becomes clear. Let's just mention a few examples. Many authors have reported the amazing facts of long forgotten languages being memorized. A man, who had left his homeland Wales, in childhood and had quite forgotten the Welsh language, in a fit of delirium seventy years later spoke Welsh fluently, but on recovery could not say a word in that language.

Farihagen observed a basket-weaver who had heard a sermon on repentance that had deeply touched him. The next night, he got out of bed still sleeping, and repeated that entire sermon with literal accuracy, pacing up and down his bedroom. When he woke up he couldn't repeat it. And even forty years later he often recited excerpts from that sermon when talking to someone.

A farmer from Rostov, Russia, being in a feverish delirium, suddenly began to say the opening words of the Gospel of John in Greek which he had accidently heard 60 years ago. Seneca mentioned a peasant woman, who being in a feverish delirium pronounced Syrian, Chaldean and Hebrew words which she heard accidentally from a scientist with whom she lived in her childhood. Even the observations made on idiots have shown not only hypermnesia, but also an amazing display of a hidden conscious life.

Maudsley in his book "Physiology and Pathology of the Soul" says: "The extraordinary memory of some idiots who, despite the limitations of their minds reproduce the longest narrations with great accuracy, provides further evidence in favour of such unconscious activity of the soul. Many idiots in a state of excitement caused, for example by great grief or other reasons (for example, the lastupsurge of fading life)show they are capable of such an inner life for which they were apparently, made forever incapable, and this indicates that many things that they can't express are perceived by them and leave impresses in their souls."

There are not enough words to express just how true it is that our consciousness does not comprise our soul. Consciousness can't give us a report on the way these impresses are formed and how exactly they are kept in our soul in a latent state.

T. Ribot tries to explain hypermnesia by an increased blood flow in the brain caused by fever. But this explanation is clearly untenable, since hypermnesia has been observed during sleep also, when the activity of the cerebral cortex is strongly suppressed. If hypermnesia occurs in two opposite states of the brain, sleep and excitement during delirious fever , they can't be its cause, but only a [suitable] occasion for it to be manifested. This is what we"ll discuss in detail in the next chapter. The correctness of our explanation of memory is shown and approved in the Scriptures. The Medium of Endor called the spirit of the Prophet Samuel on request of king Saul, and the conversation of king Saul with the ghost of the Prophet Samuel (1 Sam. 28: 13 - 15) suggests that Samuel's spirit preserved also after his death all the memories of his military life, all the abilities of the mind, will and feelings. It's definitely the same for Moses and Elijah who appeared at the Lord's Transfiguration on Mount Tabor.

How could the dead appear before their loved ones and talk to them if their spirit had not retained all the memories of their life on earth? The little brother could see and hear his dead brother, who called on him to follow him. The little girl saw and heard her dead aunt Louise, who appeared to her many times and called her to join her in the world behind the curtain. His dead father foretold Metropolitan Philaret his death on the 19th day. This list can easily be continued.

From all these facts, new and old, we conclude once again: there are in nature unknown "vibrations' which set the human intellect into motion and reveal to it facts inaccessible to man's five senses.
If we acknowledge the existence of telepathy, we must change only one word in this assumption. It is enough to say "vibrations" of human thought, instead of talking about the "unknown vibration". However, reducing cryptaesthesia to just "vibration of human thought" means an extreme confining of the concept of cryptaesthesia and, consequently, perverting it.
We have talked about it many times with people "devoid of superstition", who "believe in science only", and they always found a simple explanation for all these "new and terrible things". These are just "the waves of human thought", the "oscillations of the molecules of the human brain", which spread like waves transmitted through the wireless. Blessed are the people for whom everything is so simple and clear. They need not weary their superficial thinking with the hard work of studying and clarifying the new and the unknown. They always explain new and extraordinary facts by old and ordinary ones. Science is the only authority for them, although its axioms and hypotheses often crumble like a house of cards under the pressure of the new and unknown. They simply dismiss as superstition and old wives' tales everything that does not fit in the old scientific framework. They can accept a new vision only by getting used to it. Well, even horses stopped being frightened by cars when they got used to them. If physics recognizes the wave-like motion as the basis of material phenomena, why must it necessarily be applicable to

the phenomena of a higher order which occur in the immaterial world?

Why not admit that other laws are valid in that [immaterial] world, which are very different and unknown to us, and that the spiritual energy, the energy of love, sympathy and antipathy, may act beyond time and space, without requiring a wave-like motion?

Let us put a simple question to the "devoid of superstition": if all metaphysic phenomena, all forms of cryptaesthesia are explained by the movement of brain particles transmitted by waves through space, how should we apply this explanation to the undoubted facts of communicating with the dead , whose brain no longer exists?

The profound scientist C. Richet poses far more difficult questions. "From all these facts, both the important and the less important ones, we should make the conclusion which the petty critics cannot make. This conclusion is that the foreboding, predictions are proven facts: strange, paradoxical, absurd facts judging by their appearance, but facts that we have to recognize. So, under some not yet exactly defined conditions, certain individuals, most often (though not exclusively) easily hypnotized people, or mediums, can predict events and report on facts that have not yet happened and which can't be foreseen, in such precise detail, that these predictions cannot be explained by perspicacity, coincidence, or occasion.

We have to assume that having a special mystical knowledge, the nature and characteristics of which are not known to us, which we call cryptaesthesia may be found not only in relation to the past and the present, but also in relation to the future. In addition, the metaphysic knowledge of the present is so unusual that our knowledge of the future is even a little more amazing. A. knows that B. who is a thousand miles away has just drowned. How could A. possibly know? We have no

answers. A. predicts that B. will drown tomorrow. This is a little bit more mysterious, but not much more. In the field of metaphysic clairvoyance the oddity is so overwhelming, and the obscurity is so enigmatic that a little more obscurity and oddity should not confuse us.
I will not enter into useless speculations. I will keep to the dense area of facts. So, there are proven facts, indisputable facts of foresight. The explanation will come later (or will not come). However, there are facts, credible, irrefutable facts. Anticipation and foresight exist. Is it only due to the power of human intellect, or are there some other intellectual forces which affect our intellect? It is impossible to find out presently. Let us be satisfied with at least communicating the facts precisely as they are. It would be an unforgivable audacity for us to claim that there is a prediction if there had not been ample evidence of it." (Richet)
What do the words in the Song of Solomon mean: "I sleep but my heart awaketh?" (Song 5: 2). These are very profound words. The heart is the organ of higher cognition, the organ of communication with God and with the entire transcendental world, and it never sleeps. The most important and profound mental activity takes place beyond the threshold of our consciousness, and it never stops. A clear expression of this idea can be found in Leibniz' works.
"Our own experience shows that there is no moment of life, when there would be a pause in the designing power, and the spirit would cease forming designs. Wouldn't opponents talk about the dream world? But sleep has its images ; in fact, it is full of dreams, and we dream constantly. The so-called sleep without dreams is nothing but the deep sleep, with dreams we don't remember and have no idea of their images when waking up. But upon awakening we always have a feeling that during sleep a certain time has passed, and this feeling would be impossible if we had not dreamed, i.e. if there had

not been images during our sleep, because we always measure time by the images which took place so the same time seems longer or shorter depending on the number of images we had.

If we didn't dream at all, we would perceive the time of our sleep as non-existant; however as the passed dream seems to us a certain passed time, this experience sufficiently proves that we constantly have images. In addition, we would not wake up having ideas in our mind if we slept without any images. However, the so-called sleep without dreams is accompanied by a weak sensation of the outside world, and the stronger this feeling is, the more easily we wake up. That is why continuity of the images in our soul should not be based on our dreams only, for the dream contains also images of the outside world."

Almost all people, even the least sensitive, have the ability of higher knowledge, different from the knowledge of the five senses. The more spiritual a man is, the more pronounced is his ability for higher cognition. Under special circumstances and in highly sensitive people it is manifested by an extraordinary power of clairvoyance, premonition and prophetic anticipation as the unknown and mysterious sixth sense, through which they learn a lot about people by their belongings. We mentioned many facts of such supernatural cognition. In most of these cases the organism and first of all the brain of these people was in an abnormal state. This was the state of hypnosis, somnambulism, febrile delirium, or of a medium. But it was not always so.

In a smaller number of cases such supernatural abilities are manifested by people who are in a normal state. Speaking of the facts of transcendental abilities and somnambulism in a state of hypnosis, we have more than once compared them with similar facts from the lives of saints, or even of ordinary people.

What does this mean? This means that for the manifestation of the transcendental powers of our spirit, for the detection of super-consciousness it is necessary to "switch off", or at least significantly weaken our normal phenomenal consciousness. To express this idea Du Prel makes a good comparison: stars give light constantly, but we do not see their light when the sun is shining. It is necessary for the sun to set and the darkness of night to come, and then the light of the stars becomes visible for us.

As long as our life proceeds in the kaleidoscope and the noise of external perceptions while our phenomenal consciousness is running at full power, the never-ending activity of the super-consciousness is concealed from us. But during the normal sleep, the somnambulistic or hypnotic one, or when the brain is poisoned with opium, or hashish, or toxins of febrile illnesses, our normal brain activity and the light of phenomenal consciousness are quenched, and then the light of transcendental consciousness flares up. We also know that blindness deepens the work of thought and moral sense and widens the threshold of consciousness. The philosopher Fechner created his most profound works after having become blind. Prince Basil the Dark said to Shemyaka, the man who had blinded him: "You gave me the means to repent." It is known from the lives of many saints that long, debilitating illnesses were a great blessing for them, for they tamed their passions, deprived them of the impressions of life in the world with its noise and bustle, which distract from going deeper into the recesses of the spirit. This is well understood and deeply appreciated by the Christian and Buddhist anchorites who sought the suppression of all external impressions through a life in seclusion, the constant immersion in oneself and prayer; to subdue the flesh to the spirit by fasting and vigil, even by standing on a pillar.

In the martyrology there are many examples of the heaviest injuries to the body, the most cruel tortures causing the extinction of the phenomenal consciousness and the awakening of the inner transcendental consciousness which manifests itself by inner bliss.

Transcendental life of the spirit was well known to the ancient Indian sages and Greek philosophers, especially philosophers of the Alexandrian school. Plotinus, Porphyry, and others wrote about it. Here are the words of Plotinus: "Finally, if I dare to express freely and definitely my own belief in opposition to the opinion of other people, then, I think that in the sensual body there constantly remains not all of our soul, but only a part of it, which, being immersed in this world and therefore condensing, or rather, becoming clogged and darkened, prevents us to percept the same what the highest part of our soul perceives." In another place he says: "The souls are like amphibians: they live, according to their need, sometimes on this side of the living, and sometimes in the afterlife." Paracelsus, Van Helmont, Campanella, and many others expressed similar ideas during the Renaissance.

In "German Theology", composed by Luther and highly valued by Schopenhauer, it is said: "The created human soul has two eyes: one can contemplate the eternal, the other only temporary and created things. But these two eyes of our soul can't do their job both at once, but only so that when our soul is contemplating the eternity with its right eye, its left eye should entirely abandon all its activities and remain in idleness, like being dead. When the left eye is operating, and the soul has to deal with the temporal and made, then its right eye should give up all activity. So, who wants to watch with one eye only should get rid of the other, for no man can serve two masters".

Emmanuel Kant undoubtedly recognized the transcendental subject in an unattainable depth of his thought. His contemporaries had little interest in the magical powers of the human soul, and had a poor faith in them. But Kant, with his powerful logic, never judged anything with preconceived ideas and considered impossible only the ideas containing logical contradictions. He argued that we can prescribe nothing to the experience and should take from it all that it gives us, even if it may seem strange and unexpected to us. So when he learned about the discovery of the magical powers of Swedenborg, his contemporary, Kant not only collected accurate information about that mystic, but also bought his works. After reading them, he was struck by the similarity of Swedenborg's theory with his own, drawn from pure reason, the theory of the transcendental nature of man.

In "The Dreams of a Spirit-Seer" Kant writes: "I confess that I am very inclined to believing in the existence of immaterial beings in the world, and to reckon my own soul as such a creature." And then he adds: "Therefore, the human soul in this life should be considered simultaneously connected with both worlds, but while still being united with the body, it perceives clearly only the material world."

In "Rosencrantz" Kant expresses his thought even more clearly: "Therefore we can assume almost for granted, or easily proved, ... or, rather, it will be proved, although I do not know where or when, that the human soul in this life is in close connection with all immaterial beings of the spiritual world, that it operates alternately in the one or the other world, and perceives from those beings impressions, which it, being an earthly man, is not aware of until everything goes well (that is, until it enjoys the material world)."

Kant always kept to his teaching about "things in themselves", which he reckoned to the world of noumenons. Kant's "noumenon man" is what he called "a thing in itself", and Carl Du Prel called it a "transcendental I", and the Apostle Paul called it "the inner man." For us of course, the most important are

the words of the Apostle Paul: "Though our outward man perish, yet the inward man is renewed day by day." (2 Cor. 4:16)
When out of passion and lust the flesh weakens and fades, when its strength is weakened, and the glamour and noise of this world cloud us, then " ...put off concerning the former conversation the old man, which is corrupt according to the deceitful lusts, and renewed in the spirit of our mind, and that ye put on the new man, which after God is created in righteousness and true holiness." (Eph. 4: 22 - 24)
"The New Man" in the words of Paul, is of course the same as the "inner man".
This deep psychological change occurs in us through the action of God's Grace: "I bow my knees unto the Father of Our Lord Jesus Christ... that he would grant you according to the riches of His glory, to be strengthened with might by His Spirit in the inner man." (Eph. 3:14-16)
Then there is a co-crucifixion with Christ of our old "outer" man to destroy the body full of sin, for "henceforth we should not serve sin" (Rom. 6:6), and then "let it be the hidden man of the heart in that which is not corruptible, even the ornament of a meek and quiet spirit." (1 Pet. 3:4)
Then we even become "partakers of the Divine nature" according to the words of the Apostle Peter (2 Pet. 1:4), then "Christ may dwell in your hearts by faith; that ye, being rooted and grounded in love, may be able to comprehend with all the saints what is the breadth and length, and depth and height; and to know the love of Christ, which passeth knowledge, that ye might be filled with all the fullness of God..." (Eph. 3:17-19)
And we should add the words of Paul: "... put off the old man with its deeds and ... put on the new man, which is renewed in knowledge after the image of him that created him." (Col. 3: 9 -10)

Our inner, transcendental man, freed from the bonds of the flesh, can reach the highest knowledge of all that exists in all its "breadth and length, and depth and height", for he will be renewed and strengthened in the knowledge, even according to his creation in the image of his Creator, he will comprehend the love of Christ surpassing all earthly understanding, for Christ will dwell in him by faith. Incomprehensible to the "geometrical mind" it becomes clear to the transcendental consciousness of the inner man enlightened by Christ.

The human spirit is the breath of the Spirit of God, and for this reason only it is immortal, like all incorporeal, angelic spirits. They are numerous, as evidenced by the Scriptures, and the extent of their development, their perfection is infinite.

In the range of earthly beings man is the first and the only spiritual being, and there have been people of a very high degree of spirituality, almost liberating the spirit from the body during their life. These human-angels ascended into the air during their prayer, manifested the greatest power of the spirit over the body (the pillar ascetics, fasters); they were in a transitional stage between the spirit associated with the body and the soul (a person) and the bodiless spirit (an angel).

The whole world of living beings, even the whole of nature, demonstrates the great law of the endless gradual improvement of forms, and it is impossible to assume that the utter perfection achieved in the earth's nature, the spirituality of man, had no further development beyond the physical world. It is impossible to assume that all the innumerable worlds of stars were just vast masses of dead matter, and that the world of living beings ended with man, who is just the first stage of spiritual development. What prevents us from assuming that the heavenly bodies are the dwellings of countless sentient living beings, who possess higher forms of intelligence?

There are the usual arguments against it, namely that organic life is impossible under the physical conditions existing on the stars and the planets (except perhaps, Mars). But do the bodiless spirits need certain physical conditions of life, like organic beings? And finally, why couldn't forms of physicality exist there, very different from the ones on earth, adapted to physical conditions different from ours?

And the glowing white-hot surface of huge stars may be inhabited by fiery seraphim and cherubim: "Who maketh his angels spirits; his ministers a flaming fire." (Ps. 104:4)
If the law of development and improvement in the earth's nature is so clear, there is no reason to suppose that it is suspended outside our planet, that the spirit first manifested in man, but being evident even in the simplest creatures in the initial form, has no further development in the Universe.
The world has its origin in God's Love, and if people are given the law: "Be ye therefore perfect, even as your Father which is in heaven is perfect." (Matt.5:48), then, of course, they must be given the opportunity for this commandment to be implemented, that is the possibility of infinite perfection of the spirit. And this requires the everlasting existence of the immortal spirit and the infinite number of forms of its perfection. It couldn't be so, that the law of the infinite perfection of the spirit in its approach to the perfection of God was given to people only, not to the whole universe, not to the whole world of spiritual beings, for they are created in various degrees of perfection, far exceeding the smallest perfection of the human spirit.
If matter and energy in their physical form are perfect (indestructible), then, of course, the spiritual energy, or, in other words, the spirit of man and all living beings, should be subjected to this law. Thus, immortality is a necessary postulate of our mind.
Our Lord Jesus Christ openly testified the immortality of man: "And whosoever liveth and believeth in me shall never die." (John 11:26)
"He that heareth my word, and believeth on him that sent me, hath everlasting life." (John 5: 24)
And the Apostle James speaks of a man as of just the first stage of spirituality; he particularly says: "Of his own will begat he us with the word of truth, that we should be a kind of

firstfruits of his creatures." (James 1:18) And Paul says: "And not only they, but ourselves also, have the firstfruits of the Spirit..." (Rom. 8:23) There is no need of any other proof of immortality for us, Christians. And for the disbelievers it is useful to recall the words of Emmanuel Kant, the most profound of people, cited in the previous chapter. He believed in the existence of incorporeal and therefore immortal beings in the world, and his own soul he reckoned as one of them. C. Richet at the end of his great book, where a huge amount of undoubted metaphysic facts were gathered, discusses possible explanations, and concludes that the most probable of them should be seen in the existence of beings possessing reason other than the humans that surround us, and who can interfere in our life, in our development, although they are alien to the mechanical, physical, anatomical and chemical conditions of existence.

Why should we not acknowledge the existence of intelligent, powerful beings not belonging to the world of our senses? By what right do we, with our limited senses, our imperfect minds, our scientific development, which is just three centuries old, dare to say that man is the only being in the immeasurable space to possess reason, and that any thinking reality depends on the presence of nerve cells, irrigated with blood? The existence of intelligent beings different from humans, having an entirely different type of organization to the human, is not only possible but highly probable. It is absurd to think that the human mind is unique in the universe, and that any power which possesses reason must necessarily be organized the same way as man or animals, and have a brain as the organ of thought.

If we assume that in the universe, in the time and the space that our rudimentary psychology is subjected to, there are powers endowed with reason which sometimes interfere in our lives, we obtain a hypothesis which gives a full explanation of the facts stated in this book.

Thus acknowledging the existence of non-material forms of mystical creatures, angels or demons, spirits, who sometimes interfere in our actions, creatures who may change our matter by their will through ways completely unknown to us, direct some of our thoughts, take part in our lives; creatures who can take the bodily and psychological form of the dead to enter into communication with us (the only way for us to know them) is the easiest way to understand and explain much of the metaphysic phenomena.

This is the conclusion of a scientist used to positive thinking, obtained after the objective study of many metaphysic facts, which he assiduously collected during his lifetime.

Other prominent scientists in the fields of metapsychology, Mauer and Oliver Lodge, arrived at a similar conclusion.

The essence of this conclusion may be reduced to the statement that the human spirit communicating with the transcendental, eternal world, lives in it and belongs to eternity.

The main obstacle to unbelievers admitting the soul's immortality is the dualistic conception of body and soul, the idea of the soul as a distinct entity, just associated with the body during its lifetime.

This view is exactly what we consider being the core of disbelief and in the Scriptures we find no obstacle to understanding the relation between soul and body from the point of view of monism. We have already spoken about the necessary connection between the spirit and the form, about the fact that the spirit forms the body in the embryonic stage of its growth. Spiritual energy is inherent in all the cells of the body as they are alive and the life comes from the Spirit.

There is, of course, the two-way causal relationship between all the functions of the body and psychic activity, interpreted by psychologists.

But this concerns only that part of our threefold being, which could be called the lower, animal soul; it is that part of our spiritual essence which is embraced by our consciousness: it is, so to speak, our "phenomenal" soul. And "The Spirit projects outside the brain from all sides" (Bergson); the spirit is the sum of our soul and its part lying outside the boundaries of our consciousness.

There is constant communication and interaction between the body and the spirit. Everything that happens in the soul of a man throughout his life has its meaning and is necessary only because the whole life of our soul and body, every thought, feeling, volition, having started in sensory perceptions, is closely connected with the life of the spirit. All acts of the body and the soul are imprinted in the spirit, and are kept in it, and form it. The life of the spirit and its orientation towards good or evil evolves under their formative influence. The life of the brain and the heart, and the entire, marvellously coordinated life of all organs of the body are necessary only for the formation of the spirit and they cease when its formation is completed, or its direction has become definite. The life of the body and the soul can be compared with the life of a bunch of grapes full of beauty and charm. The moment comes when it is no longer fed with the juice of the vine and the dew from heaven sprinkled on the gentle bloom of the grapes ceases, and they become just pressed skins, doomed to decay; but the life of grapes goes on in the wine obtained from them. All that was valuable, beautiful and fragrant, that has been produced in living grapes under the beneficial influence of light and heat from the sun, goes into the wine. And just as wine does not spoil, but continues to live its own life after the death of the grapes, becoming better and more valuable the longer it lives, so the eternal life goes on in the immortal human spirit as well as an infinite development in the way of

good or evil after the death of the body, brain and heart, when the soul is not longer active.
We understand the eternal bliss of the righteous and the eternal torment of sinners to mean that the immortal spirit of the righteous, enlightened and powerfully amplified after the liberation from the body, has the possibility of an infinite development towards God's goodness and love, in constant communication with God and all His bodiless powers. While the dark spirit of the wicked and those opposing God, which is in constant communication with the devil and his angels, will be eternally tormented because it is alienated from God, Whose sanctity will finally become evident to it, and because of the unbearable poison full of evil and hate, which infinitely augment through the constant communication with the centre and source of evil, Satan.
It is certainly impossible to blame God for the eternal torment of sinners, thinking of Him as infinitely vindictive, punishing with eternal torment the sins of a short-lived life. Each person receives and possesses the breath of the Holy Spirit. No one is born of the spirit of Satan. But as the dark clouds obscure the sun and absorb light, so the evil acts of the mind, of the will and of the feelings by their constant repetition and prevalence darken constantly the light of Christ in the soul of an evil man, and his mind becomes more and more determined by the influence of the spirit of the devil. They who loved evil rather than good, have prepared for themselves eternal torment in the life everlasting.
But here we are confronted with an ancient dispute on free will and determinism.
Only the great Kant gave a profound solution to this dispute. Freedom can't be attributed to man as a phenomenon of the material world, because he is subjected to causality in this world. Like everything in nature, he has his own empirical character, which determines his reactions to external stimuli.

But in his spirit man belongs to the noumenal, transcendental world, thus his reaction to external stimuli is determined not only by his empirical character, but also by his spirit. Kant says, "Thus, freedom and necessity each in its full meaning can co-exist and do not contradict each other in the same act, as our every act is the product of noumenal and sensible reasons."

Simply put, the human spirit is free, "blows where it wishes..." And its lower, perceptible soul obeys the laws of causality.

We have still spoken only of the immortality of the spirit, but also our bodies will be resurrected to eternal life and they will be partaking in the endless blissfulness of the righteous or the interminable torment of sinners according to the clear testimony of Revelation.

This is also a stumbling block to unbelievers and a great mystery for the faithful. The unbelievers find impossible the restoration and resurrection of the bodies, which were completely destroyed by decay or burnt, turned to dust and gases, decomposed to atoms. But if during the life of the body the spirit was intimately connected with it, with all its organs and tissues, penetrating all the molecules and atoms of the body, if it was its organizing source, why should the relationship disappear forever after the death of the body? Why is it unthinkable that this relationship is preserved for ever after death, and at the universal resurrection at the sound of the archangel's trumpet the immortal spirit will be reconnected with all the physical and chemical elements of the rotted body, and the organizing power of the spirit, creating forms, will manifest itself again? Nothing disappears, it is just modified.

Another difficult issue, which is a mystery to believers, is understanding the purpose of the resurrection of the dead in their earthly integrity. The immortality of the spirit freed from the bondage of the body is clearer for us. Why is it necessary for the whole man to participate in eternal life, not only the spirit, but also his body and soul?

Of course, we can't clearly comprehend the mystery of God ruling over His creation, yet the Scriptures give us the opportunity to lift the veil. The Holy Apostles Peter and John to some extent explain the secrets of the resurrection of human bodies. They speak clearly about the end of the world, the great and terrible catastrophe which will occur in the universe at the time of the second coming of the Lord Jesus Christ.

"But the day of the Lord will come as a thief in the night; in the which the heavens shall pass away with a great noise, and the elements shall melt with fervent heat, the earth also and the works that are therein shall be burned up." (2 Pet. 3:10)

And the Apostle John in his Revelation clearly depicts this world catastrophe divided into phases.

What will be next? What is the purpose of this cataclysm?

"Nevertheless we, according to His promise, look for new heavens and a new earth, wherein dwelleth righteousness." (2 Pet. 3:13)

"And I saw a new heaven and a new earth: for the first heaven and the first earth were passed away; and there was no more sea. And I, John, saw the holy city, new Jerusalem, coming down from God out of heaven, prepared as a bride adorned for her husband. And I heard a great voice out of heaven, saying, Behold, the tabernacle of God is with men, and he will dwell with them, and they shall be his people, and God himself shall be with them, and be their God. And God shall wipe away all tears from their eyes; and there shall be no more death, neither sorrow, nor crying, neither shall there be any more pain: for the former things are passed away. And he that sat upon the throne said, Behold, I make all things new." (Rev. 21:1-5)

"I make all things new". There will be a time for the new creation, the new earth and the new heavens. Everything will be completely different, and our new life will take place in a totally new environment, and we should have our full nature in that life and perceive a quite new experience by our renewed and enlightened senses. Consequently, the activity of the part of our spirit, that we now call the lower, physiological soul, will be necessary.

Our thoughts will work hard to discover the new world, where our spirit, freed from the power of the earth, from our sinful flesh, will be built and strive to come closer to God. The mind is a part of our spirit, and therefore our brains have to be immortal, too. The immortal heart will be the focus of new, pure and deep feelings.

The eternal life is not only the life of the spirit, freed from the body and soul, but the life in the New Jerusalem, described so vividly by St. John the Theologian in his "Revelation".

The immortality of the body, not just the spirit, perhaps has one more purpose, full of justice and truth, i.e. the purpose of honouring the body of the saints as the great instrument of the spirit, which worked hard and suffered during the formation and perfection of the spirit during the earthly life. And the bodies of grave sinners, which were the principal instruments of sin, of course deserve punishment. The powerful and terrible pictures of Dante's "Inferno" are probably not just a figment of poetic imagination.

St. Paul reveals to a significant extent the mystery of the resurrection of the bodies in his first epistle to the Corinthians: "How are the dead raised up? And with what body do they come? Thou fool, that which thou sowest is not quickened, except it die: And that which thou sowest, thou sowest not that body that shall be, but bare grain, it may chance of wheat, or of some other grain: But God giveth it a body as it hath pleased him, and to every seed his own body... So also is the

resurrection of the dead. It is sown in corruption; it is raised in incorruption: it is sown in dishonour; it is raised in glory: it is sown in weakness; it is raised in power: it is sown a natural body; it is raised a supernatural body. There is a natural body, and there is a spiritual body." (Cor. 15: 35-44)
Sown, buried in the ground, the grain as if perishes; it ceases to exist as a seed, but out of it there grows something much larger than it was, incomparably better in its complexity and form, a new plant. God gives it shape and beauty, and life full of utility and delights.
A human body is buried in the ground, and it ceases to exist as a body. But from the elements in which it is decomposed, as from cells of grains of wheat, the power of God will raise a new body, and not a destroyed, weak and powerless corpse, but a new spiritual body, full of power, incorruption and glory.
"The first man is of the earth, earthy: the second man is the Lord from heaven. As is the earthy, such are they also that are earthy: and as is the heavenly, such are they also that are heavenly. And as we have borne the image of the earthy, we shall also bear the image of the heavenly." (1 Cor. 15:47-49)
During life our body is of dust, of soul, as the body of Adam. After the resurrection it will be different, spiritual, like the heavenly body of the second Adam, Jesus Christ, the one he had after His glorious Resurrection.
We do not know all the properties of the resurrected body of Jesus Christ. We only know that it passed through locked doors, could suddenly disappear from sight. (Luke 24:36; John 20: 19).
He was not immediately recognized by the apostles and the myrrh-bearing women. The Lord ascended into heaven in this glorious body. But it was a real body, which the apostles could touch, and it could have the usual functions of the human body (Luke 24:43). Our body will be similar to the body of Christ after our resurrection to eternal life.

Does only man possess immortality? The great words: "Behold, I make all things new" are certainly related not to man only, but to all creation, to every creature. We have already said that the spirit of animals, even the smallest part of it, the spirit of life, cannot be mortal, for it is from the Holy Spirit. And the spirit of animals is connected with the body, like the spirit of man, and therefore there is all reason to expect that their bodies will exist in the new nature also, in the new universe which comes after the end of this world. This is evidenced by the Apostle Paul: "For the earnest expectation of the creature waiteth for the manifestation of the sons of God. For the creature was made subject to vanity, not willingly, but by reason of him who hath subjected the same in hope, because the creature itself also shall be delivered from the bondage of corruption into the glorious liberty of the children of God. For we know that the whole creation groaneth and travaileth in pain together until now." (Rom. 8:19-22)

All the creatures would have lived in light and joy, if the sinful Adam's fall had not changed the destinies of the world. The fate of the world became sad, and according to the sinful will of Adam, to whom God had subjected it, it has become subjected to vanity, discord and suffering. But there is hope for it, that on the day of glorifying all the righteous, redeemed by Christ from the bondage of corruption, it will be freed from suffering and decay, i.e. will become imperishable.

In the new Jerusalem, the new universe, there will be a place for the animals, too; there will be nothing unclean there, and the new creature will regain the old justification and sanctification by God's Word: "And God saw every thing that he had made, and behold, it was very good." (Gen. 1:31)

Certainly, immortality will not be of the same value for animals as for man. Its primitive spirit can't grow infinitely and be morally improved. Eternal life for the low creatures will be only quiet happiness and enjoyment of the new radiant

nature full of light, in communication with man, who will no longer torture and exterminate it. Man will be whole and harmonious in the future new universe, and there will be a place for every creature in it.
God bless! God bless!

www.ingramcontent.com/pod-product-compliance
Ingram Content Group UK Ltd.
Pitfield, Milton Keynes, MK11 3LW, UK
UKHW042017290726
14061UKWH00001BB/41